华章IT

HZBOOKS | Information Technology

Istio in Practice

Istio入门与实战

毛广献 编著

机械工业出版社
China Machine Press

图书在版编目（CIP）数据

Istio 入门与实战 / 毛广献编著 . 一北京：机械工业出版社，2019.4
（实战）

ISBN 978-7-111-62524-7

I. I…　II. 毛…　III. 互联网络 – 网络服务器　IV. TP368.5

中国版本图书馆 CIP 数据核字（2019）第 068126 号

Istio 入门与实战

出版发行：机械工业出版社（北京市西城区百万庄大街 22 号　邮政编码：100037）

责任编辑：吴　怡　　责任校对：殷　虹

印　　刷：三河市宏图印务有限公司　　版　　次：2019 年 5 月第 1 版第 1 次印刷

开　　本：186mm×240mm　1/16　　印　　张：20

书　　号：ISBN 978-7-111-62524-7　　定　　价：79.00 元

凡购本书，如有缺页、倒页、脱页，由本社发行部调换

客服热线：（010）88379426　88361066　　投稿热线：（010）88379604

购书热线：（010）68326294　　读者信箱：hzit@hzbook.com

Preface 前　　言

近几年来，容器技术的飞速发展使得微服务技术更容易落地，微服务架构在业界逐渐流行起来。但是微服务架构对基础设施要求较高，微服务依赖持续集成、服务注册、服务发现、负载均衡、健康检测、配置管理、服务路由、服务容错、日志收集、指标收集、调用链追踪等，而构建这一套基础设施的成本巨大。因此，微服务相关的框架逐渐露出水面，比如Java语言的Spring Cloud框架。虽然这些微服务框架为我们提供了很多便利，但由于这些框架是与编程语言绑定的，使得我们应用的技术栈受到了限制。即使后来有其他的微服务框架也支持多编程语言的技术栈，但由于这些微服务框架代码对服务代码入侵严重，给后续服务框架的bug修复和版本升级带来了一定的困难。所以，服务网格的概念一经提出，便得到了很多人的支持，人们对这项技术抱有很大的期望，希望能解决当前微服务所遇到的问题。

当我第一次接触服务网格技术的时候，就觉得服务网格将来可能会像IaaS、PaaS一样成为业界的主流技术，会得到广泛应用。其实，服务网格并没有提供什么新的概念和功能，它只是把原来服务框架所做的功能完全独立出来，整合了一个服务网格的基础设施层。这个改变看似很小，但是能使服务与服务治理功能实现完全解耦，这个影响是巨大的。

2017年5月，谷歌、IBM、Lyft等公司共同努力实现的开源服务网格Istio正式发布了第一个版本。而后又一个服务网格开源实现Conduit开始启航，服务网格进入了更多技术人员的视野，经过努力，Istio在2018年7月正式发布了1.0.0版本。

2018年是服务网格快速发展的一年，Istio发布的1.0.0版本标志着Istio已经成熟到可以接受生产流量的考验。2019年服务网格将会持续保持热度，作为一名技术人员，现在是时候了解一下服务网格技术了。而在所有的开源服务网格实现中，最成熟的肯定是Istio。因此，要学习和了解服务网格，首先应该学习使用Istio。

由于Istio官方文档使用英文编写，而且只简单演示了Istio提供的功能，没有基于一般的使用场景，我在学习Istio时就很迷茫。于是，我结合自己的学习经验和方法编写了这本如何在实战中学习 Istio 的书，以便技术人员能以最简单的方式上手 Istio，理解 Istio，并能在生产环境中应用 Istio 服务网格。

由于本书是一本实战类型的书，书中没有大篇的理论知识。对于 Istio 提供的功能，本书只简单地描述其作用和使用方式，然后使用实验来演示效果。我相信，只要你跟着书中的实验操作，并理解这些实验的目的，学习完本书后，你就一定能熟练使用 Istio，并在以后的 Istio 使用中得心应手。

对服务网格感兴趣的人都可以阅读本书。如果你想了解服务网格，想知道服务网格提供了哪些功能，能解决什么问题，本书将是一个不错的选择。如果你想了解 Istio，体验 Istio，将来有可能将 Istio 应用于生产环境，那么阅读本书将是一个不错的开始。

本书主要内容如下：

第 1 章介绍服务网格的由来以及服务网格能给我们带来什么，接着介绍开源的服务网格实现 Istio 的主要功能特性及其架构设计。

第 2 章说明本书后续实验相关的环境和实验中的注意事项，并详细介绍后续实验中将会使用的微服务架构应用，以及微服务的容器化构建。

第 3 章简单介绍 Vagrant 及其使用方法。使用 Vagrant 可以快速创建实验环境，这对于我们后续实验环境的准备提供了非常大的便利，接着对创建实验环境的场景进行模拟，帮助读者熟悉 Vagrant 的使用流程。

第 4 章介绍如何使用 Kubeadm 快速创建一个多节点的 Kubernetes 集群。Kubernetes 集群是后续部署 Istio 的基础。

第 5 章介绍如何以官方示例的方式部署一个包含完整功能的 Istio 集群，以及如何以最小资源的方式部署一个能满足大部分实验场景要求的 Istio 集群。此外，还简单介绍了 Istio 中常用的资源，以及常用的 istioctl、kubectl 命令。

第 6 章介绍微服务架构应用如何部署在 Kubernetes 集群中，以及如何访问部署在 Kubernetes 集群中的服务，还简单介绍了如何在 Istio 集群中部署和对外暴露服务。

第 7 章介绍 Istio 提供的服务流量管理功能，包括管理网格的入口和出口流量，根据请求进行流量拆分，如何借助 Istio 实现 A/B 测试和灰度发布等功能。

第 8 章介绍如何让部署在 Istio 中的服务更具弹性，包括负载均衡、连接池、服务健康检测、服务熔断、服务超时、服务重试、服务限流功能的配置。

第 9 章介绍服务故障注入的相关功能，提前给服务注入故障，可以测试服务在故障中的稳定性，提前发现问题并修复问题。

第 10 章介绍服务间通信加密和服务间访问权限的控制。Istio 提供了双向 TLS 进行服务间的通信加密，使用 RBAC 来实现细粒度的服务访问权限控制。

第 11 章介绍如何提升服务的可观测性。在 Istio 中通过简单的配置，就可以实现服务的指标收集、日志收集。通过传递指定的服务请求头，就可以轻松实现服务的调用链追踪功能，这不但增强了服务的可观测性，还大大减轻了运维人员和开发人员的负担。

第 12 章介绍 Istio 部署后的维护工作。通过部署开源的第三方仪表板工具，我们可以更方便地管理 Istio。接着介绍如何在不停机的情况下升级 Istio，如何使用 Helm 以定制化的方式部署 Istio，以及当 Istio 出现故障时应该如何排查并解决问题。最后介绍了在 Istio 中一个请求从发出到响应的整个流程。

第 13 章介绍一些不适合放在其他章节的 Istio 功能，包括跨域、跳转、TCP 路由、TLS 路由等，以及如何在 Gateway 上启用 HTTPS，如何为部署在 Istio 中的服务开启健康检查功能，如何使用 Envoy 代理 Ingressgateway 来实现把集群内的服务暴露给集群外部使用。最后还简单介绍了 Mixer 和 Adapter 模型。

本书源码

本书所有示例代码都放在 GitHub 上，地址为 https://github.com/mgxian/istio-lab，读者可以查看或下载。由于部署 Kubernetes 集群和 Istio 的过程中会涉及比较多的命令，我也把相关的命令放在了源码的 cmd 目录下。

由于作者水平和时间有限，书中难免会有一些纰漏和错误，欢迎读者及时指正。非常希望和大家一起学习和讨论服务网格与 Istio，并共同推动服务网格和 Istio 在国内的发展。可以通过电子邮件 will835559313@163.com 联系到我。

致谢

感谢所有在本书撰写、出版过程中给予过帮助的人。这里要特别感谢机械工业出版社的吴怡编辑，没有她的鼎力相助，就没有本书。同时也要感谢我的家人和朋友，没有他们的支持和理解，我不可能在有限的时间里完成本书。最后要感谢阅读本书的读者，非常感谢大家的支持！

目　录 *Contents*

前言

第1章　服务网格与Istio……1

1.1　服务网格简介……1

1.1.1　服务网格的概念与特点……2

1.1.2　服务网格的优势……3

1.2　Istio简介……4

1.3　Istio的架构设计……5

1.3.1　数据平面……6

1.3.2　控制平面……7

1.4　Istio的功能特性……9

1.5　本章小结……10

第2章　实验说明……11

2.1　实验的环境……11

2.1.1　基础环境……11

2.1.2　命令说明……12

2.1.3　问题及解决方案……13

2.2　实验的应用……16

2.2.1　应用架构说明……16

2.2.2　应用详细说明……17

2.3　应用的构建……26

2.4　本章小结……31

第3章　使用Vagrant管理虚拟机……32

3.1　Vagrant简介……32

3.2　Vagrant常用命令……33

3.3　模拟实验时的场景……38

3.4　本章小结……47

第4章　创建Kubernetes集群……48

4.1　安装Docker……48

4.2　安装Kubeadm……51

4.3　配置基础环境……52

4.4　创建Kubernetes集群的步骤……55

4.5　测试集群的正确性……61

4.6　注意事项与技巧……65

4.7　本章小结……67

第5章　Istio部署与常用命令……68

5.1　部署Istio……68

5.2　常用资源类型……77

5.2.1 流量控制 ······ 77
5.2.2 请求配额 ······ 80
5.2.3 mTLS 认证策略 ······ 81
5.2.4 RBAC 访问权限 ······ 81
5.3 常用的 kubectl 命令 ······ 83
5.4 常用的 istioctl 命令 ······ 83
5.4.1 通用参数说明 ······ 84
5.4.2 常用命令 ······ 84
5.5 注意事项与技巧 ······ 85
5.6 本章小结 ······ 89

第 6 章 微服务应用的部署 ······ 90

6.1 微服务应用架构 ······ 90
6.2 部署服务 ······ 94
6.3 访问服务 ······ 98
6.4 在 Istio 中部署微服务 ······ 102
6.5 本章小结 ······ 105

第 7 章 让服务流量控制更简单 ······ 106

7.1 整体介绍 ······ 106
7.2 管理集群的入口流量 ······ 110
7.3 把请求路由到服务的指定版本 ······ 111
7.4 根据服务版本权重拆分流量 ······ 113
7.5 根据请求信息路由到服务的不同版本 ······ 114
7.6 流量镜像 ······ 115
7.7 管理集群的出口流量 ······ 117
7.8 实现服务 A/B 测试 ······ 126
7.9 实现服务灰度发布 ······ 128
7.10 灰度发布与 A/B 测试结合 ······ 132
7.11 本章小结 ······ 135

第 8 章 让服务更具弹性 ······ 136

8.1 整体介绍 ······ 136
8.2 负载均衡 ······ 138
8.3 连接池 ······ 141
8.4 健康检测 ······ 144
8.5 熔断 ······ 145
8.6 超时 ······ 149
8.7 重试 ······ 151
8.8 限流 ······ 153
8.9 本章小结 ······ 165

第 9 章 让服务故障检测更容易 ······ 166

9.1 整体介绍 ······ 166
9.2 给服务增加时延 ······ 168
9.3 给服务注入错误 ······ 169
9.4 时延与错误配合使用 ······ 171
9.5 本章小结 ······ 173

第 10 章 让服务通信更安全可控 ······ 174

10.1 整体介绍 ······ 174
10.2 Denier 适配器 ······ 176
10.3 黑白名单 ······ 177
10.4 服务与身份认证 ······ 180
10.5 RBAC 访问控制 ······ 194
10.6 本章小结 ······ 205

第 11 章 让服务更易观测与监控 ······ 206

11.1 整体介绍 ······ 206

11.2 指标收集……209
11.3 日志收集……216
11.4 调用链追踪……224
11.5 服务指标可视化……230
11.6 服务调用树……235
11.7 本章小结……239

第12章 Istio维护……240

12.1 整体介绍……240
12.2 Istio服务网格仪表板……241
12.3 升级Istio……245
12.4 使用Helm定制部署Istio……253
12.5 故障排除……257
12.6 一个请求的完整过程分析……272
12.7 本章小结……282

第13章 杂项……283

13.1 CORS……284
13.2 URL重定向……287
13.3 URL重写……289
13.4 TCP路由……290
13.5 TLS路由……292
13.6 mTLS迁移……295
13.7 EnvoyFilter……297
13.8 添加请求头……299
13.9 在Gateway上使用HTTPS……300
13.10 在HTTPS服务上开启mTLS……304
13.11 网格中的服务健康检查……306
13.12 Envoy代理Ingressgateway……308
13.13 Mixer与Adapter模型……311
13.14 本章小结……312

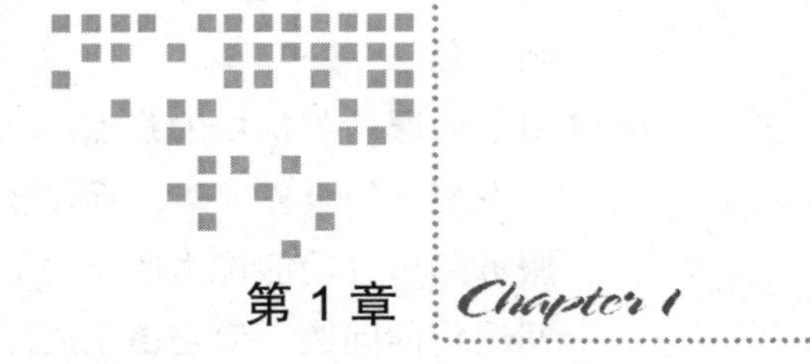

第 1 章 Chapter 1

服务网格与 Istio

本章介绍服务网格的由来，以及服务网格给服务开发部署带来什么样的变化，并介绍一个成熟的开源服务网格实现——Istio，这也是本书主要学习的服务网格。通过本章，我们可以了解 Istio 的架构设计，了解 Istio 实现了哪些服务网格功能，从而为后面的学习和实验打下基础。

1.1 服务网格简介

服务网格出现的大环境如下：

- **容器技术的广泛应用**。由于 Docker 的出现，容器技术得到更广泛的认可和应用，各种服务于容器的工具如雨后春笋般涌现，出现了众多容器部署、容器集群、容器编排等平台，例如 Swarm、Mesos、Kubernetes。由于容器具备轻量级、启动速度快、性能损失小、扩容缩容快、开发与生产环境统一等特性，越来越多的公司开始尝试使用容器来部署服务，容器技术的飞速发展也大大加速了微服务的应用。
- **微服务的快速流行**。随着近几年云计算的飞速发展，公有云也越来越成熟，微服务架构模式在大公司的兴起，特别是在 Netflix、亚马逊等公司的大规模实践，使得越来越多的公司开始尝试使用微服务架构来重构应用。当微服务的服务数量越来越大时，微服务间的服务通信也越来越重要，我们所看到的一个应用，有可能背后需要协调成百上千个微服务来处理用户的请求。随着服务数和服务实例数的不断增长，服务可能上线下线，服务实例也可能出现上线下线和宕机的情况，服务之间的通信变得异常复杂，每个服务都需要自己处理复杂的服务间通信。

- **目前微服务架构中的痛点**。面对复杂的服务间通信问题，一般的解决方案是为服务开发统一的服务框架，所有服务依赖于服务框架开发，所有服务间通信、服务注册、服务路由等功能都由底层服务框架来实现，这样做固然可以在某种程度上解决服务间通信的问题，但是由于底层服务框架的限制，业务人员可能无法基于实际情况选择合适的技术栈；由于所有服务都依赖于底层的服务框架代码库，当框架代码需要更新时，业务开发人员可能并不能立即更新服务框架，导致服务框架整体升级困难。后来 Netflix 开源了自己的微服务间通信组件，之后被 Spring Cloud 集成到了一起，组成了 Java 语言的通用微服务技术栈，而其他编程语言可能并没有如此强大功能的开源组件，只能继续饱受微服务间通信的各种痛。

基于以上服务间通信出现的问题，有人开始思考：能不能把服务间的复杂通信分层并下沉到基础设施层，让应用无感知呢？答案是肯定的。于是服务网格开始渐渐浮出水面，越来越多的人看到了服务网格的价值，尝试把服务网格应用于微服务实践中。

1.1.1 服务网格的概念与特点

服务网格（service mesh）这个概念来源于 Buoyant 公司的 CEO Willian Morgan 的文章“What’s a service mesh？ And why do I need one？”。服务网格是一个专注于处理服务间通信的基础设施层，它负责在现代云原生应用组成的复杂服务拓扑中可靠地传递请求。在实践中，服务网格通常是一组随着应用代码部署的轻量级网络代理，而应用不用感知它的存在。

服务网格的特点如下：

- 轻量级的网络代理。
- 应用无感知。
- 应用之间的流量由服务网格接管。
- 把服务间调用可能出现的超时、重试、监控、追踪等工作下沉到服务网格层处理。

服务网格的原理大致如图 1-1 所示，深色部分代表应用，浅灰色部分代表服务网格中轻量级的网络代理，代理之间可以相互通信，而应用之间的通信完全由代理来进行。如果只看代理部分，可以看到一个网状结构，服务网格由此得名。

服务网格一般由数据平面（data plane）和控制平面（control plane）组成，数据平面负责在服务中部署一个称为“边车”(sidecar) 的请求代理，控制平面负责请求代理之间的交互，以及用户与请求代理的交互。服务网格的基本架构如图 1-2 所示。

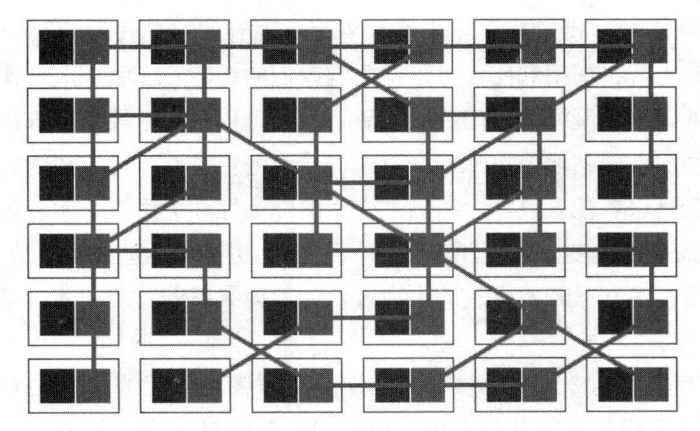

图 1-1　服务网格原理（图片来源：Pattern：Service Mesh）

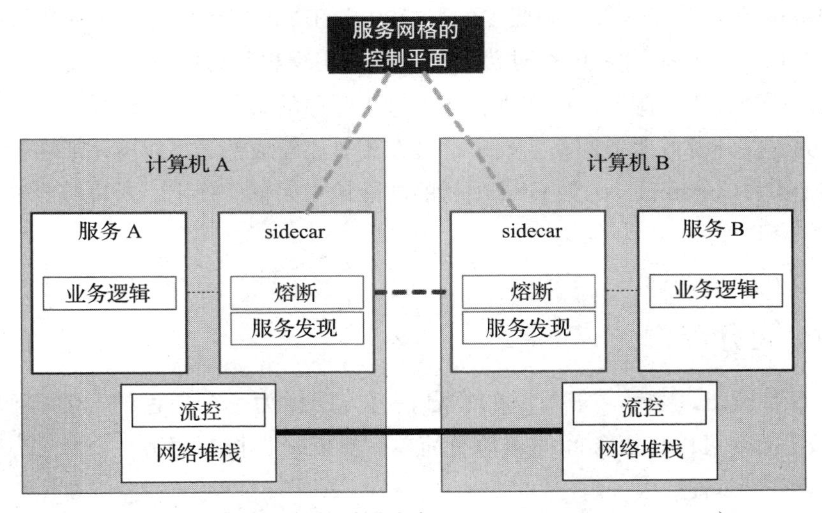

图 1-2　服务网格架构（图片来源：Pattern：Service Mesh）

1.1.2　服务网格的优势

微服务架构流行以后，服务的数量在不断增长。在不使用服务网格的情况下，每个服务都需要自己管理复杂的服务间网络通信，开发人员不得不使用各种库和框架来更好地处理服务间的复杂网络通信问题，这导致代码中包含很多与业务逻辑完全不相关的代码，稍有不慎就有可能给业务带来额外的复杂度和 bug。

当服务规模逐渐变大，复杂度增加，服务间的通信也变得越来越难理解和管理，这就要求服务治理包含很多功能，例如：服务发现、负载均衡、故障转移、服务度量指标收集和监控等。

在使用服务网格时，我们甚至完全不需要改动现有的服务代码，服务开发完全可以使用不同的语言和技术栈，框架和库再也不是限制我们的绊脚石。服务应用代码中将不再需要那些用于处理服务间复杂网络通信的底层代码，我们可以更好地控制请求的流量，对服务进行更好的路由，使服务间的通信更加安全可靠，让服务更具有弹性，还能让我们更好地观测服务，并可以提前给服务注入故障，以测试应用的健壮性和可用性。而拥有这些功能只需要我们的服务做出微小的改变，甚至不需要改变。以上提到的这些功能，在中小规模的公司中，使用服务网格技术，只需要少量的人力投入就能拥有以前大公司才具备的高级服务治理能力。

在云原生大行其道的今天，容器和Kubernetes增强了应用的伸缩能力，开发者可以基于此快速地创造出依赖关系复杂的应用；而使用服务网格，开发者不用关心应用的服务管理，只需要专注于业务逻辑的开发，这将赋予开发者更多的创造性。

既然服务网格能给我们带来如此多的好处，我们应该如何快速上手使用服务网格呢？从头开发一个服务网格平台不仅费时费力，最后的实现也很可能与开源实现的功能基本一致，不如直接选择开源的服务网格实现，或者基于开源的服务网格实现做二次开发以适应自己公司的业务。在开源的服务网格实现中，现在相对比较成熟、可以应用于生产环境的只有Istio和Linkerd2（Conduit）。因为Istio项目的功能最为完整、稳定，所以综合来看，选择使用Istio更为合理一些。

1.2 Istio简介

Istio出自希腊语，表示“航行”的意思，官方图标为一个白色的小帆船。使用Istio可以让服务间的通信更简单、更安全，控制服务更容易，观测服务更方便。

Istio是由Google、IBM、Lyft公司主导开发的影响力最大的开源服务网格实现，使用Go语言编码，由于Go语言的性能较好，使得Istio性能不错。这个项目由众多代码贡献者完成。

Istio能有效减少部署的复杂性，可以方便地将Istio应用到已有的分布式应用系统架构中，Istio包含的API可以方便地集成任何日志平台、监控平台、数据收集平台和访问策略系统。Istio有丰富的功能，可以帮助开发者更加高效地运行一个分布式微服务架构，并提供一个统一的方式来保护、连接、监控微服务。Istio提供了一个完整的解决方案，可以满足多样化的分布式微服务的需求。

1. Istio的主要功能特性

Istio可以让你轻松部署一个服务网格，而不需要在服务代码中做任何改变。只需要在你的环境中部署一个特殊的代理用来拦截所有微服务间的网络通信，就可以通过控制平面配置和管理 Istio。Istio 的功能特性如下：

- HTTP、gRPC、WebSocket、TCP 流量的自动负载均衡。
- 细粒度的流量路由控制，包含丰富的路由控制、重试、故障转移和故障注入。
- 可插拔的访问控制策略层，支持 ACL、请求速率限制和请求配额。
- 集群内度量指标，日志和调用链的自动收集，管理集群的入口、出口流量。
- 使用基于身份的认证和授权方式来管理服务间通信的安全。

由于 Istio 提供了足够多的可扩展性，这也使得 Istio 能满足多样化的需求。基于 Istio 你完全可以搭建出一套适合自己公司基础设施层的服务网格。

2. Istio 的设计目标

Istio 的架构设计中有几个关键目标，这些目标对于系统应对大规模流量和高性能地进行服务处理至关重要：

- **最大化透明：**为了让 Istio 被更广泛采用，运维和开发人员只需要付出很少的代价就可以从中受益。为此 Istio 将自己自动注入所有的网络路径的服务中。Istio 使用 Sidecar 代理来捕获流量，不需要对已部署的应用程序代码进行任何改动。在 Kubernetes 中，代理被注入 Pod 中，通过编写 iptables 规则来捕获流量。注入 Sidecar 代理到 Pod 中并且修改路由规则后，Istio 就能够拦截所有流量。这个原则也适用于性能，所有组件和 API 在设计时都必须考虑性能和规模。
- **增量：**随着运维人员和开发人员越来越依赖 Istio，系统必然会一起成长。Istio 会继续添加新功能，但是最重要的是扩展策略系统的能力，集成其他策略和控制来源，并将网格行为信号传播到其他系统进行分析。策略运行时支持标准扩展机制以便插入其他服务中。
- **可移植性：**Istio 必须支持以最小的代价在任何云和本机环境上运行。将 Istio 上的服务进行迁移也是可行的。
- **策略一致性：**在服务间的 API 调用中，策略的应用使得可以对网格间行为进行全面的控制，但对于不需要在 API 级别表达的资源来说，对资源应用策略也同样重要。例如，将配额应用到 ML 训练任务消耗的 CPU 数量上，比将配额应用到启动这个工作的调用上更为有用。因此，策略系统作为独特的服务来维护，具有自己的 API，而不是将其放到 Sidecar 代理中，这容许服务根据需要直接与其集成。

1.3　Istio 的架构设计

Istio 的架构设计在逻辑上分为数据平面和控制平面：

- **数据平面**由一系列称为“边车”(sidecar)的智能代理组成，这些代理通过 Mixer 来控制所有微服务间的网络通信，Mixer 是一个通用的策略和遥测中心。
- **控制平面**负责管理和配置代理来路由流量，另外，控制平面通过配置 Mixer 来实施策略与遥测数据收集。

Istio 的数据平面主要负责流量转发、策略实施与遥测数据上报；Istio 的控制平面主要负责接收用户配置生成路由规则、分发路由规则到代理、分发策略与遥测数据收集。

图 1-3 展示了 Istio 的架构。

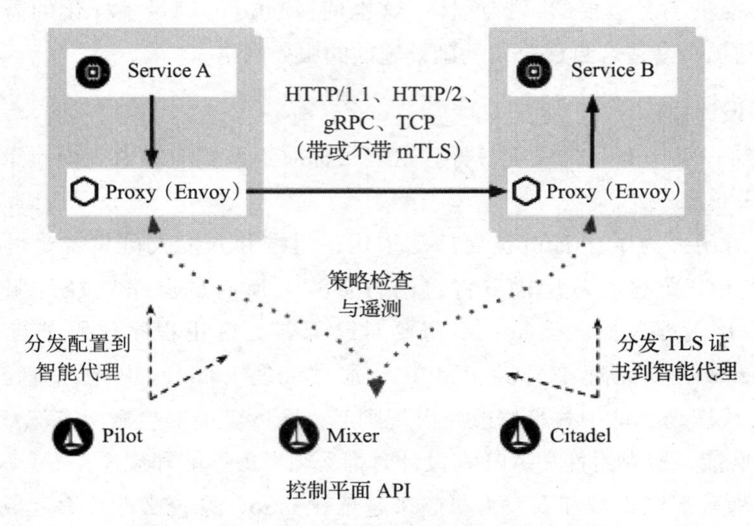

图 1-3　Istio 架构（图片来源：Istio 官方网站）

用户通过控制平面提供的 API 提交路由配置规则、策略配置规则与遥测数据收集的配置规则。Pilot 把用户提交的配置规则转换成智能代理需要的配置形式，推送给智能代理。智能代理根据用户的配置来执行服务路由、遥测数据收集与服务访问策略。智能代理拦截服务所有流量，并与其他智能代理通信。

1.3.1　数据平面

Istio 在数据平面中使用一个 Envoy 代理的扩展版本。Envoy 是使用 C++ 语言开发的高性能代理，它能拦截服务网络中所有服务的入口和出口流量。Istio 利用了众多 Envoy 内置的功能特性，例如：

- 动态服务发现
- 负载均衡
- TLS 终止

- HTTP/2 和 gRPC 代理
- 熔断器
- 健康检查
- 基于百分比流量分隔的灰度发布
- 故障注入
- 丰富的度量指标

Envoy 作为一个边车与对应的服务部署在同一个 Kubernetes Pod 中。这种部署方式使得 Istio 能提取丰富的流量行为信号作为属性。Istio 又可以反过来使用这些数据在 Mixer 中进行策略决策，并发送这些数据到监控系统中，提供整个网络中的行为信息。

采用边车部署方式，可以把 Istio 的功能添加到一个已经存在的部署中，并且不需要重新构建或者重新编写代码。

1.3.2　控制平面

控制平面中主要包括 Mixer、Pilot、Citadel 部件。

1. Mixer

Mixer 是一个与平台无关的组件。Mixer 负责在服务网络中实施访问控制和策略，并负责从 Envoy 代理和其他服务上收集遥测数据。代理提取请求级别的属性并发送到 Mixer 用于评估，评估请求是否能放行。

Mixer 有一个灵活的插件模型。这个模型使得 Istio 可以与多种主机环境和后端基础设施对接。因此，Istio 从这些细节中抽象了 Envoy 代理和 Istio 管理的服务。

Mixer 架构如图 1-4 所示。

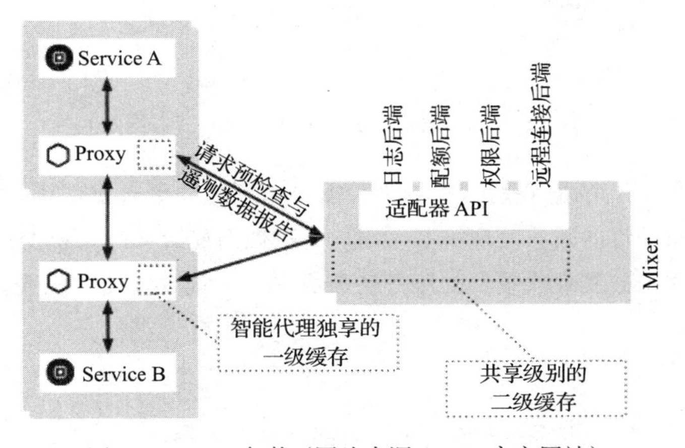

图 1-4　Mixer 架构（图片来源：Istio 官方网站）

在每一个请求过程中，Envoy代理会在请求之前调用Mixer组件进行前置条件检查，在请求结束之后上报遥测数据给Mixer组件。为了提高性能，每个Envoy代理都会提前缓存大量前置条件检查规则，当需要进行前置条件检查时，直接在缓存中检查规则。如果本地缓存中没有需要的规则，再去调用Mixer组件获取规则。Mixer组件也有自己的缓存，以加速前置条件检查。需要上报的遥测数据也会被Envoy代理暂时缓存起来，等待时机上报Mixer组件，从而减少上报数据的调用次数。

2. Pilot

Pilot为Envoy代理提供服务发现功能，并提供智能路由功能（例如：A/B测试、金丝雀发布等）和弹性功能（例如：超时、重试、熔断器等）。

Pilot将高级别的控制流量行为的路由策略转换为Envoy格式的配置形式，并在运行时分发给Envoy代理。Pilot抽象了平台相关的服务发现机制，并转换成Envoy数据平面支持的标准格式。这种松耦合设计使得Istio能运行在多平台环境，并保持一致的流量管理接口。

Pilot架构如图1-5所示。

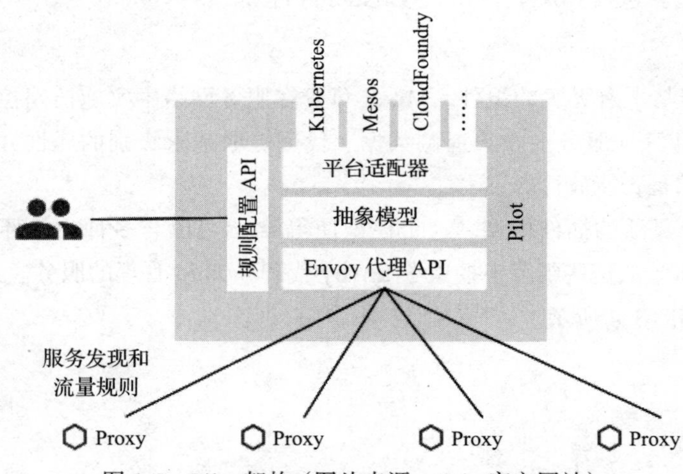

图1-5 Pilot架构（图片来源：Istio官方网站）

Pilot抽象不同平台的服务发现机制，只需要为不同的平台实现统一的抽象模型接口，即可实现对接不同平台的服务发现机制。用户通过规则配置API来提交配置规则，Pilot把用户配置的规则和服务发现收集到的服务转换成Envoy代理需要的配置格式，推送给每个Envoy代理。

3. Citadel

Citadel内置有身份和凭证管理，提供了强大的服务间和终端用户的认证。Citadel可以把不加密的通信升级为加密的通信。运维人员可以使用Citadel实施基于服务身份的策略而不用在网络层控制。现在，Istio还支持基于角色的访问控制，用于控制谁能够访问服务。

1.4　Istio的功能特性

1. 强大的流量管理

Istio提供了简单的规则配置和流量路由功能，用于控制服务间的流量流动和API调用。Istio简化了服务级别特性的配置，如熔断器、超时、重试，并且能更简单地配置一些复杂的功能，如A/B测试、金丝雀发布、基于百分比进行流量分隔的灰度发布等。

Istio提供了开箱即用的故障恢复功能，你可以在问题出现之前找到它，使服务调用更可靠，网络更加健壮。

2. 安全可靠

Istio提供了强大的安全功能，使得开发者不必再过分关注应用级别的安全问题。Istio提供了底层的安全通信，并且管理认证、授权和服务间通信加密。Istio对服务间通信默认加密以保证安全，在不同的协议和运行环境中也可使用统一的策略，只需要很少的修改或者完全不需要应用做出修改。

虽然Istio是平台无关的，但是当你在Kubernetes上使用Istio时，结合Kubernetes的网络策略，就可以在网络层实现Pod级别安全隔离，使用Istio提供的安全功能实现服务级别的安全通信。

3. 便捷的观测能力

Istio提供了强大的调用链跟踪、指标收集、日志收集和监控功能，使用者可以更深入地了解服务网格的部署情况和运行状态，可以真正了解服务性能如何影响上游和下游的功能，可以自定义仪表板对所有服务性能进行可视化管理，并了解该性能如何影响其他的应用程序。

所有这些功能都可以帮助你更高效地观测服务，增强服务的SLO（服务等级目标），最重要的是，可以帮助你更高效地找到服务的问题并修复。

4. 多平台支持

Istio的设计目标是平台无关，可以运行在多种环境，包括跨云环境、裸机环境、Kubernetes、Mesos等。到目前为止，官方支持如下的部署方式：

- 服务部署在Kubernetes上。
- 服务注册在Consul上。

❑ 服务运行在独立的虚拟机上。

虽然官方支持多种部署方式，但是目前有许多功能是基于 Kubernetes 的原生功能来实现的，Istio 在 Kubernetes 上实现的功能也是最多的，比如自动注入代理。综合来看，在 Kubernetes 上尝试使用 Istio 是目前最好的选择。

5. 易于集成和定制

Istio 具备扩展和定制功能，并可以与当前已经存在的解决方案集成，包括访问控制（ACL）、日志、监控、配额、认证等功能。

1.5 本章小结

Istio 是一个生产已经可用的开源服务网格平台实现，它提供了丰富的功能。虽然 Istio 是与平台无关的，但是由于许多功能是基于 Kubernetes 实现的，目前还是建议在 Kubernetes 上使用 Istio。

由于 Istio 实用性很强，直接上手实验是学习 Istio 的最好途径，所以在后续章节中，我们将搭建实验环境，逐一尝试 Istio 的强大功能。

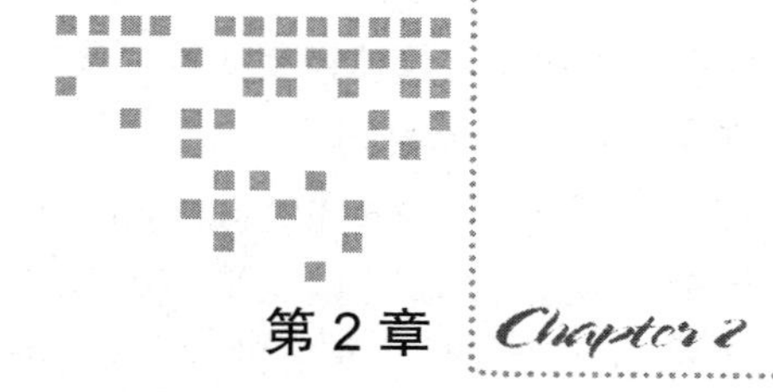

第 2 章 Chapter 2

实验说明

为了能更好地演示和学习 Istio，我们搭建了一个 Istio 的实验环境，以便后续章节进行实验。本章主要介绍实验的环境，以及在创建实验环境时需要注意的事项。实验中使用的应用是我用多种编程语言编写的简单样例程序，在本章中也会做详细的介绍，最后简单介绍一下应用的容器化镜像构建方法。

2.1 实验的环境

2.1.1 基础环境

在没有特殊说明的情况下，本书的实验是在 Windows 10 系统下进行的，在 MacOS 和 Linux 系统下也可以进行实验，并无太大差别。

考虑到不是每位读者都有足够的云主机或者物理机，本实验使用 3 台虚拟机来进行实验。虚拟机管理软件使用开源的 Virtualbox，为了方便进行重复的快速实验，我们会使用 Vagrant 配合 Virtualbox 来进行虚拟机的管理，包括创建、启动、关闭虚拟机以及虚拟机的快照保存恢复等操作。

本实验对硬件有一些基本要求，否则可能会出现实验无法成功的现象。对 CPU 并没有太多要求，由于我们实验时会创建 3 台虚拟机，每台虚拟机会分配至少 2G 内存，所以这就要求至少有 6G 的空闲内存，加上系统本身的内存占用，所以推荐在进行实验时，电脑至少有 8G 内存。一般的机械硬盘就能满足实验条件，当然如果使用 SSD 硬盘会更好一些。硬盘推荐至少保留30G以上可用空间，虚拟机并不会占用这么多存储空间，但是由于我们会保存多个快照来快速恢复实验环境，这些快照需要占用大量的存储空间，所以综合起来可能需要30G的存储空间。

关于Virtualbox和Vagrant的安装不再详细描述，具体安装细节可以参考官方文档或者通过搜索引擎获取安装文档。本书实验时所使用的Virtualbox和Vagrant软件安装包以及后面会使用的Vagrant box文件可以到如下地址下载：https://pan.baidu.com/s/1Q8s4mnhj2ROnUzW1ZTn44Q。

由于在Kubernetes上部署Istio更加方便，并且能体验到最全功能，所以本书的实验会依赖Kubernetes环境。后面的章节会详细介绍如何通过Kubeadm部署Kubernetes集群。

实验中还会涉及Git的基本使用，所以也需要提前在系统上安装好对应操作系统的Git软件。由于Windows系统下默认的命令行终端CMD并不好用，所以在Windows系统上我会使用Git Bash作为默认的命令行终端。Git Bash是Windows系统上安装好Git软件后就自带的。当然，你也可以根据自己的喜好选择终端软件。

实验时使用Xshell软件来应用SSH远程登录到实验环境的Linux虚拟机，免费家用版本提供在一个窗口中最多可以同时打开4个会话终端，已经能满足我们的实验需求。Xshell只提供了Windows系统版本的软件，安装时一直点击“下一步”按钮就可以完成安装，详细安装步骤请查阅相关文档。MacOS和Linux系统的用户可以直接使用系统自带的终端来应用SSH远程登录到实验环境的Linux虚拟机。

注意 默认情况下使用Virtualbox启动虚拟机时，虚拟机目录会存放在用户的主目录，如果Windows下C盘剩余空间过小，可能会由于硬盘空间不够，导致创建虚拟机失败。如果有需要，可以在命令行终端设置虚拟机的存放目录，先创建D:\virtualbox目录。如果提示找不到VBoxManage命令，可能需要把Virtualbox的安装目录添加到系统的PATH环境变量中。使用如下命令设置：

```
$ VBoxManage setproperty machinefolder D:\virtualbox
$ VBoxManage list systemproperties | grep machine
Default machine folder:          D:virtualbox
```

2.1.2 命令说明

由于实验使用Vagrant和Virtualbox来管理虚拟机集群环境。使用Vagrant创建的虚拟机一般情况下登录用户名为vagrant，当部署kubernetes集群或者其他安装软件的操作时，由于需要root权限，我们有时会直接切换到root用户来执行接下来的操作。当以vagrant用户登录虚拟机环境的Linux主机后，使用sudo su - root命令可临时切换到root用户来执行接下来的操作，这也可以从执行命令的提示符中看出，当以 vagrant 用户操作时，命令会以 $ 符号来开头；而当以 root 用户操作时，命令会以 # 符号来开头。如下所示：

```
$ whoami
vagrant

# whoami
root
```

本书所有实验中的命令执行操作如果没有特殊说明，普通用户的操作均是以 vagrant 用户执行。

本书实验内容的大部分代码和使用的 Kubernetes 以及 Istio 的 yaml 文件，都在 https://github.com/mgxian/istio-lab 仓库中。可使用如下方式获取源码：

```
$ git clone https://github.com/mgxian/istio-lab
```

如果没有安装 Git，需要提前安装 Git，可使用如下命令安装：

```
$ sudo yum install -y git
```

本书后面的大部分实验在执行 kubectl 和 istioctl 命令时，使用到的命令都是以仓库的根目录作为当前工作目录的，也就是说，当执行这两个命令时会先进入 istio-lab 目录，然后再执行相应的命令。

第 3 章之后（不包含第 3 章）的大部分实验操作，在没有特殊说明的情况下，都是在使用 Xshell 登录到实验环境虚拟机中后进行的操作，安装 Git 和下载实验中使用的源代码只需要在一台虚拟机上进行即可，一般会选择第一台虚拟机。只有管理虚拟机相关的操作命令比如：启动、停止、暂停虚拟机，才需要在 Windows 系统宿主机上进行操作。

2.1.3　问题及解决方案

由于实验虚拟机集群资源与性能的问题，可能会出现实验结果和书中展示的结果有所不同的情况，本小节主要介绍实验中可能会遇到的问题及解决方法。

1. 路由不生效

在进行服务路由等功能验证时，很有可能会出现路由不生效的问题。当创建路由后，由于机器性能问题，导致服务路由信息传播速度慢，如果立即访问测试，就很有可能会出现与书中展示的结果不同的现象，此时可以稍等片刻，让服务路由信息传播完成，再进行访问测试。或者删除路由规则，再重新创建。当然，你也可以重启 Pilot 组件，这通常是最快的解决方法。另外，也可以通过如下介绍的方式深入地排查问题。

创建路由后，可以通过如下命令查看路由的分发情况：

```
$ istioctl proxy-status
PROXY                                                       CDS         LDS
EDS               RDS          PILOT                        VERSION
dns-test.default                                            SYNCED      SYNCED
SYNCED (100%)     SYNCED       istio-pilot-64958c46fc-jsn48  1.0.2
istio-egressgateway-7dc5cbbc56-fvjxm.istio-system           SYNCED      SYNCED
SYNCED (100%)     NOT SENT     istio-pilot-64958c46fc-jsn48  1.0.2
istio-ingressgateway-7958d776b5-wwmx2.istio-system          SYNCED      SYNCED
SYNCED (100%)     SYNCED       istio-pilot-64958c46fc-jsn48  1.0.2
service-go-v1-7cc5c6f574-m7xtn.default                      SYNCED      SYNCED
SYNCED (100%)     SYNCED       istio-pilot-64958c46fc-jsn48  1.0.2
service-go-v2-7656dcc478-dk2sz.default                      SYNCED      SYNCED
SYNCED (100%)     SYNCED       istio-pilot-64958c46fc-jsn48  1.0.2
service-js-v1-55756d577-4vxsh.default                       SYNCED      SYNCED
SYNCED (100%)     SYNCED       istio-pilot-64958c46fc-jsn48  1.0.2
service-js-v2-86bdfc86d9-9gpcj.default                      SYNCED      SYNCED
SYNCED (100%)     SYNCED       istio-pilot-64958c46fc-jsn48  1.0.2
service-lua-v1-5c9bcb7778-zkgdm.default                     SYNCED      SYNCED
SYNCED (100%)     SYNCED       istio-pilot-64958c46fc-jsn48  1.0.2
service-lua-v2-75cb5cdf8-hzns6.default                      SYNCED      SYNCED
SYNCED (100%)     SYNCED       istio-pilot-64958c46fc-jsn48  1.0.2
service-node-v1-d44b9bf7b-4glm7.default                     SYNCED      SYNCED
SYNCED (100%)     SYNCED       istio-pilot-64958c46fc-jsn48  1.0.2
service-node-v2-86545d9796-nml56.default                    SYNCED      SYNCED
SYNCED (100%)     SYNCED       istio-pilot-64958c46fc-jsn48  1.0.2
service-python-v1-79fc5849fd-p7cgn.default                  SYNCED      SYNCED
SYNCED (100%)     SYNCED       istio-pilot-64958c46fc-jsn48  1.0.2
service-python-v2-7b6864b96b-7w6jw.default                  SYNCED      SYNCED
SYNCED (100%)     SYNCED       istio-pilot-64958c46fc-jsn48  1.0.2
```

当路由中的状态都为SYNCED，不存在Stale状态时，表明路由已经分发完成。此时再进行访问测试，一般情况下不会再出现问题。istio-egressgateway和istio-ingressgateway会有部分状态为NOT SENT，这是正常的，因为如果没有创建过Gateway，就不会发送RDS给istio-egressgateway和istio-ingressgateway，此时的状态就为NOT SENT。

此外，还可以通过查看Pod日志观察Envoy有无接收到最新的路由规则，可以通过如下方式查看日志：

```
$ INGRESS_GATEWAY_POD=$(kubectl get pod -n istio-system | grep istio-ingressgateway | awk '{print $1}')
$ kubectl logs -f $INGRESS_GATEWAY_POD -n istio-system
```

有时候也可能会出现使用istioctl proxy-status不能获取全部Pod的路由同步状态的情况，或者使用istioctl proxy-status获取到的状态都是正常状态，但路由仍然没有生效，此时可能是由于Pilot处于异常状态。可以使用如下的方式重启Pilot实例：

```
$ kubectl delete pod -n istio-system $(kubectl get pod -l app=pilot -n istio-system -o jsonpath='{.items[*].metadata.name}')
```

```
$ kubectl get pod -l app=pilot -n istio-system
NAME                          READY   STATUS    RESTARTS   AGE
istio-pilot-5fb59666cb-7jnt6  2/2     Running   0          37s
$ kubectl logs -f -n istio-system $(kubectl get pod -l app=pilot -n istio-system -o jsonpath='{.items[*].metadata.name}') discovery
```

当然，你也可以参考本书第 12 章中关于路由不生效的排错步骤来进行问题排查。

2. 应用路由规则时出现超时错误

在实验中创建路由规则时，无法成功创建或更新，出现如下的超时错误信息：

```
Error from server (Timeout): error when applying patch:
...
for: "istio/route/virtual-service-go-canary.yaml": Timeout: request did not complete within requested timeout 30s
```

这一般是由于 Istio 中的 Galley 组件出现了问题，使用如下命令重启 Galley 组件即可解决：

```
$ kubectl delete pod -n istio-system $(kubectl get pod -l istio=galley -n istio-system -o jsonpath='{.items[*].metadata.name}')
$ kubectl get pod -l istio=galley -n istio-system
NAME                            READY   STATUS    RESTARTS   AGE
istio-galley-545b6b8f5b-kkdgk   1/1     Running   0          9s
```

3. 自动注入失败

创建 Pod 时，提示如下的错误信息：

```
Error from server (InternalError): error when creating "kubernetes/dns-test.yaml": Internal error occurred: failed calling admission webhook "sidecar-injector.istio.io": Post https://istio-sidecar-injector.istio-system.svc:443/inject?timeout=30s: dial tcp 10.97.113.82:443: connect: connection refused
```

创建或扩容 Deployment 时，没有创建出对应数量的 Pod，查看 Deployment 对应的 ReplicaSet 信息，可以看到如下所示的错误信息：

```
$ kubectl describe rs $(kubectl get rs -l app=service-go,version=v1 -o jsonpath='{.items[*].metadata.name}')
Name:         service-go-v1-7cc5c6f574
Namespace:    default
Selector:     app=service-go,pod-template-hash=7cc5c6f574,version=v1
...
Events:
  Type     Reason        Age                  From                    Message
  ----     ------        ----                 ----                    -------
  Warning  FailedCreate  28s (x4 over 119s)   replicaset-controller   Error creating: Internal error occurred: failed calling admission webhook "sidecar-injector.istio.io": Post https://istio-sidecar-injector.istio-system.svc:443/inject?timeout=30s: net/http: request canceled (Client.Timeout exceeded while
```

```
awaiting headers)
        Warning   FailedCreate       17s                     replicaset-controller
Error creating: Internal error occurred: failed calling admission webhook
"sidecar-injector.istio.io": Post https://istio-sidecar-injector.istio-system.
svc:443/inject?timeout=30s: unexpected EOF
        Warning   FailedCreate        12s (x5 over 17s)     replicaset-controller
Error creating: Internal error occurred: failed calling admission webhook
"sidecar-injector.istio.io": Post https://istio-sidecar-injector.istio-system.
svc:443/inject?timeout=30s: dial tcp 10.97.113.82:443: connect: connection refused
```

这一般是由于Istio中的Sidecar-injector组件出现了问题，使用如下命令重启Sidecar-injector组件即可解决：

```
$ kubectl delete pod -n istio-system $(kubectl get pod -l istio=sidecar-
injector -n istio-system -o jsonpath='{.items[*].metadata.name}')
kubectl get pod -l istio=sidecar-injector -n istio-system
NAME                                      READY   STATUS    RESTARTS   AGE
istio-sidecar-injector-99b476b7b-sc8k5    1/1     Running   0          13s
```

注意 有时也可能碰到其他异常问题，比如：拉取镜像失败，可能是由于Virtualbox的nat网络出了问题，这些问题一般都无法快速解决，甚至没有办法解决，不用浪费太多时间在这些异常问题上。可以尝试使用虚拟机的快照功能，直接恢复虚拟机环境到创建好Istio集群的初始状态，再重新进行实验。

2.2 实验的应用

2.2.1 应用架构说明

为了充分展示Istio的功能，我们使用不同的语言来模拟数个微服务，服务之间存在相应的调用关系，服务之间通过HTTP协议通信。我们并没有写一个实际的综合应用，例如：购物网站、论坛等，来模拟生产环境的情况，我们只是简单地模拟服务间的调用关系来进行Istio相关的功能实验，目的是通过演示Istio相关功能来学习Istio。每个服务以其使用的编程语言为服务名，例如：使用Python语言编写的服务命名为service-python。各服务的调用关系如图2-1所示。

service-js服务是一个由Vue/React实现的前端应用，当用户访问前端Web页面时，用户会看到一个静态页面。当用户点击相应的按钮时，前端页面会通过浏览器异步请求后端service-python服务提供的API接口，service-python调用后端service-lua服务和service-node服务，而service-node服务又会调用service-go服务，最终，所有服务配合来完成用户的请求，并把结果合并处理之后发送给前端浏览器。当前端页面收到请求的响应数据时会渲染出新的页面呈现给用户。

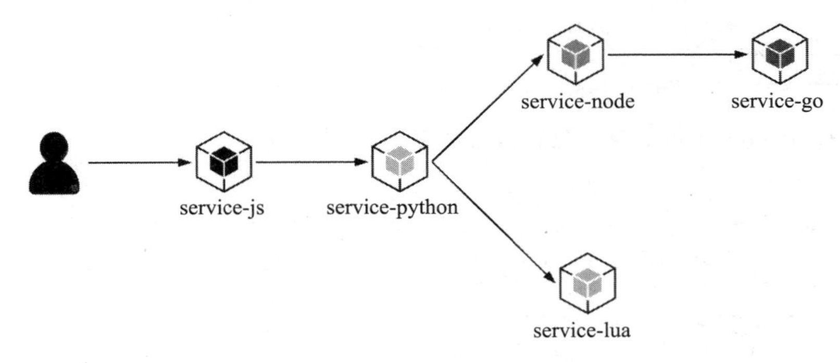

图 2-1　实验应用架构

应用架构说明：

- 本应用采用当前比较流行的前后端分离架构。
- 前端项目使用 Vue/React 实现。
- 前端调用 Python 实现的 API 接口。
- Python 服务调用后端 Node 实现的服务和 Lua 实现的服务。
- Node 服务调用 Go 实现的服务。

2.2.2　应用详细说明

1. service-js 服务

service-js 服务分别使用 Vue 和 React 各实现一套 Web 界面，主要用于服务路由中的 A/B 测试，可以让不同的终端用户看到不同的前端 Web 界面。service-js 服务主要负责根据 service-python 服务的响应数据，使用 ECharts 图表库在浏览器上展示出后端服务的具体调用关系和各个服务的调用耗时，具体的代码在实验源码根目录的 service/js 目录下。

v1 版本使用 React 框架实现，源码目录如下：

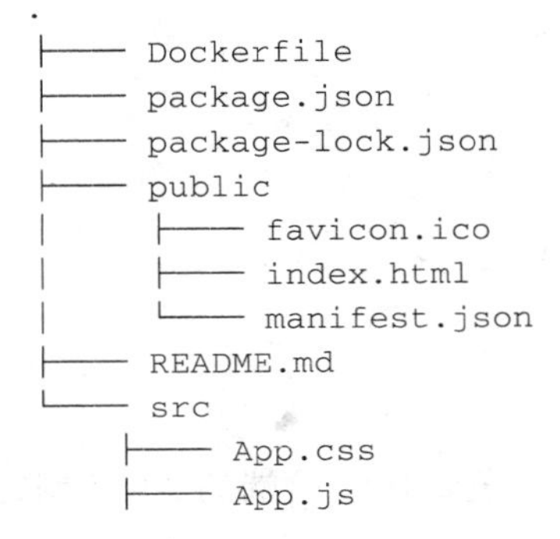

```
.
├── Dockerfile
├── package.json
├── package-lock.json
├── public
│   ├── favicon.ico
│   ├── index.html
│   └── manifest.json
├── README.md
└── src
    ├── App.css
    ├── App.js
```

```
├── App.test.js
├── index.css
├── index.js
├── logo.svg
└── registerServiceWorker.js
```

v2 版本使用 Vue 框架实现，源码目录如下：

```
.
├── build
│   ├── build.js
│   ├── check-versions.js
│   ├── logo.png
│   ├── utils.js
│   ├── vue-loader.conf.js
│   ├── webpack.base.conf.js
│   ├── webpack.dev.conf.js
│   └── webpack.prod.conf.js
├── config
│   ├── dev.env.js
│   ├── index.js
│   └── prod.env.js
├── Dockerfile
├── index.html
├── package.json
├── package-lock.json
├── README.md
├── src
│   ├── App.vue
│   ├── assets
│   │   └── logo.png
│   ├── components
│   │   └── HelloWorld.vue
│   └── main.js
└── static
```

用于容器化的 Dockerfile 文件如下所示：

```
FROM node:8-alpine
LABEL maintainer="will835559313@163.com"
COPY . /app
WORKDIR /app
RUN npm i && npm run build \
    && rm -rf ./node_modules \
    && npm install -g serve
EXPOSE 80
CMD ["serve", "-s", "build", "-p", "80"]
```

2. service-python 服务

service-python 服务是一个用 Python 编写的 API 服务，负责接收前端的 API 请求，调用整合后端其他服务的响应数据，返回给前端使用。service-python 服务分别使用 Python2 和 Python3 实现了两个版本的服务，具体代码在实验源码根目录的 service/python 目录下。service-python 服务使用 Flask 框架实现，具体源码如下：

```
 1 import time
 2 import requests
 3 from functools import partial
 4 from flask import Flask, jsonify, g, request
 5 from multiprocessing.dummy import Pool as ThreadPool
 6
 7
 8 app = Flask(__name__)
 9
10
11 def getForwardHeaders(request):
12     headers = {}
13     incoming_headers = [
14         'x-request-id',
15         'x-b3-traceid',
16         'x-b3-spanid',
17         'x-b3-parentspanid',
18         'x-b3-sampled',
19         'x-b3-flags',
20         'x-ot-span-context'
21     ]
22
23     for ihdr in incoming_headers:
24         val = request.headers.get(ihdr)
25         if val is not None:
26             headers[ihdr] = val
27
28     return headers
29
30
31 @app.before_request
32 def before_request():
33     g.forwardHeaders = getForwardHeaders(request)
34
35
36 def get_url_response(url, headers={}):
37     try:
38         start = time.time()
39         resp = requests.get(url, headers=headers, timeout=20)
40         response_time = round(time.time() - start, 2)
41         data = resp.json()
42         data['response_time'] = response_time
43     except Exception as e:
44         print(e)
45         data = None
46     return data
```

```
47
48
49 @app.route("/env")
50 def env():
51     service_lua_url = 'http://' + 'service-lua' + '/env'
52     service_node_url = 'http://' + 'service-node' + '/env'
53
54     services_url = [service_lua_url, service_node_url]
55     pool = ThreadPool(2)
56     wrap_get_url_response = partial(get_url_response, headers=g.forwardHeaders)
57     results = pool.map(wrap_get_url_response, services_url)
58     upstream = [r for r in results if r]
59
60     return jsonify({
61         "message": 'python v1',
62         "upstream": upstream
63     })
64
65
66 @app.route("/status")
67 def status():
68     return "ok"
69
70
71 if __name__ == '__main__':
72     app.run(host='0.0.0.0', port=80)
```

第 11 ～ 28 行定义的 getForwardHeaders 函数是为了从请求中提取出用于 Istio 调用链追踪的头信息，用于传递给 service-python 要调用的其他后端服务。

第 31 ～ 33 行表示每个请求在被处理前，调用 getForwardHeaders 函数，从请求中提取出 Istio 调用链追踪的头信息，并保存到全局对象 g 的 forwardHeaders 变量中。

第 36 ～ 46 行定义了用于获取后端服务响应数据的 get_url_response 函数。

第 49 ～ 63 行定义了真正的业务路由，使用线程池的方式并发请求后端服务，并把后端服务的响应数据组合处理后返回给调用方。

第 66 ～ 68 行定义了用于服务健康检查的路由。

第 71 ～ 72 行表示服务启动在 0.0.0.0 地址的 80 端口。

v1 版本和 v2 版本源码只有第 61 行有略微差别，在 v2 版本中 "python v1" 修改为 "python v2"。

用于容器化的 Dockerfile 文件如下所示：

```
FROM python:2-alpine
LABEL maintainer="will835559313@163.com"
COPY . /app
WORKDIR /app
```

```
RUN pip install -r requirements.txt
CMD [ "python", "main.py" ]
```

v1 版本和 v2 版本的 Dockerfile 只有使用的基础镜像版本不同，其他保持一致。

3. service-lua 服务

service-lua 服务使用 OpenResty 的不同版本用 Lua 语言分别实现了两个版本的服务。具体的代码在实验源码根目录的 service/lua 目录下。源代码如下：

```
 1 worker_processes  4;
 2 error_log logs/error.log;
 3 events {
 4     worker_connections 10240;
 5 }
 6 http {
 7     server {
 8         listen 80;
 9         location / {
10             default_type text/html;
11             content_by_lua '
12                 ngx.say("hello, world")
13             ';
14         }
15
16         location = /status {
17             default_type text/html;
18             content_by_lua '
19                 ngx.say("ok")
20             ';
21         }
22
23         location = /env {
24             charset utf-8;
25             charset_types application/json;
26             default_type application/json;
27             content_by_lua '
28                 json = require "cjson"
29                 ngx.status = ngx.HTTP_OK
30                 version = "lua v1"
31                 data = {
32                     message = version
33                 }
34                 ngx.say(json.encode(data))
35                 return ngx.exit(ngx.HTTP_OK)
36             ';
37         }
38     }
39 }
```

第 8 行表示服务启动在 0.0.0.0 地址的 80 端口。

第 9 ～ 14 行定义了访问服务的 / 链接时返回 "hello world"。

第 16 ～ 21 行定义了用于服务健康检查的 /status 链接。

第 23 ～ 37 行定义了真正的业务逻辑，当访问 /env 链接时，响应服务的版本信息。

v1 版本和 v2 版本源码只有第 30 行有略微差别，在 v2 版本中 "lua v1" 修改为 "lua v2"。

用于容器化的 Dockerfile 文件如下所示：

```
FROM openresty/openresty:1.11.2.5-alpine
LABEL maintainer="will835559313@163.com"
COPY . /app
WORKDIR /app
EXPOSE 80
ENTRYPOINT ["/usr/local/openresty/bin/openresty", "-c", "/app/nginx.conf",
"-g", "daemon off;"]
```

v1 版本和 v2 版本的 Dockerfile 只有使用的基础镜像版本不同，其他保持一致。

4. service-node 服务

service-node 服务使用 Node 的不同版本分别实现了两个版本的服务。具体的代码在实验源码根目录的 service/node 目录下。源代码如下：

```
 1 const Koa = require('koa');
 2 const Router = require('koa-router');
 3 const axios = require('axios')
 4 const app = new Koa();
 5 const router = new Router();
 6
 7
 8 function getForwardHeaders(request) {
 9     headers = {}
10     incoming_headers = [
11         'x-request-id',
12         'x-b3-traceid',
13         'x-b3-spanid',
14         'x-b3-parentspanid',
15         'x-b3-sampled',
16         'x-b3-flags',
17         'x-ot-span-context'
18     ]
19
20     for (idx in incoming_headers) {
21         ihdr = incoming_headers[idx]
22         val = request.headers[ihdr]
23         if (val !== undefined && val !== '') {
24             headers[ihdr] = val
25         }
26     }
27     return headers
28 }
```

```
29
30
31
32 router.get('/status', async (ctx, next) => {
33     ctx.body = 'ok';
34 })
35
36 router.get('/env', async (ctx, next) => {
37     forwardHeaders = getForwardHeaders(ctx.request)
38     service_go_url = 'http://' + 'service-go' + '/env'
39     upstream_ret = null
40     try {
41         let start = Date.now()
42         const response = await axios.get(service_go_url, {
43             headers: forwardHeaders,
44             timeout: 20000
45         });
46         response_time = ((Date.now() - start) / 1000).toFixed(2)
47         upstream_ret = response.data
48         upstream_ret.response_time = response_time
49     } catch (error) {
50         console.error('error');
51     }
52     if (upstream_ret) {
53         ctx.body = {
54             'message': 'node v1',
55             'upstream': [upstream_ret]
56         };
57     } else {
58         ctx.body = {
59             'message': 'node v1',
60             'upstream': []
61         }
62     }
63 })
64
65 app.use(router.routes()).use(router.allowedMethods());
66 app.listen(80);
```

第 8 ～ 28 行定义的 getForwardHeaders 函数是为了从请求中提取出用于 Istio 调用链追踪的头信息，用于传递给 service-node 要调用的其他后端服务。

第 32 ～ 34 行定义了用于服务简单健康检查的路由。

第 36 ～ 63 行定义了真正的业务路由，请求后端服务，并把后端服务的响应数据组合处理后返回给调用方。

第 65 ～ 66 行表示服务启动在 0.0.0.0 地址的 80 端口。

v1 版本和 v2 版本源码只有第 54 和 59 行有略微差别，在 v2 版本中 "node v1" 修改 "node v2"。

用于容器化的Dockerfile文件如下：

```
FROM node:8-alpine
LABEL maintainer="will835559313@163.com"
COPY . /app
WORKDIR /app
RUN npm i
EXPOSE 80
CMD ["node", "main.js"]
```

v1版本和v2版本的Dockerfile只有使用的基础镜像版本不同，其他保持一致。

5. service-go服务

service-go服务使用Go语言的不同版本分别实现了两个版本的服务，具体的代码在实验源码根目录的service/go目录下。源代码如下：

```
 1 package main
 2
 3 import (
 4     "github.com/gin-gonic/gin"
 5 )
 6
 7 func main() {
 8     r := gin.Default()
 9
10     r.GET("/env", func(c *gin.Context) {
11         c.JSON(200, gin.H{
12             "message": "go v1",
13         })
14     })
15
16     r.GET("/status", func(c *gin.Context) {
17         c.String(200, "ok")
18     })
19
20     r.Run(":80")
21 }
```

第10～14行定义了真正的业务路由，以及返回服务的版本信息。

第16～18行定义了用于服务健康检查的路由。

第20行表示服务启动在0.0.0.0地址的80端口。

v1版本和v2版本源码只有第12行有略微差别，在v2版本中"go v1"修改为"go v2"。

用于容器化的Dockerfile文件如下：

```
FROM golang:1.10-alpine as builder
LABEL maintainer="will835559313@163.com"
COPY . /app
```

```
WORKDIR /app
RUN apk update && apk add git \
    && go get github.com/gin-gonic/gin \
    && go build

FROM alpine:latest
WORKDIR /app
COPY --from=builder /app/app .
EXPOSE 80
CMD ["./app"]
```

你可能已经注意到，上面的 Dockerfile 代码使用了两次 FROM 关键词和一个不太一样的 COPY 用法，这体现了 Docker 镜像的多阶段构建功能，可以在一个镜像中编译代码，然后复制编译后的产物到另一个镜像中，这样可以非常有效地减小应用的 Docker 镜像大小，特别是 Go、Java 这类编译型静态语言，因为这类语言编译之后就不再需要原来编译时的依赖库，可以把编译后的产物直接放在一个极小运行环境中启动运行。由于 Go 语言可以编译为在操作系统上直接运行的二进制文件，所以可以把编译后的文件直接复制到 alpine 这类极简的操作系统镜像中，这种优化使得 service-go 服务编译构建后的镜像体积可缩小到 10M 级别，进而使服务镜像的分发效率大幅度提升。

v1 版本和 v2 版本的 Dockerfile 只有使用的基础镜像版本不同，其他保持一致。

6. service-redis 服务

service-redis 服务是使用 Go 语言实现的服务，用于从 Redis 服务器获取信息，具体的代码在实验源码根目录的 service/redis 目录下。源代码如下：

```
 1 package main
 2
 3 import (
 4     "log"
 5
 6     "github.com/gin-gonic/gin"
 7     "github.com/go-redis/redis"
 8 )
 9
10 // NewClient new redis client
11 func NewClient() *redis.Client {
12     client := redis.NewClient(&redis.Options{
13         Addr:     "redis:6379",
14         Password: "",
15         DB:       0,
16     })
17     return client
18 }
19
20 func main() {
```

```
21     r := gin.Default()
22     client := NewClient()
23     r.GET("/env", func(c *gin.Context) {
24         val, err := client.Info().Result()
25         if err != nil {
26             log.Print(err)
27         }
28         c.String(200, val)
29     })
30     r.GET("/status", func(c *gin.Context) {
31         c.String(200, "ok")
32     })
33     r.Run(":80")
34 }
```

第23～29行定义了真正的业务路由，以及返回Redis服务器的信息。

第30～32行定义了用于服务健康检查的路由。

第33行表示服务启动在0.0.0.0地址的80端口。

用于容器化的Dockerfile文件如下：

```
FROM golang:1.11-alpine as builder
LABEL maintainer="will835559313@163.com"
COPY . /app
WORKDIR /app
RUN apk update && apk add git \
    && go get github.com/gin-gonic/gin \
    && go build

FROM alpine:latest
WORKDIR /app
COPY --from=builder /app/app .
EXPOSE 80
CMD ["./app"]
```

7. httpbin服务

httpbin服务是一个用于HTTP测试的开源服务[⊖]。它既提供了在线的测试服务，也可以通过源码或者使用Docker镜像在本地部署运行，这两种使用方式在后续章节的实验中都有涉及。

2.3 应用的构建

如果只是跟着本书做实验，本节可以跳过，本节属于应用镜像构建部分，不会影响后面的Istio实验。了解本节内容需要掌握Docker的基础知识。

由于本书的实验重点在于如何使用Istio，如何使用本地Docker私有镜像仓库部署服务

⊖ 详细使用方法可参考官方文档https://httpbin.org/。

并不是我们关注的重点，所以本书实验所使用的镜像均采用阿里云镜像服务免费提供的镜像构建功能。当然，你也可以使用 Docker Hub 提供的镜像构建功能。下面以阿里云镜像服务为例实现镜像构建，具体步骤如下。

1. 构建应用镜像

（1）上传代码到 GitHub

此步骤不做详细说明，请参考相关文档了解 Git 和 GitHub 的基本使用。

（2）在阿里云上的镜像构建

阿里云镜像服务地址：https://cr.console.aliyun.com/，如下操作均在这个链接的 Web 上进行。

1）创建命名空间。

命名空间不能重复，请注意修改命名空间名称，如图 2-2 所示，可创建名为 istio-lab 的命名空间。

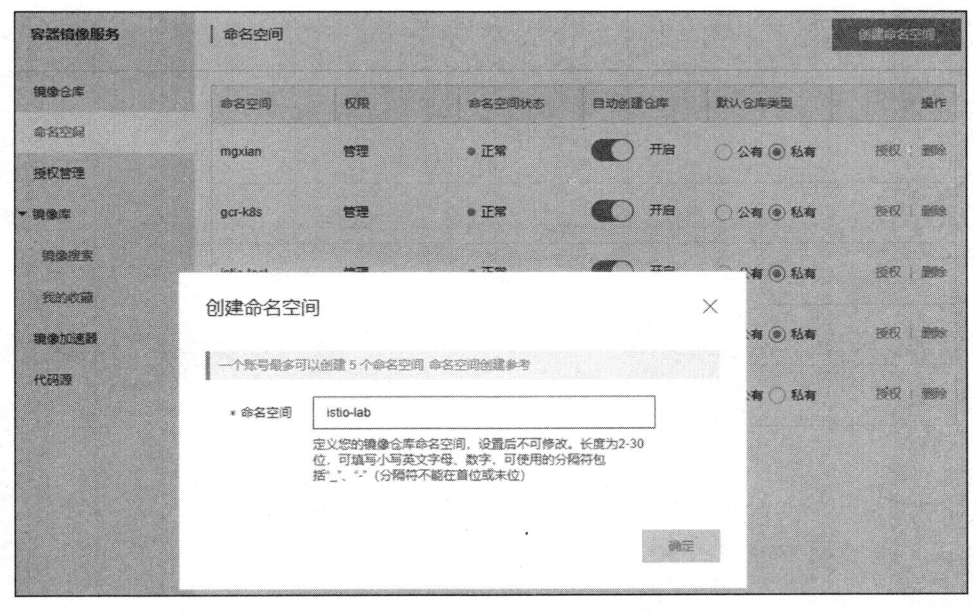

图 2-2　创建命名空间

2）创建镜像仓库。

选择命名空间为上个步骤中创建的命名空间，填写要创建的镜像仓库名和简介，如 service-go。注意选择仓库类型为公开，如图 2-3 所示。

选择已经绑定的 GitHub 账号和要构建的源码仓库，由于构建时需要访问国外资源，因此勾选使用海外机器构建，并取消“选择代码变更时自动构建镜像”（如果首次使用需要先绑定 GitHub 账号），如图 2-4 所示。

图 2-3 创建镜像仓库（一）

图 2-4 创建镜像仓库（二）

3）添加镜像构建规则。

选择要构建规则的镜像仓库，并点击“管理”按钮，出现如图 2-5 所示的界面。

容器镜像服务 | 镜像仓库 | 设置Registry登录密码 | 创建镜像仓库

镜像仓库　命名空间　授权管理　镜像库　镜像搜索　我的收藏　镜像加速器　代码源

istio-lab　仓库名称

容器镜像服务已停止控制台授权功能，请迁移至RAM控制台设置子账号权限，具体参考《RAM权限管理文档》。

仓库名称	命名空间	仓库状态	仓库类型	权限	仓库地址	创建时间	操作
apiextensions-ca-helper	istio-lab	正常	公开	管理		2019-01-11 16:01:04	管理 | 删除
cert-manager-controller	istio-lab	正常	公开	管理		2019-01-11 17:32:56	管理 | 删除
cert-manager-webhook	istio-lab	正常	公开	管理		2019-01-11 16:00:34	管理 | 删除
envoy	istio-lab	正常	公开	管理		2019-01-09 09:51:30	管理 | 删除
service-go	istio-lab	正常	公开	管理		2018-11-25 16:43:09	管理 | 删除

图 2-5　添加镜像构建规则（一）

选择“构建”选项卡上的“添加规则”按钮，添加镜像构建规则，如图 2-6 所示。

图 2-6　添加镜像构建规则（二）

添加如图 2-7 所示的版本构建规则，实验中会使用两个版本，因此需要创建两个版本的构建规则。

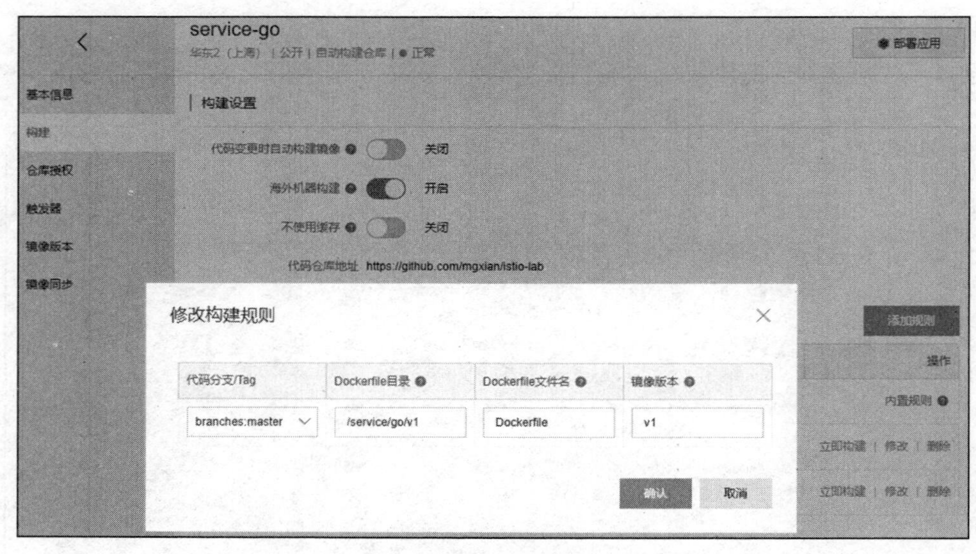

图 2-7　添加镜像构建规则（三）

4）构建镜像。

点击“立即构建”按钮开始构建，如果构建失败，可以通过点击日志链接查看构建日志，找出失败原因，如图 2-8 所示。

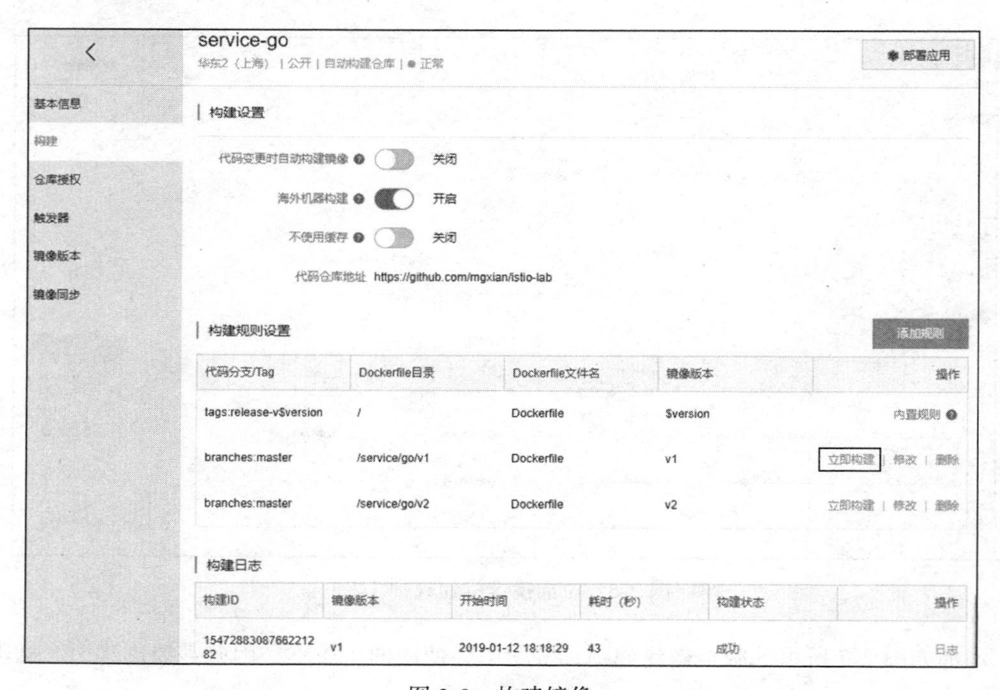

图 2-8　构建镜像

2. 本地拉取镜像验证

镜像构建完成后，拉取到本地测试镜像是否能正常工作。具体步骤如下。

1）拉取镜像：

```
$ sudo docker pull registry.cn-shanghai.aliyuncs.com/istio-lab/service-go:v1
v1: Pulling from istio-lab/service-go
4fe2ade4980c: Already exists
7eb00a8eb80c: Already exists
1ed92905e9ab: Pull complete
Digest: sha256:fffe0f892ed657952afcc4bd07216fe5e30e27fae014a3c81220368cccb
bd161
Status: Downloaded newer image for registry.cn-shanghai.aliyuncs.com/istio-
lab/service-go:v1
```

2）启动容器：

```
$ sudo docker run -d --name service-go-v1 \
-p 8000:80 registry.cn-shanghai.aliyuncs.com/istio-lab/service-go:v1
e936daf6ccdcdcca811369fe19f6163559a4e796f6971c063cc83a05e386c2fe

$ sudo docker ps | grep service-go-v1
e936daf6ccdc        registry.cn-shanghai.aliyuncs.com/istio-lab/service-go:v1
"./app"             13 seconds ago      Up 12 seconds       0.0.0.0:8000->80/
tcp   service-go-v1
```

3）访问测试：

```
$ curl 127.0.0.1:8000/env
{"message":"go v1"}
```

4）清理：

```
$ sudo docker stop service-go-v1
service-go-v1

$ sudo docker rm service-go-v1
service-go-v1
```

2.4 本章小结

要掌握 Istio，我们需要一步一步地实验 Istio 的功能特性，所以我们先创建一个实验环境，而实验时最有可能遇到的问题就是实验环境的机器性能不足，进而出现 Istio 路由功能生效缓慢，甚至导致后续实验失败的问题，这也是我们需要特别关注的。我也在本章中说明了这种情况，并介绍了几种解决方法。为了更好地模拟生产环境中多服务调用的场景，我使用多种编程语言实现了多个服务的不同版本，这也能帮助我们在后续章节更好地展示 Istio 的相关功能特性。

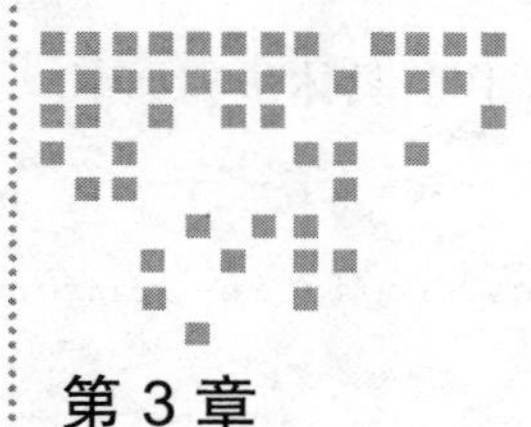

Chapter 3 第3章

使用 Vagrant 管理虚拟机

本章主要介绍如何使用 Vagrant 管理虚拟机，为了保持实验环境的一致性，并能快速创建恢复实验环境，本章会介绍 Vagrant 的简单使用方式，并模拟实验时的操作步骤来使用 Vagrant。借助 Vagrant，我们可以更方便地管理 Istio 实验环境，这种快速创建实验环境的能力大大地减少了我们实验前的环境准备时间。

3.1 Vagrant 简介

Vagrant 是一款用于构建及配置管理虚拟机环境的开源软件，使用 Ruby 语言开发，主要以命令行的方式运行。Vagrant 使用 Oracle 的开源 VirtualBox 虚拟化系统，与 Chef、Salt、Puppet 等环境配置管理软件搭配使用，使用方便且自动化程度很高。使用 Vagrant 可以快速构建出虚拟机环境，这将大大降低开发者创建新开发环境的难度，并减少开发时间。Vagrant 还支持创建虚拟机集群，可以快速搭建出一个用于开发和实验的虚拟机集群环境。虽然 Vagrant 可用于云环境的虚拟机管理，但是生产环境应用不多，仍以开发环境使用场景居多。

使用 Vagrant 很简单，只需要配置好文件，就可以快速重建出一个可移植的工作环境，而这个功能正好可以帮助我们快速重建出一个虚拟机实现环境，帮助我们快速上手 Istio，而不用于陷于实验前期的环境准备过程中可能遇到的各种棘手问题。基于这个理由，本书选择了使用 Vagrant 和 Virtualbox 快速搭建实验环境。当然，如果你对 Vagrant 感兴趣，想进一步了解 Vagrant 的使用方法，可以查阅 Vagrant 的官方文档。

3.2　Vagrant 常用命令

注意事项： 由于默认情况下，Vagrant 会把临时文件和 Box 文件存放在用户主目录里。如果 Box 文件过大，会导致占用过大空间，可以通过设置环境变量 VAGRANT_HOME 来设定 Vagrant 的主目录路径。Vagrant 默认设置的主目录为用户主目录的 .vagrant.d 文件夹。本次实验时把此环境变量设置为 D:\vagrant\home，关于各个操作系统的环境变量的设置，请查阅相关文档。

1. 基本命令

基础命令总览：

- box add：导入 box
- box list：查看 box
- box remove：删除 box
- init：初始化
- up：启动
- status：查看状态
- ssh：SSH 连接
- reload：重载
- halt：关闭
- suspend：暂停
- destroy：删除

下面举例说明这些命令的使用方法。

（1）导入 Box

导入已经下载的 Box 命令如下：

```
$ vagrant box add --name centos-7.4-base /f/vagrant/box/centos-7.4-base.box
==> box: Box file was not detected as metadata. Adding it directly...
==> box: Adding box 'centos-7.4-base' (v0) for provider:
    box: Unpacking necessary files from: file://F:/vagrant/box/centos-7.4-base.box
    box:
==> box: Successfully added box 'centos-7.4-base' (v0) for 'virtualbox'!
```

/f/vagrant/box/centos-7.4-base.box 表示 Box 的路径地址，这是由于使用了 Git Bash 的路径表示方法。如果使用 CMD 命令行，使用 F:/vagrant/box/centos-7.4-base.box 即可。此处使用的 Box 从第 2 章“实验说明”中提供的百度云盘链接上下载。

（2）查看 Box

查看可用的 Box 如下所示：

```
$ vagrant box list
centos-7.4-base      (virtualbox, 0)
```

（3）删除 Box

删除不再使用的 Box 如下所示：

```
$ vagrant box remove centos-7.4-base
Removing box 'centos-7.4-base' (v0) with provider 'virtualbox'...
```

（4）初始化

初始化虚拟机如下所示：

```
$ mkdir test
$ cd test
$ vagrant init centos-7.4-base
A 'Vagrantfile' has been placed in this directory. You are now
ready to 'vagrant up' your first virtual environment! Please read
the comments in the Vagrantfile as well as documentation on
'vagrantup.com' for more information on using Vagrant.
```

查看未被注释配置文件内容：

```
$ cat Vagrantfile | tr -d '\r' | egrep -v '^#|^ +#|^$'
Vagrant.configure("2") do |config|
    config.vm.box = "centos-7.4-base"
end
```

（5）启动

启动虚拟机如下所示：

```
$ vagrant up
Bringing machine 'default' up with 'virtualbox' provider...
==> default: Importing base box 'centos-7.4-base'...
==> default: Matching MAC address for NAT networking...
==> default: Setting the name of the VM: test_default_1543216915133_85015
==> default: Clearing any previously set network interfaces...
==> default: Preparing network interfaces based on configuration...
    default: Adapter 1: nat
==> default: Forwarding ports...
    default: 22 (guest) => 2222 (host) (adapter 1)
==> default: Booting VM...
==> default: Waiting for machine to boot. This may take a few minutes...
    default: SSH address: 127.0.0.1:2222
    default: SSH username: vagrant
    default: SSH auth method: private key
    default:
    default: Vagrant insecure key detected. Vagrant will automatically replace
    default: this with a newly generated keypair for better security.
    default:
```

```
    default: Inserting generated public key within guest...
    default: Removing insecure key from the guest if it's present...
    default: Key inserted! Disconnecting and reconnecting using new SSH key...
==> default: Machine booted and ready!
==> default: Checking for guest additions in VM...
    default: No guest additions were detected on the base box for this VM! Guest
    default: additions are required for forwarded ports, shared folders, host only
    default: networking, and more. If SSH fails on this machine, please install
    default: the guest additions and repackage the box to continue.
    default:
    default: This is not an error message; everything may continue to work properly,
    default: in which case you may ignore this message.
```

（6）查看状态

查看虚拟机状态如下所示：

```
$ vagrant status
Current machine states:

default                   running (virtualbox)

The VM is running. To stop this VM, you can run 'vagrant halt' to
shut it down forcefully, or you can run 'vagrant suspend' to simply
suspend the virtual machine. In either case, to restart it again,
simply run 'vagrant up'.
```

（7）SSH连接

注意，如果Windows下使用Git Bash时无法使用SSH连接虚拟机，可以尝试使用系统自己带的命令行工具CMD连接虚拟机：

```
$ vagrant ssh
Last login: Tue Dec 12 16:01:23 2017 from 10.0.2.2
[vagrant@localhost ~]$ hostname
localhost.localdomain
[vagrant@localhost ~]$ ip a
1: lo: <LOOPBACK,UP,LOWER_UP> mtu 65536 qdisc noqueue state UNKNOWN qlen 1
    link/loopback 00:00:00:00:00:00 brd 00:00:00:00:00:00
    inet 127.0.0.1/8 scope host lo
       valid_lft forever preferred_lft forever
    inet6 ::1/128 scope host
       valid_lft forever preferred_lft forever
2: eth0: <BROADCAST,MULTICAST,UP,LOWER_UP> mtu 1500 qdisc pfifo_fast state UP
qlen 1000
    link/ether 52:54:00:ca:e4:8b brd ff:ff:ff:ff:ff:ff
    inet 10.0.2.15/24 brd 10.0.2.255 scope global dynamic eth0
       valid_lft 86194sec preferred_lft 86194sec
    inet6 fe80::5054:ff:feca:e48b/64 scope link
       valid_lft forever preferred_lft forever
[vagrant@localhost ~]$ exit;
```

```
logout
Connection to 127.0.0.1 closed.
```

（8）重新

当我们编辑当前文件夹下的虚拟机配置文件 Vagrantfile 后，可以使用 reload 命令重载虚拟机，使配置生效。比如添加设置主机名的配置：

```
...
config.vm.box = "centos-7.4-base"
config.vm.hostname = "istio"
...
```

重载使配置文件生效：

```
$ vagrant reload
==> default: Attempting graceful shutdown of VM...
Connection to 127.0.0.1 closed by remote host.
==> default: Clearing any previously set forwarded ports...
==> default: Clearing any previously set network interfaces...
==> default: Preparing network interfaces based on configuration...
    default: Adapter 1: nat
==> default: Forwarding ports...
    default: 22 (guest) => 2222 (host) (adapter 1)
==> default: Booting VM...
==> default: Waiting for machine to boot. This may take a few minutes...
    default: SSH address: 127.0.0.1:2222
    default: SSH username: vagrant
    default: SSH auth method: private key
==> default: Machine booted and ready!
==> default: Checking for guest additions in VM...
    default: No guest additions were detected on the base box for this VM! Guest
    default: additions are required for forwarded ports, shared folders, host only
    default: networking, and more. If SSH fails on this machine, please install
    default: the guest additions and repackage the box to continue.
    default:
    default: This is not an error message; everything may continue to work properly,
    default: in which case you may ignore this message.
==> default: Setting hostname...
==> default: Machine already provisioned. Run 'vagrant provision' or use the
'--provision'
==> default: flag to force provisioning. Provisioners marked to run always
will still run.

$ vagrant ssh
Last login: Tue Dec 12 16:11:12 2017 from 10.0.2.2
[vagrant@istio ~]$ hostname
istio
[vagrant@istio ~]$ exit;
logout
```

```
Connection to 127.0.0.1 closed.
```

（9）关闭

关闭虚拟机如下所示：

```
$ vagrant halt
==> default: Attempting graceful shutdown of VM...
```

（10）暂停

由于上一步骤关闭虚拟机，执行本步骤时需要先启动虚拟机，然后再暂停虚拟机：

```
$ vagrant suspend
==> default: Saving VM state and suspending execution...
```

（11）删除

删除虚拟机的命令如下所示：

```
$ vagrant destroy
    default: Are you sure you want to destroy the 'default' VM? [y/N] y
==> default: Discarding saved state of VM...
==> default: Destroying VM and associated drives...
```

2. 使用虚拟机快照命令

虚拟机快照命令如下：

- save：保存虚拟机快照。
- list：查看虚拟机快照。
- restore：用快照恢复虚拟机。
- delete：删除虚拟机快照。

进行如下快照的相关操作时，需要先创建虚拟机并启动虚拟机。

保存虚拟机快照示例：

```
$ vagrant snapshot save istio
==> default: Snapshotting the machine as 'istio'...
==> default: Snapshot saved! You can restore the snapshot at any time by
==> default: using 'vagrant snapshot restore'. You can delete it using
==> default: 'vagrant snapshot delete'.
```

查看虚拟机快照示例：

```
$ vagrant snapshot list
istio
```

用快照恢复虚拟机示例：

```
$ vagrant snapshot restore istio
```

```
==> default: Forcing shutdown of VM...
==> default: Restoring the snapshot 'istio'...
==> default: Resuming suspended VM...
==> default: Booting VM...
==> default: Waiting for machine to boot. This may take a few minutes...
    default: SSH address: 127.0.0.1:2222
    default: SSH username: vagrant
    default: SSH auth method: private key
==> default: Machine booted and ready!
```

删除虚拟机快照示例：

```
$ vagrant snapshot delete istio
==> default: Deleting the snapshot 'istio'...
==> default: Snapshot deleted!
```

3.3 模拟实验时的场景

由于后续实验时使用三台虚拟机进行实验，在实验时，不可避免地会使用Vagrant管理虚拟机环境，包括初始化创建虚拟机，启动虚拟机，连接并登录到虚拟机环境，保存实验环境，以及快速恢复实验环境等。本节会模拟后续实验场景用到的步骤，方便你熟悉整个实验环境管理的流程。

1. 初始化虚拟机集群

（1）创建虚拟机配置文件

创建名为istio的目录，并把如下的配置文件写入istio目录的Vagrantfile文件：

```
 1 # -*- mode: ruby -*-
 2 # vi: set ft=ruby :
 3
 4 Vagrant.configure("2") do |config|
 5     (1..3).each do |i|
 6         config.vm.define "lab#{i}" do |node|
 7             node.vm.box = "centos-7.4-base"
 8
 9             node.ssh.insert_key = false
10             node.vm.hostname = "lab#{i}"
11
12             node.vm.network "private_network", ip: "11.11.11.11#{i}"
13
14             node.vm.provision "shell", run: "always",
15                 inline: "ntpdate ntp.api.bz"
16
17             node.vm.provision "shell", run: "always",
18                 inline: "echo hello from lab#{i}"
19
```

```
20              node.vm.provider "virtualbox" do |v|
21                  v.cpus = 2
22                  v.customize ["modifyvm", :id, "--name", "lab#{i}", "--memory",
                          "2048"]
23              end
24          end
25      end
26 end
```

第5行表示创建三台虚拟机。

第7行表示使用名为"centos-7.4-base"的box。

第9行表示不自动生成新的ssh key，使用Vagrant默认的ssh key注入到虚拟机中，这么做主要是为了方便登录。不用为每台虚拟机设置ssh key登录。

第10行设置3台虚拟机主机名分别为lab1、lab2、lab3。

第12行设置3台虚拟机的私有网络为11.11.11.111、11.11.11.112、11.11.11.113。

第14～15行表示当虚拟机启动完成之后，执行inline中配置的shell命令，此命令用于开机之后的时间同步。

第17～18行表示当虚拟机启动完成之后，执行inline中配置的shell命令，此命令用于输出测试字符串。

第21～22行设置虚拟机的CPU核心数和内存大小，本次实验设置为每台虚拟机2核2G内存，可以根据自己电脑的实际情况适当加大。

上面的虚拟机集群配置文件，也是我们后续实验时所使用的虚拟机环境配置文件。

（2）启动虚拟机集群

如果启动不成功，请调整上一步骤中配置文件时CPU和Memery相关的配置。还需要确保虚拟机目录D:\virtualbox中没有和此次实验中创建的同名的虚拟机目录（lab1、lab2、lab3）存在。代码如下：

```
$ vagrant up
Bringing machine 'lab1' up with 'virtualbox' provider...
Bringing machine 'lab2' up with 'virtualbox' provider...
Bringing machine 'lab3' up with 'virtualbox' provider...
==> lab1: Importing base box 'centos-7.4-base'...
==> lab1: Matching MAC address for NAT networking...
==> lab1: Setting the name of the VM: istio_lab1_1543225240637_26922
==> lab1: Clearing any previously set network interfaces...
==> lab1: Preparing network interfaces based on configuration...
    lab1: Adapter 1: nat
    lab1: Adapter 2: hostonly
==> lab1: Forwarding ports...
    lab1: 22 (guest) => 2222 (host) (adapter 1)
==> lab1: Running 'pre-boot' VM customizations...
```

```
==> lab1: Booting VM...
...
==> lab1: Running provisioner: shell...
    lab1: Running: inline script
    lab1: hello from lab1
==> lab2: Importing base box 'centos-7.4-base'...
...
==> lab2: Running provisioner: shell...
    lab2: Running: inline script
    lab2: hello from lab2
==> lab3: Importing base box 'centos-7.4-base'...
==> lab3: Matching MAC address for NAT networking...
...
==> lab3: Setting hostname...
==> lab3: Configuring and enabling network interfaces...
    lab3: SSH address: 127.0.0.1:2201
    lab3: SSH username: vagrant
    lab3: SSH auth method: private key
==> lab3: Running provisioner: shell...
    lab3: Running: inline script
    lab3: hello from lab3
```

（3）查看虚拟机状态

代码如下：

```
$ vagrant status
Current machine states:

lab1                      running (virtualbox)
lab2                      running (virtualbox)
lab3                      running (virtualbox)

This environment represents multiple VMs. The VMs are all listed
above with their current state. For more information about a specific
VM, run 'vagrant status NAME'.
```

2. 连接虚拟机集群

1）使用Xshell导入vagrant的密钥。密钥存储在VAGRANT_HOME环境变量里指定目录的insecure_private_key文件中，添加该密钥到Xshell中。选择顶部菜单中“工具”菜单的“用户密钥管理者”，在弹出的对话框中点击“导入”按钮，选择insecure_private_key文件即可完成密钥的导入，如图3-1所示。

2）使用Xshell创建新的会话。

选择顶部菜单中“文件”菜单的“新建”，创建连接到三台虚拟机的会话。实验中配置的三台虚拟机IP地址分别为11.11.11.111、11.11.11.112、11.11.11.113，ssh端口为22，如图3-2所示。

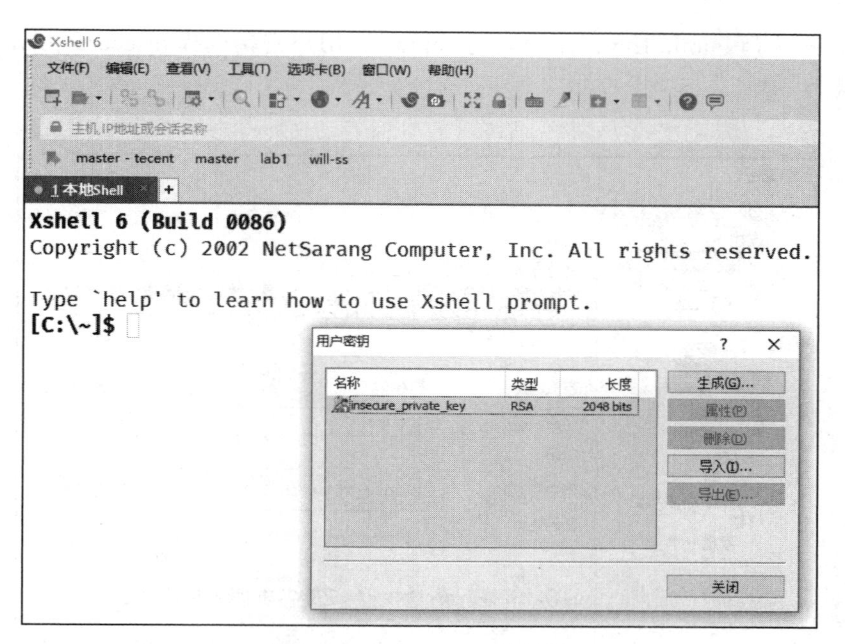

图 3-1　导入 vagrant 的密钥

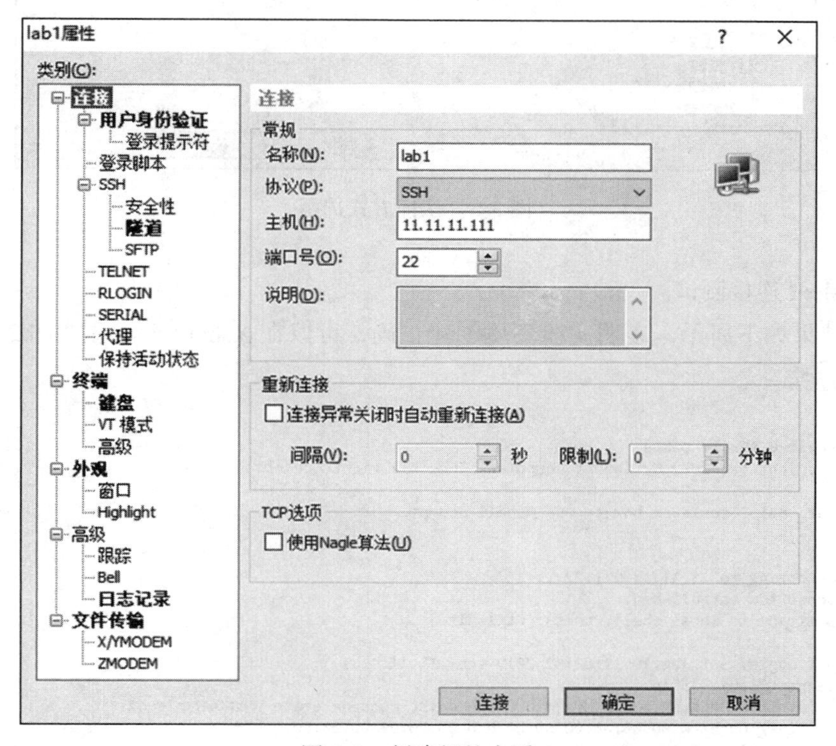

图 3-2　创建新的会话

认证方式选择Public Key，用户名为vagrant，用户密钥选择insecure_private_key，如图3-3所示。

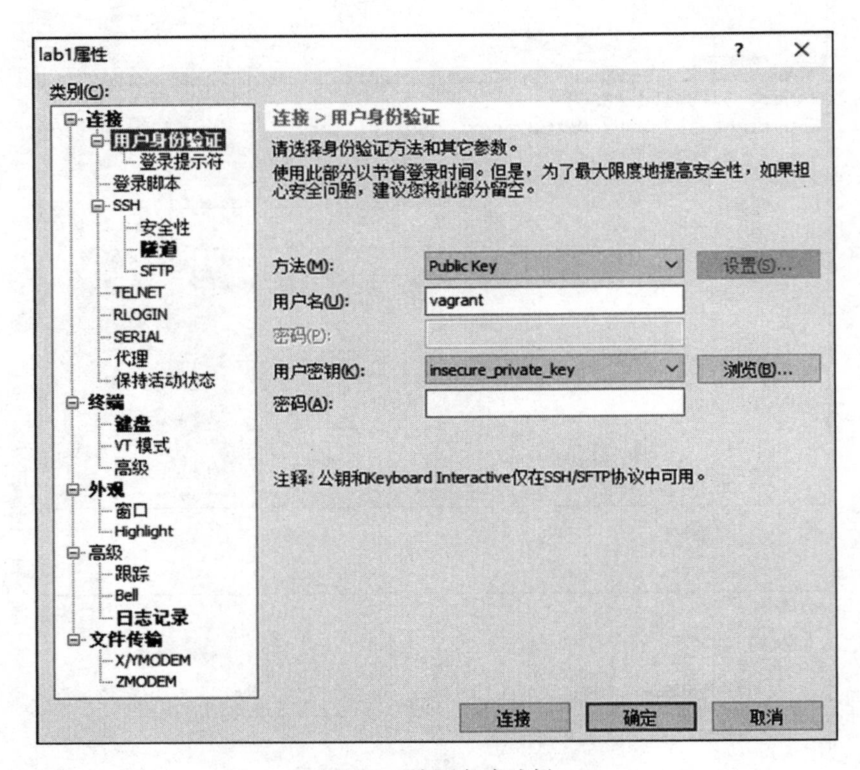

图3-3　认证方式选择

3）Xshell连接测试。

测试结果如下所示，表明实验环境创建正确，可以在Xshell继续添加lab2、lab3虚拟机的SSH连接。

```
Xshell 6 (Build 0086)
Copyright (c) 2002 NetSarang Computer, Inc. All rights reserved.

Type `help' to learn how to use Xshell prompt.
[C:\~]$

Connecting to 11.11.11.111:22...
Connection established.
To escape to local shell, press 'Ctrl+Alt+]'.

Last login: Sat Jan 19 11:02:02 2019 from 11.11.11.1
[vagrant@lab1 ~]$ ip a
1: lo: <LOOPBACK,UP,LOWER_UP> mtu 65536 qdisc noqueue state UNKNOWN qlen 1
    link/loopback 00:00:00:00:00:00 brd 00:00:00:00:00:00
    inet 127.0.0.1/8 scope host lo
```

```
        valid_lft forever preferred_lft forever
    inet6 ::1/128 scope host
        valid_lft forever preferred_lft forever
2: eth0: <BROADCAST,MULTICAST,UP,LOWER_UP> mtu 1500 qdisc pfifo_fast state UP qlen 1000
    link/ether 52:54:00:ca:e4:8b brd ff:ff:ff:ff:ff:ff
    inet 10.0.2.15/24 brd 10.0.2.255 scope global dynamic eth0
        valid_lft 86318sec preferred_lft 86318sec
    inet6 fe80::5054:ff:feca:e48b/64 scope link
        valid_lft forever preferred_lft forever
3: eth1: <BROADCAST,MULTICAST,UP,LOWER_UP> mtu 1500 qdisc pfifo_fast state UP qlen 1000
    link/ether 08:00:27:11:b0:fb brd ff:ff:ff:ff:ff:ff
    inet 11.11.11.111/24 brd 11.11.11.255 scope global eth1
        valid_lft forever preferred_lft forever
```

3. 暂停实验环境虚拟机

当我们的实验进行到一定步骤后，可能需要暂停，换个时间再次进行实验。这个时候我们可以直接暂停整个实验环境中的所有虚拟机，等下次再进行实验时，直接恢复之前的环境即可，非常方便。

暂停集群中所有虚拟机如下所示：

```
$ vagrant suspend
==> lab1: Saving VM state and suspending execution...
==> lab2: Saving VM state and suspending execution...
==> lab3: Saving VM state and suspending execution...
```

暂停集群中单个虚拟机如下所示：

```
$ vagrant suspend lab1
==> lab1: Saving VM state and suspending execution...
```

恢复集群中所有虚拟机如下所示：

```
$ vagrant resume
==> lab1: Resuming suspended VM...
==> lab1: Booting VM...
...
==> lab2: Resuming suspended VM...
==> lab2: Booting VM...
...
==> lab3: Resuming suspended VM...
==> lab3: Booting VM...
==> lab3: Waiting for machine to boot. This may take a few minutes...
    lab3: SSH address: 127.0.0.1:2201
    lab3: SSH username: vagrant
    lab3: SSH auth method: private key
==> lab3: Machine booted and ready!
==> lab3: Machine already provisioned. Run 'vagrant provision' or use the '--provision'
==> lab3: flag to force provisioning. Provisioners marked to run always will
still run.
==> lab3: Running provisioner: shell...
    lab3: Running: inline script
```

```
    lab3: 16 Dec 13:30:58 ntpdate[2983]: step time server 114.118.7.161 offset
25.563654 sec
==> lab3: Running provisioner: shell...
    lab3: Running: inline script
    lab3: hello from lab3
```

恢复集群中单个虚拟机，使用resume和up都能恢复暂停的虚拟机：

```
$ vagrant resume lab1
==> lab1: Resuming suspended VM...
==> lab1: Booting VM...
==> lab1: Waiting for machine to boot. This may take a few minutes...
    lab1: SSH address: 127.0.0.1:2222
    lab1: SSH username: vagrant
    lab1: SSH auth method: private key
==> lab1: Machine booted and ready!
==> lab1: Machine already provisioned. Run 'vagrant provision' or use the '--provision'
==> lab1: flag to force provisioning. Provisioners marked to run always will still run.
==> lab1: Running provisioner: shell...
    lab1: Running: inline script
    lab1: 16 Dec 13:40:35 ntpdate[2983]: step time server 114.118.7.161 offset
25.563654 sec
==> lab1: Running provisioner: shell...
    lab1: Running: inline script
lab1: hello from lab1

$ vagrant up lab1
Bringing machine 'lab1' up with 'virtualbox' provider...
==> lab1: Resuming suspended VM...
==> lab1: Booting VM...
==> lab1: Waiting for machine to boot. This may take a few minutes...
    lab1: SSH address: 127.0.0.1:2222
    lab1: SSH username: vagrant
    lab1: SSH auth method: private key
==> lab1: Machine booted and ready!
==> lab1: Machine already provisioned. Run 'vagrant provision' or use the '--provision'
==> lab1: flag to force provisioning. Provisioners marked to run always will
still run.
==> lab1: Running provisioner: shell...
    lab1: Running: inline script
    lab1: 16 Dec 13:45:55 ntpdate[2983]: step time server 114.118.7.161 offset
25.563654 sec
==> lab1: Running provisioner: shell...
    lab1: Running: inline script
        lab1: hello from lab1
```

4. 保存与恢复实验环境

完成前面的步骤后，我们的实验虚拟机环境就已经基本搭建完成了。之后我们可以安装 Docker、Git、Wget 等基础软件。安装之后保存实验环境，之后实验不成功或者实验环境被污染，可以快速恢复到当前的实验环境。

保存集群中所有虚拟机快照：

```
$ vagrant snapshot save base
==> lab1: Snapshotting the machine as 'base'...
==> lab1: Snapshot saved! You can restore the snapshot at any time by
==> lab1: using 'vagrant snapshot restore'. You can delete it using
==> lab1: 'vagrant snapshot delete'.
==> lab2: Snapshotting the machine as 'base'...
==> lab2: Snapshot saved! You can restore the snapshot at any time by
==> lab2: using 'vagrant snapshot restore'. You can delete it using
==> lab2: 'vagrant snapshot delete'.
==> lab3: Snapshotting the machine as 'base'...
==> lab3: Snapshot saved! You can restore the snapshot at any time by
==> lab3: using 'vagrant snapshot restore'. You can delete it using
==> lab3: 'vagrant snapshot delete'.
```

保存集群中单个虚拟机快照：

```
$ vagrant snapshot save lab1 base
==> lab1: Snapshotting the machine as 'base'...
==> lab1: Snapshot saved! You can restore the snapshot at any time by
==> lab1: using 'vagrant snapshot restore'. You can delete it using
==> lab1: 'vagrant snapshot delete'.
```

删除集群中所有虚拟机快照：

```
$ vagrant snapshot delete base
==> lab1: Deleting the snapshot 'base'...
==> lab1: Snapshot deleted!
==> lab2: Deleting the snapshot 'base'...
==> lab2: Snapshot deleted!
==> lab3: Deleting the snapshot 'base'...
==> lab3: Snapshot deleted!
```

删除集群中单个虚拟机快照：

```
$ vagrant snapshot delete lab1 base
==> lab1: Deleting the snapshot 'base'...
==> lab1: Snapshot deleted!
```

查看集群中所有虚拟机快照：

```
$ vagrant snapshot list
base
base
base
```

查看集群中单个虚拟机快照：

```
$ vagrant snapshot list lab1
base
```

恢复集群中所有虚拟机快照：

```
$ vagrant snapshot restore base
==> lab1: Forcing shutdown of VM...
==> lab1: Restoring the snapshot 'base'...
==> lab1: Resuming suspended VM...
==> lab1: Booting VM...
...
==> lab2: Forcing shutdown of VM...
==> lab2: Restoring the snapshot 'base'...
==> lab2: Resuming suspended VM...
==> lab2: Booting VM...
==> lab2: Waiting for machine to boot. This may take a few minutes...
...
==> lab3: Forcing shutdown of VM...
==> lab3: Restoring the snapshot 'base'...
==> lab3: Resuming suspended VM...
==> lab3: Booting VM...
...
==> lab3: Machine booted and ready!
==> lab3: Running provisioner: shell...
    lab3: Running: inline script
    lab3: 16 Dec 13:50:38 ntpdate[2983]: step time server 114.118.7.161 offset
55.563654 sec
==> lab3: Running provisioner: shell...
    lab3: Running: inline script
    lab3: hello from lab3
```

恢复集群中单个虚拟机快照：

```
$ vagrant snapshot restore lab1 base
==> lab1: Forcing shutdown of VM...
==> lab1: Restoring the snapshot 'base'...
==> lab1: Resuming suspended VM...
==> lab1: Booting VM...
==> lab1: Waiting for machine to boot. This may take a few minutes...
    lab1: SSH address: 127.0.0.1:2222
    lab1: SSH username: vagrant
    lab1: SSH auth method: private key
==> lab1: Machine booted and ready!
==> lab1: Running provisioner: shell...
    lab1: Running: inline script
    lab1: 16 Dec 14:10:29 ntpdate[2983]: step time server 114.118.7.161 offset
25.563654 sec
```

```
==> lab1: Running provisioner: shell...
    lab1: Running: inline script
    lab1: hello from lab1
```

注意事项

在使用 Vagrant 时，可能会出现偶发的异常错误，大部分情况下，重启电脑即可解决。

3.4 本章小结

通过使用 Vagrant，我们可以更快速地创建 Istio 的实验环境，这将大大节省我们创建实验环境的时间。当我们的实验环境被污染时，也可以使用 Vagrant 快速恢复到之前的实验环境，这也使得我们可以重复验证 Istio 功能，并节省了重复准备实验环境的时间，大大提升了实验效率。

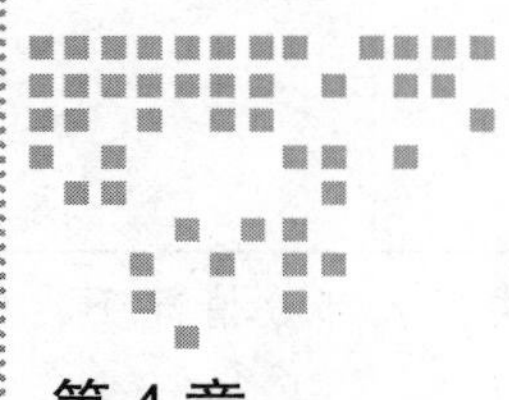

Chapter 4　第4章

创建 Kubernetes 集群

由于Istio在Kubernetes上实现的功能最为完整，本书后续章节的实验都是基于Kubernetes集群的，因此本章介绍如何创建一个Kubernetes集群，对实验来说本章内容至关重要。本章将介绍如何安装Docker、安装Kubeadm、创建Kubernetes集群并配置所需要的基础环境，以及创建Kubernetes集群之后如何测试检查集群是否能正常工作。

请确保跟着本章实验创建Kubernetes集群，并完成后面的集群测试；否则可能导致后续Istio相关的实验无法顺利进行，或者出现实验结果与书中所示不一致的情况。

4.1 安装 Docker

由于目前Docker与Kubernetes集成最为方便，所以选择使用Docker作为集群使用的容器。下面介绍在集群中所有的节点上安装Docker。

1）配置阿里云Docker CE镜像源。

由于Docker官方网站在国外，阿里云有相关的镜像源，可以更快下载，所以这里就用阿里云的镜像源。

添加阿里云软件源信息：

```
$ sudo yum-config-manager --add-repo http://mirrors.aliyun.com/docker-ce/linux/centos/docker-ce.repo
Loaded plugins: fastestmirror
adding repo from: http://mirrors.aliyun.com/docker-ce/linux/centos/docker-ce.repo
```

```
grabbing file http://mirrors.aliyun.com/docker-ce/linux/centos/docker-ce.repo
to /etc/yum.repos.d/docker-ce.repo
repo saved to /etc/yum.repos.d/docker-ce.repo
```

更新软件源信息：

```
$ sudo yum makecache fast
Loaded plugins: fastestmirror
...
Determining fastest mirrors
  * base: mirrors.aliyun.com
  * epel: mirrors.aliyun.com
  * extras: mirrors.aliyun.com
  * updates: mirrors.aliyun.com
epel                                                            12719/12719
Metadata Cache Created
```

2）安装 Docker CE。

查看可安装版本：

```
$ sudo yum list docker-ce --showduplicates | sort -r
  * updates: mirrors.aliyun.com
Loading mirror speeds from cached hostfile
Loaded plugins: fastestmirror
  * extras: mirrors.aliyun.com
  * epel: mirrors.aliyun.com
docker-ce.x86_64      3:18.09.0-3.el7               docker-ce-stable
docker-ce.x86_64      18.06.1.ce-3.el7              docker-ce-stable
docker-ce.x86_64      18.06.0.ce-3.el7              docker-ce-stable
docker-ce.x86_64      18.03.1.ce-1.el7.centos       docker-ce-stable
docker-ce.x86_64      18.03.0.ce-1.el7.centos       docker-ce-stable
docker-ce.x86_64      17.12.1.ce-1.el7.centos       docker-ce-stable
docker-ce.x86_64      17.12.0.ce-1.el7.centos       docker-ce-stable
docker-ce.x86_64      17.09.1.ce-1.el7.centos       docker-ce-stable
docker-ce.x86_64      17.09.0.ce-1.el7.centos       docker-ce-stable
docker-ce.x86_64      17.06.2.ce-1.el7.centos       docker-ce-stable
docker-ce.x86_64      17.06.1.ce-1.el7.centos       docker-ce-stable
docker-ce.x86_64      17.06.0.ce-1.el7.centos       docker-ce-stable
docker-ce.x86_64      17.03.3.ce-1.el7              docker-ce-stable
docker-ce.x86_64      17.03.2.ce-1.el7.centos       docker-ce-stable
docker-ce.x86_64      17.03.1.ce-1.el7.centos       docker-ce-stable
docker-ce.x86_64      17.03.0.ce-1.el7.centos       docker-ce-stable
  * base: mirrors.aliyun.com
Available Packages
```

安装指定版本 Docker CE（Kubernetes 1.12 版本官方推荐使用 Docker CE 18.06 版本）：

```
# 安装 Docker 依赖
$ sudo yum install -y yum-utils device-mapper-persistent-data lvm2
```

```
# 安装 Docker CE
$ sudo yum install -y docker-ce-18.06.1.ce
...
Installed:
  docker-ce.x86_64 0:18.06.1.ce-3.el7
...
Dependency Installed:
...
Complete!
```

3）运行测试。

启动 Docker CE：

```
# 启动 Docker
$ sudo systemctl start docker

# 运行 Docker 命令，测试是否正常启动
$ sudo docker ps
CONTAINER ID    IMAGE           COMMAND         CREATED         STATUS
PORTS           NAMES
```

运行测试：

```
$ sudo docker run --rm alpine echo 'hello world !'
Unable to find image 'alpine:latest' locally
latest: Pulling from library/alpine
4fe2ade4980c: Pull complete
Digest: sha256:621c2f39f8133acb8e64023a94dbdf0d5ca81896102b9e57c0dc184cadaf5528
Status: Downloaded newer image for alpine:latest
hello world !
```

4）配置镜像加速。

有时 Docker 官方镜像可能会出现故障，出现拉取不到镜像的问题，也可以尝试使用阿里云提供的免费镜像加速功能[⊖]。

配置加速。配置国内 Docker 官方镜像拉取加速：

```
$ sudo mkdir -p /etc/docker
$ sudo tee /etc/docker/daemon.json <<-'EOF'
{
  "registry-mirrors": ["https://registry.docker-cn.com"]
}
EOF

# 查看配置
$ sudo cat /etc/docker/daemon.json
{
```

⊖ 参考链接地址：https://cr.console.aliyun.com/cn-shanghai/mirrors。

```
  "registry-mirrors": ["https://registry.docker-cn.com"]
}
```

重启 Docker：

```
# 重载配置
$ sudo systemctl daemon-reload

# 重启 Docker
$ sudo systemctl restart docker
```

拉取镜像测试。此时会感觉镜像拉取速度明显加快。拉取镜像测试如下：

```
$ sudo docker pull centos:7
7: Pulling from library/centos
a02a4930cb5d: Pull complete
Digest: sha256:184e5f35598e333bfa7de10d8fb1cebb5ee4df5bc0f970bf2b1e7c7345136426
Status: Downloaded newer image for centos:7

$ sudo docker rmi centos:7
Untagged: centos:7
Untagged: centos@sha256:184e5f35598e333bfa7de10d8fb1cebb5ee4df5bc0f970bf2b1e
7c7345136426
Deleted: sha256:1e1148e4cc2c148c6890a18e3b2d2dde41a6745ceb4e5fe94a923d811bf82ddb
Deleted: sha256:071d8bd765171080d01682844524be57ac9883e53079b6ac66707e192ea25956
```

4.2　安装 Kubeadm

Kubeadm 用于创建 Kubernetes 集群，下面介绍如何在集群中的所有节点上安装 Kubeadm 相关软件。

1）配置阿里云 Kubernetes 镜像源：

```
$ sudo tee /etc/yum.repos.d/kubernetes.repo <<-'EOF'
[kubernetes]
name=Kubernetes
baseurl=https://mirrors.aliyun.com/kubernetes/yum/repos/kubernetes-el7-x86_64
enabled=1
gpgcheck=1
repo_gpgcheck=1
gpgkey=https://mirrors.aliyun.com/kubernetes/yum/doc/yum-key.gpg
https://mirrors.aliyun.com/kubernetes/yum/doc/rpm-package-key.gpg
EOF

# 查看 yum 源配置
$ sudo cat /etc/yum.repos.d/kubernetes.repo
[kubernetes]
name=Kubernetes
baseurl=https://mirrors.aliyun.com/kubernetes/yum/repos/kubernetes-el7-x86_64
```

```
enabled=1
gpgcheck=1
repo_gpgcheck=1
gpgkey=https://mirrors.aliyun.com/kubernetes/yum/doc/yum-key.gpg
https://mirrors.aliyun.com/kubernetes/yum/doc/rpm-package-key.gpg
```

2）安装 Kubeadm 相关软件：

```
$ sudo yum install -y kubelet-1.12.4 kubeadm-1.12.4 kubectl-1.12.4 ipvsadm
...
Installed:
  ipvsadm.x86_64 0:1.27-7.el7                    kubeadm.x86_64 0:1.12.4-0
  kubectl.x86_64 0:1.12.4-0                      kubelet.x86_64 0:1.12.4-0

Dependency Installed:
  cri-tools.x86_64 0:1.12.0-0                    kubernetes-cni.x86_64 0:0.6.0-0
  socat.x86_64 0:1.7.3.2-2.el7

Complete!
```

4.3 配置基础环境

由于 Kubernetes 集群对基础环境有要求，下面介绍如何在集群中的所有节点上完成 Kubernetes 集群的配置。

1）关闭防火墙和 SELinux：

```
# 关闭防火墙
$ sudo systemctl stop firewalld
$ sudo systemctl disable firewalld

# 临时关闭 SELinux
$ sudo setenforce 0

# 关闭开机启用 SELinux
$ sudo sed -i 's/SELINUX=permissive/SELINUX=disabled/' /etc/sysconfig/selinux
```

2）关闭 Swap：

```
# 临时关闭 Swap
$ sudo swapoff -a

# 关闭开机自动挂载 Swap 分区
$ sudo sed -ri 's@(^/.*swap.*)@#\1@g' /etc/fstab

# 查看内存情况，Swap 为 0 表示已经关闭成功
$ free -m
         total    used    free    shared    buff/cache    available
Mem:     1839     162     653     16        1022          1453
Swap:    0        0       0
```

3）加载 IPVS 相关内核模块：

```
$ sudo modprobe ip_vs
$ sudo modprobe ip_vs_rr
$ sudo modprobe ip_vs_wrr
$ sudo modprobe ip_vs_sh
$ sudo modprobe nf_conntrack_ipv4

# 查看 IPVS 相关内核模块是否导入成功
$ sudo lsmod | grep ip_vs
ip_vs_sh        12688   0
ip_vs_wrr       12697   0
ip_vs_rr        12600   0
ip_vs           141092  6 ip_vs_rr,ip_vs_sh,ip_vs_wrr
nf_conntrack    133387  7 ip_vs,nf_nat,nf_nat_ipv4,xt_conntrack,nf_nat_
    masquerade_ipv4,nf_conntrack_netlink,nf_conntrack_ipv4
libcrc32c       12644  4 xfs,ip_vs,nf_nat,nf_conntrack

# 配置开机自动导入 IPVS 相关内核模块
$ sudo tee /etc/modules-load.d/k8s-ipvs.conf <<-'EOF'
ip_vs
ip_vs_rr
ip_vs_wrr
ip_vs_sh
nf_conntrack_ipv4
EOF

# 查看 IPVS 相关内核模块配置
$ sudo cat /etc/modules-load.d/k8s-ipvs.conf
ip_vs
ip_vs_rr
ip_vs_wrr
ip_vs_sh
nf_conntrack_ipv4
```

4）RHEL/CentOS 7 需要的特殊配置：

```
$ sudo tee /etc/sysctl.d/k8s.conf <<-'EOF'
net.bridge.bridge-nf-call-ip6tables = 1
net.bridge.bridge-nf-call-iptables = 1
vm.swappiness=0
EOF

# 使配置生效
$ sudo sysctl --system
...
* Applying /etc/sysctl.d/k8s.conf ...
net.bridge.bridge-nf-call-ip6tables = 1
net.bridge.bridge-nf-call-iptables = 1
vm.swappiness = 0
* Applying /etc/sysctl.conf ...
```

5）开启 Forward。Docker 从 1.13 版本开始调整了默认的防火墙规则，禁用了 iptables filter 表中 FOWARD 链，这可能会引起 Kubernetes 集群中跨 Node 的 Pod 无法正常通信，本次实验并没执行此步骤，如果碰到上述问题，可以使用如下方式开启：

```
$ sudo iptables -P FORWARD ACCEPT
$ sudo sed -i '/ExecStart/a ExecStartPost=/sbin/iptables -P FORWARD ACCEPT' /
    usr/lib/systemd/system/docker.service
$ sudo systemctl daemon-reload
```

6）配置 Hosts 解析：

```
# 配置 Hosts 解析
$ sudo tee /etc/hosts <<-'EOF'
127.0.0.1   localhost localhost.localdomain localhost4 localhost4.localdomain4
::1         localhost localhost.localdomain localhost6 localhost6.localdomain6
11.11.11.111 lab1
11.11.11.112 lab2
11.11.11.113 lab3
EOF

# 解析测试
$ ping -c2 lab1
PING lab1 (11.11.11.111) 56(84) bytes of data.
64 bytes from lab1 (11.11.11.111): icmp_seq=1 ttl=64 time=0.034 ms
64 bytes from lab1 (11.11.11.111): icmp_seq=2 ttl=64 time=0.068 ms

--- lab1 ping statistics ---
2 packets transmitted, 2 received, 0% packet loss, time 999ms
rtt min/avg/max/mdev = 0.034/0.051/0.068/0.017 ms

$ ping -c2 lab2
PING lab2 (11.11.11.112) 56(84) bytes of data.
64 bytes from lab2 (11.11.11.112): icmp_seq=1 ttl=64 time=2.00 ms
64 bytes from lab2 (11.11.11.112): icmp_seq=2 ttl=64 time=1.09 ms

--- lab2 ping statistics ---
2 packets transmitted, 2 received, 0% packet loss, time 1002ms
rtt min/avg/max/mdev = 1.094/1.549/2.005/0.457 ms

$ ping -c2 lab3
PING lab3 (11.11.11.113) 56(84) bytes of data.
64 bytes from lab3 (11.11.11.113): icmp_seq=1 ttl=64 time=0.100 ms
64 bytes from lab3 (11.11.11.113): icmp_seq=2 ttl=64 time=0.031 ms

--- lab3 ping statistics ---
2 packets transmitted, 2 received, 0% packet loss, time 1000ms
```

7）配置 Kubelet：

```
$ DOCKER_CGROUPS=$(sudo docker info | grep 'Cgroup' | cut -d' ' -f3)
```

```
$ echo $DOCKER_CGROUPS
$ sudo tee /etc/sysconfig/kubelet <<-EOF
KUBELET_EXTRA_ARGS="--cgroup-driver=$DOCKER_CGROUPS --pod-infra-container-
    image=registry.cn-hangzhou.aliyuncs.com/google_containers/pause-amd64:3.1"
EOF

# 查看配置
$ sudo cat /etc/sysconfig/kubelet
KUBELET_EXTRA_ARGS="--cgroup-driver=cgroupfs --pod-infra-container-image=registry.
    cn-hangzhou.aliyuncs.com/google_containers/pause-amd64:3.1"

# 重新加载配置
$ sudo systemctl daemon-reload
```

4.4　创建Kubernetes集群的步骤

经过上面的基础步骤后，下面介绍如何使用Kubeadm创建Kubernetes集群的步骤，并使安装Flannel网格插件实现容器间通信。

1. 集群架构说明

集群由三台虚拟机组成，主机名分别为lab1、lab2、lab3，操作系统为CentOS 7.4版本。安装Kubernetes版本为1.12.4版本，网络插件使用Flannel来实现跨节点通信，每台虚拟机有两张网卡，分别为eth0、eth1。eth0用于连接外网，安装下载软件，eth1用来集群内部通信。实验采用1个主节点两个工作节点，lab1作为master节点，lab2、lab3作为node工作节点，如下所示：

主机名	集群中的角色	集群通信IP地址
lab1	Master	11.11.11.111/24
lab2	Node	11.11.11.112/24
lab3	node	11.11.11.113/24

2. 初始化Master节点

下面对lab1节点进行初始化。

1）配置Docker和Kubelet开机启动：

```
# 设置Docker开机启动
$ sudo systemctl enable docker.service

# 设置Kubelet开机启动
$ sudo systemctl enable kubelet.service
```

2）生成配置文件。

根据实际情况修改如下的配置文件，11.11.11.111为master节点的IP地址，默认情况下会使用gcr.io的镜像，由于无法拉取镜像，会导致集群创建失败，所以应配置imageRepository，使用国内阿里云的镜像，kube-proxy使用了ipvs模式。使用如下命令写入配置文件：

```
$ cat >kubeadm-master.config<<EOF
apiVersion: kubeadm.k8s.io/v1alpha2
kind: MasterConfiguration
kubernetesVersion: v1.12.4
imageRepository: registry.cn-hangzhou.aliyuncs.com/google_containers
api:
  advertiseAddress: 11.11.11.111

controllerManagerExtraArgs:
  node-monitor-grace-period: 10s
  pod-eviction-timeout: 10s

networking:
  podSubnet: 10.244.0.0/16

kubeProxy:
  config:
    mode: ipvs
EOF
```

3）提前拉取镜像。

如果执行失败可以多次执行，使用如下命令提前拉取镜像：

```
$ sudo kubeadm config images pull --config kubeadm-master.config
[config/images] Pulled registry.cn-hangzhou.aliyuncs.com/google_containers/
    kube-apiserver:v1.12.4
[config/images] Pulled registry.cn-hangzhou.aliyuncs.com/google_containers/
    kube-controller-manager:v1.12.4
[config/images] Pulled registry.cn-hangzhou.aliyuncs.com/google_containers/
    kube-scheduler:v1.12.4
[config/images] Pulled registry.cn-hangzhou.aliyuncs.com/google_containers/
    kube-proxy:v1.12.4
[config/images] Pulled registry.cn-hangzhou.aliyuncs.com/google_containers/
    pause:3.1
[config/images] Pulled registry.cn-hangzhou.aliyuncs.com/google_containers/
    etcd:3.2.24
[config/images] Pulled registry.cn-hangzhou.aliyuncs.com/google_containers/
    coredns:1.2.2
```

4）执行初始化。

初始化完成之后保留命令输出的join相关命令，用于后面步骤node节点加入集群中。使用如下命令执行初始化：

```
$ sudo kubeadm init --config kubeadm-master.config
[init] using Kubernetes version: v1.12.4
[preflight] running pre-flight checks
[preflight/images] Pulling images required for setting up a Kubernetes cluster
[preflight/images] This might take a minute or two, depending on the speed of
  your internet connection
[preflight/images] You can also perform this action in beforehand using
  'kubeadm config images pull'
[kubelet] Writing kubelet environment file with flags to file "/var/lib/
  kubelet/kubeadm-flags.env"
[kubelet] Writing kubelet configuration to file "/var/lib/kubelet/config.yaml"
[preflight] Activating the kubelet service
...
[bootstraptoken] creating the "cluster-info" ConfigMap in the "kube-public"
  namespace
[addons] Applied essential addon: CoreDNS
[addons] Applied essential addon: kube-proxy

Your Kubernetes master has initialized successfully!

To start using your cluster, you need to run the following as a regular user:

  mkdir -p $HOME/.kube
  sudo cp -i /etc/kubernetes/admin.conf $HOME/.kube/config
  sudo chown $(id -u):$(id -g) $HOME/.kube/config

You should now deploy a pod network to the cluster.
Run "kubectl apply -f [podnetwork].yaml" with one of the options listed at:
  https://kubernetes.io/docs/concepts/cluster-administration/addons/

You can now join any number of machines by running the following on each node
as root:

   kubeadm join 11.11.11.111:6443 --token xlivmy.jj73qkeobqoyfs0r --discovery-
     token-ca-cert-hash sha256:c6290480baaef64f3f1c6b6861de8685ca74d5a4a7c2703
     eace18a2e2117e87d
```

5）配置使用 kubectl。

另外打开一个主机 lab1 终端会话，以 vagrant 用户执行如下操作：

```
$ rm -rf $HOME/.kube
$ mkdir -p $HOME/.kube
$ sudo cp -i /etc/kubernetes/admin.conf $HOME/.kube/config
$ sudo chown $(id -u):$(id -g) $HOME/.kube/config
```

使用 kubectl 获取集群中节点信息，测试 kubectl 配置是否正确：

```
$ kubectl get node
NAME          STATUS        ROLES         AGE        VERSION
lab1          NotReady      master        56s        v1.12.4
```

6）配置 master 节点接收负载。

由于实验环境虚拟机内存可能不足，导致后面的 Istio 实验无法成功，可以让 master 节点和其他 node 节点一样接收负载，分担其他 node 节点的压力，但也可能使 master 节点压力过大，导致 master 上 kubernetes 组件出现异常，实验无法成功。只有当不配置此步骤实验就不能继续进行时，才推荐执行此步骤，本书每台虚拟机 2G 内存的实验环境中并不需要此步骤。使用如下命令设置：

```
$ kubectl taint nodes lab1 node-role.kubernetes.io/master-
node/lab1 untainted
```

3. 添加 Node 节点

1）配置 Docker 和 Kubelet 开机启动。

在 lab2、lab3 分别执行如下命令：

```
# 设置 Docker 开机启动
$ sudo systemctl enable docker.service

# 设置 Kubelet 开机启动
$ sudo systemctl enable kubelet.service
```

2）加入到集群。

在 lab2、lab3 分别执行如下命令，此处的命令为初始化 master 完成时，输出的 join 命令，注意替换为你在实验初始化时的输出命令：

```
    $ sudo kubeadm join 11.11.11.111:6443 --token xlivmy.jj73qkeobqoyfs0r
--discovery-token-ca-cert-hash sha256:c6290480baaef64f3f1c6b6861de8685ca74d5a4a7c2
703eace18a2e2117e87d
    [preflight] running pre-flight checks
    [discovery] Trying to connect to API Server "11.11.11.111:6443"
    [discovery] Created cluster-info discovery client, requesting info from
"https://11.11.11.111:6443"
    [discovery] Requesting info from "https://11.11.11.111:6443" again to validate
TLS against the pinned public key
    [discovery] Cluster info signature and contents are valid and TLS certificate
validates against pinned roots, will use API Server "11.11.11.111:6443"
    [discovery] Successfully established connection with API Server
"11.11.11.111:6443"
    [kubelet] Downloading configuration for the kubelet from the "kubelet-
config-1.12" ConfigMap in the kube-system namespace
    [kubelet] Writing kubelet configuration to file "/var/lib/kubelet/config.yaml"
    [kubelet] Writing kubelet environment file with flags to file "/var/lib/
kubelet/kubeadm-flags.env"
    [preflight] Activating the kubelet service
    [tlsbootstrap] Waiting for the kubelet to perform the TLS Bootstrap...
    [patchnode] Uploading the CRI Socket information "/var/run/dockershim.sock" to
```

```
the Node API object "lab2" as an annotation

    This node has joined the cluster:
    * Certificate signing request was sent to apiserver and a response was
received.
    * The Kubelet was informed of the new secure connection details.

    Run 'kubectl get nodes' on the master to see this node join the cluster.
```

3）查看集群中的节点。

在lab1执行如下命令：

```
$ kubectl get node
NAME          STATUS         ROLES         AGE         VERSION
lab1          NotReady       master        2m38s       v1.12.4
lab2          NotReady       <none>        12s         v1.12.4
lab3          NotReady       <none>        5s          v1.12.4
```

4. 部署Kubernetes网络插件

使用Kubeadm创建Kubernetes集群时，如果没有部署网络插件，所有的节点状态都会是NotReady。当网络插件部署完成后，状态会更新为Ready，并会启动CoreDNS，用于集群中的服务发现。

以下步骤在lab1上操作。

1）下载yaml文件：

```
$ curl -s https://raw.githubusercontent.com/coreos/flannel/v0.10.0/
   Documentation/kube-flannel.yml -o kube-flannel.yml
```

2）修改配置文件。修改kube-flannel.yml文件如下位置的Network参数为"10.244.0.0/16"，此参数与初始化master节点时的配置文件中的podSubnet值保持一致：

```
net-conf.json: |
  {
    "Network": "10.244.0.0/16",
    "Backend": {
      "Type": "vxlan"
    }
  }
```

修改kube-flannel.yml文件中的image字段：

```
image: registry.cn-shanghai.aliyuncs.com/gcr-k8s/flannel:v0.10.0-amd64
```

修改如下位置的tolerations配置，详见：https://github.com/coreos/flannel/issues/1044

```
# 修改前
```

```
hostNetwork: true
nodeSelector:
  beta.kubernetes.io/arch: amd64
tolerations:
- key: node-role.kubernetes.io/master
  operator: Exists
  effect: NoSchedule

# 修改后
hostNetwork: true
nodeSelector:
  beta.kubernetes.io/arch: amd64
tolerations:
- operator: Exists
  effect: NoSchedule
```

在如下的位置添加指定 kube-flannel 容器启动时的网卡名称，此网卡用于 flannel 内部通信（详见 https://github.com/kubernetes/kubernetes/issues/39701）：

```
# 修改前
containers:
- name: kube-flannel
  image: registry.cn-shanghai.aliyuncs.com/gcr-k8s/flannel:v0.10.0-amd64
  command:
  - /opt/bin/flanneld
  args:
  - --ip-masq
  - --kube-subnet-mgr
  resources:
    requests:
      cpu: "100m"
      memory: "50Mi"
    limits:
      cpu: "100m"
      memory: "50Mi"

# 修改后
containers:
- name: kube-flannel
  image: registry.cn-shanghai.aliyuncs.com/gcr-k8s/flannel:v0.10.0-amd64
  command:
  - /opt/bin/flanneld
  args:
  - --ip-masq
  - --kube-subnet-mgr
  - --iface=eth1
  resources:
    requests:
      cpu: "100m"
      memory: "50Mi"
```

```
    limits:
      cpu: "100m"
      memory: "50Mi"
```

3）部署 Flannel：

```
$ kubectl apply -f kube-flannel.yml
clusterrole.rbac.authorization.k8s.io/flannel created
clusterrolebinding.rbac.authorization.k8s.io/flannel created
serviceaccount/flannel created
configmap/kube-flannel-cfg created
daemonset.extensions/kube-flannel-ds created
```

4）查看 Flannel 状态。

在 lab1 执行如下命令：

```
$ kubectl get pod -n kube-system -o wide
NAME          READY   STATUS    RESTARTS   AGE     IP             NODE   NOMINATED NODE
coredns-6c66ffc55b-5tpjm       1/1   Running   0   9m12s   10.244.1.3     lab2   <none>
coredns-6c66ffc55b-mztsx       1/1   Running   0   9m12s   10.244.1.2     lab2   <none>
etcd-lab1                      1/1   Running   0   8m50s   11.11.11.111   lab1   <none>
kube-apiserver-lab1            1/1   Running   0   8m44s   11.11.11.111   lab1   <none>
kube-controller-manager-lab1   1/1   Running   0   8m55s   11.11.11.111   lab1   <none>
kube-flannel-ds-bkz59          1/1   Running   0   82s     11.11.11.111   lab1   <none>
kube-flannel-ds-dk466          1/1   Running   0   82s     11.11.11.113   lab3   <none>
kube-flannel-ds-gmh51          1/1   Running   0   82s     11.11.11.112   lab2   <none>
kube-proxy-82mtm               1/1   Running   0   7m7s    11.11.11.112   lab2   <none>
kube-proxy-vtv89               1/1   Running   0   7m      11.11.11.113   lab3   <none>
kube-proxy-xw66f               1/1   Running   0   9m13s   11.11.11.111   lab1   <none>
kube-scheduler-lab1            1/1   Running   0   8m46s   11.11.11.111   lab1   <none>

$ kubectl get svc -n kube-system
NAME       TYPE        CLUSTER-IP   EXTERNAL-IP   PORT(S)          AGE
kube-dns   ClusterIP   10.96.0.10   <none>        53/UDP,53/TCP    9m56s
```

5）查看集群节点状态。

在 lab1 执行如下命令，部署完成 Flannel 之后，集群中的节点全部变为 Ready 状态：

```
$ kubectl get node
NAME         STATUS         ROLES          AGE         VERSION
lab1         Ready          master         42m         v1.12.4
lab2         Ready          <none>         40m         v1.12.4
lab3         Ready          <none>         39m         v1.12.4
```

4.5　测试集群的正确性

在 lab1 节点上执行如下的步骤，验证 Kubernetes 集群部署是否正确可用。

1）部署 Nginx 应用：

```
$ kubectl create deployment nginx --image=nginx:alpine
deployment.apps/nginx created
```

2）暴露 Nginx 应用：

```
$ kubectl expose deployment nginx --name=nginx-service --port=80 --target-
    port=80
service/nginx-service exposed

$ kubectl expose deployment nginx --type=NodePort --name=nginx-service-nodeport
    --port=80 --target-port=80
service/nginx-service-nodeport exposed
```

3）查看 Nginx 应用状态：

```
$ kubectl get deploy
NAME      DESIRED    CURRENT    UP-TO-DATE     AVAILABLE     AGE
nginx     1          1          1              1             2m9s

$ kubectl get pod
NAME                         READY      STATUS      RESTARTS      AGE
nginx-65d5c4f7cc-ztkkf       1/1        Running     0             2m9s

$ kubectl get svc
NAME                      TYPE        CLUSTER-IP       EXTERNAL-IP   PORT(S)        AGE
kubernetes                ClusterIP   10.96.0.1        <none>        443/TCP        3h5m
nginx-service             ClusterIP   10.106.97.98     <none>        80/TCP         96s
nginx-service-nodeport    NodePort    10.106.177.223   <none>        80:31994/TCP   44s
```

4）通过 Pod IP 访问 Nginx 应用。

查看 Nginx 应用的 Pod IP：

```
$ kubectl get pod -o wide
NAME                     READY   STATUS    RESTARTS   AGE    IP           NODE   NOMINATED NODE
nginx-65d5c4f7cc-ztkkf   1/1     Running   0          6m4s   10.244.1.3   lab2   <none>
```

访问 Nginx 应用，10.244.1.3 为上一步骤中查看到的 Pod IP：

```
curl -I http://10.244.1.3/
HTTP/1.1 200 OK
Server: nginx/1.15.8
Date: Sun, 13 Jan 2019 04:54:30 GMT
Content-Type: text/html
Content-Length: 612
Last-Modified: Wed, 26 Dec 2018 23:21:49 GMT
Connection: keep-alive
ETag: "5c240d0d-264"
Accept-Ranges: bytes
```

5）通过 Cluster IP 访问 Nginx 应用。

查看 Nginx 应用的 Cluster IP：

```
$ kubectl get svc nginx-service
NAME            TYPE        CLUSTER-IP     EXTERNAL-IP   PORT(S)   AGE
nginx-service   ClusterIP   10.106.97.98   <none>        80/TCP    22m
```

访问 Nginx 应用，10.106.97.98 为上一步骤中查看到的 Cluster IP：

```
curl -I http://10.106.97.98/
HTTP/1.1 200 OK
Server: nginx/1.15.8
Date: Sun, 13 Jan 2019 04:55:03 GMT
Content-Type: text/html
Content-Length: 612
Last-Modified: Wed, 26 Dec 2018 23:21:49 GMT
Connection: keep-alive
ETag: "5c240d0d-264"
Accept-Ranges: bytes
```

6）通过 NodePort 访问 Nginx 应用。

查看 Nginx 应用的 Cluster IP：

```
$ kubectl get svc nginx-service-nodeport
NAME                     TYPE       CLUSTER-IP       EXTERNAL-IP   PORT(S)        AGE
nginx-service-nodeport   NodePort   10.105.227.107   <none>        80:31499/TCP   110s
```

访问 Nginx 应用，31499 为上一步骤中查看到的 NodePort：

```
curl -I http://lab1:31499/
HTTP/1.1 200 OK
Server: nginx/1.15.8
Date: Sun, 13 Jan 2019 04:55:58 GMT
Content-Type: text/html
Content-Length: 612
Last-Modified: Wed, 26 Dec 2018 23:21:49 GMT
Connection: keep-alive
ETag: "5c240d0d-264"
Accept-Ranges: bytes

curl -I http://lab2:31499/
HTTP/1.1 200 OK
Server: nginx/1.15.8
Date: Sun, 13 Jan 2019 04:56:37 GMT
Content-Type: text/html
Content-Length: 612
Last-Modified: Wed, 26 Dec 2018 23:21:49 GMT
Connection: keep-alive
ETag: "5c240d0d-264"
```

```
Accept-Ranges: bytes

curl -I http://lab3:31499/
HTTP/1.1 200 OK
Server: nginx/1.15.8
Date: Sun, 13 Jan 2019 04:56:54 GMT
Content-Type: text/html
Content-Length: 612
Last-Modified: Wed, 26 Dec 2018 23:21:49 GMT
Connection: keep-alive
ETag: "5c240d0d-264"
Accept-Ranges: bytes
```

7）DNS 解析测试。

部署测试 Pod：

```
# 生成部署文件
$ cat >dns-test.yaml<<EOF
apiVersion: v1
kind: Pod
metadata:
  name: dns-test
spec:
  containers:
  - image: radial/busyboxplus:curl
    name: dns-test
    stdin: true
    tty: true
    resources:
      requests:
        cpu: 50m
        memory: 50Mi
      limits:
        cpu: 100m
        memory: 100Mi
EOF

# 部署
$ kubectl apply -f dns-test.yaml

# 查看状态
$ kubectl get pod dns-test
NAME          READY      STATUS      RESTARTS     AGE
dns-test      1/1        Running     0            84s
```

DNS 解析测试：

```
$ kubectl exec dns-test -- nslookup kubernetes
Server:    10.96.0.10
Address 1: 10.96.0.10 kube-dns.kube-system.svc.cluster.local
```

```
Name:      kubernetes
Address 1: 10.96.0.1 kubernetes.default.svc.cluster.local

$ kubectl exec dns-test -- nslookup nginx-service
Server:    10.96.0.10
Address 1: 10.96.0.10 kube-dns.kube-system.svc.cluster.local

Name:      nginx-service
Address 1: 10.106.97.98 nginx-service.default.svc.cluster.local

$ kubectl exec dns-test -- nslookup www.baidu.com
Server:    10.96.0.10
Address 1: 10.96.0.10 kube-dns.kube-system.svc.cluster.local

Name:      www.baidu.com
Address 1: 115.239.210.27
Address 2: 115.239.211.112
```

8）通过DNS访问Nginx应用：

```
$ kubectl exec dns-test -- curl -s -I http://nginx-service/
HTTP/1.1 200 OK
Server: nginx/1.15.8
Date: Sun, 13 Jan 2019 05:12:22 GMT
Content-Type: text/html
Content-Length: 612
Last-Modified: Wed, 26 Dec 2018 23:21:49 GMT
Connection: keep-alive
ETag: "5c240d0d-264"
Accept-Ranges: bytes
```

9）清理：

```
$ kubectl delete service nginx-service-nodeport
$ kubectl delete service nginx-service
$ kubectl delete deployment nginx
$ kubectl delete -f dns-test.yaml
```

4.6 注意事项与技巧

注意事项：

- 实验时三台虚拟机要保持时间同步，如果你用vagrant完成实验，那么基本不会出现这种问题。
- 要注意当前的操作用户，安装软件时使用的是sudo模拟root用户，在操作kubernetes集群时，使用的是vagrant普通用户。

1. Pod 无法启动的排错步骤

1）查看 Pod 日志：

```
$ kubectl logs pod-name
```

2）查看 Pod 详细信息。

重点查看事件（events）信息：

```
$ kubectl describe pod-name
```

3）查看 Docker 日志。

通过下面的命令找到容器运行的节点：

```
$ kubectl get pod -o wide
```

登录到容器运行的节点，找到出错容器：

```
$ sudo docker ps -a
```

查看容器日志：

```
$ sudo docker logs container-id
```

4）查看 Kubelet 及其他组件日志：

```
$ sudo journalctl -u kubelet
```

2. 创建集群失败后如何快速重来

当创建集群失败后，可以通过如下的命令快速恢复到没有初始化集群的状态，如下命令需要在 master 节点和已经加入集群的 node 节点上执行：

```
$ sudo kubeadm reset
```

3. 重新获取集群的 join 命令

可以通过以下两种方法重新获取 join 命令。

获取一个有过期时间的 token：

```
$ sudo kubeadm token create --print-join-command
kubeadm join 11.11.11.111:6443 --token uhuqes.c6u9p96s0dfipj3h --discovery-token-ca-cert-hash sha256:c6290480baaef64f3f1c6b6861de8685ca74d5a4a7c2703eace18a2e2117e87d
```

获取一个永不过期的 token：

```
$ sudo kubeadm token create --print-join-command --ttl=0
kubeadm join 11.11.11.111:6443 --token k6j5ic.xh3b29r8unfsbvax --discovery-token-ca-cert-hash sha256:c6290480baaef64f3f1c6b6861de8685ca74d5a4a7c2703eace18a2e2117e87d
```

4. 让 masterl 不再接收负载

使用如下的命令来停止 master 再接收普通 pod 负载，但是已经调度到 master 节点上的 pod 不受影响：

```
$ kubectl taint nodes lab1 node-role.kubernetes.io/master=true:NoSchedule
node/lab1 tainted
```

4.7　本章小结

官方有编译好的对应平台的二进制文件，但使用二进制文件的方式部署一个 Kubernetes 集群对于新手来说难度较大，而且这对于学习 Istio 并没有太大的作用，所以本书选择使用 Kubeadm 来快速创建一个 Kubernetes 集群。使用什么方式部署 Kubernetes 集群并不会影响我们后续的实验，读者如果对于使用二进制文件的方式部署 Kubernetes 集群感兴趣，可以自行参考相关的资料。由于创建 Kubernetes 集群时，需要修改的配置项比较多，所以实验时需要格外注意，有时一个参数配置错误，都有可能导致整个 Kubernetes 集群无法成功创建。

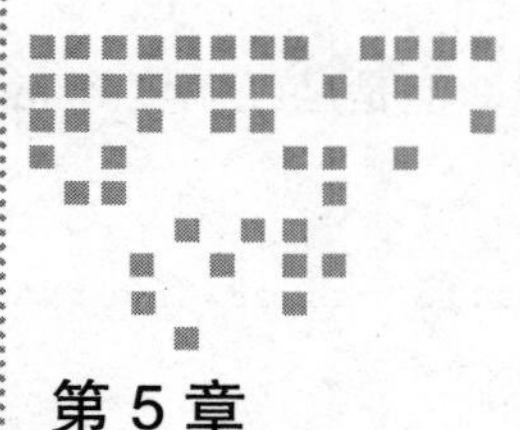

Chapter 5 第 5 章

Istio 部署与常用命令

本章开始进入 Istio 主题，介绍如何在 Kubernetes 集群上部署 Istio，包括 Istio 中常用的资源类型，操作 Istio 集群时常用的命令，以及本章实验时的一些注意事项。由于实验环境机器可能出现性能不足的情况，本章也介绍了实验时用到的 Istio 基础功能的最小化部署，这样，可使 Istio 集群占用的资源大大减少，有助于更流畅地进行实验。

5.1 部署 Istio

本节主要介绍部署 Istio 的方法，包括使用官方示例的方式来部署 Istio，定制最小化部署 Istio。后续的大部分实验都选择定制最小化部署 Istio 的方式，使用官方示例的部署方式时会有特殊说明。

1. 下载安装 Istio

到 Istio 的发布页面 https://github.com/istio/istio/releases，查找当前可用版本的对应操作系统的包，由于这个压缩包在国外服务器上，可能由于网络问题不能顺利下载，可以自己先在网络正常的机器上下载完成后，再上传到实验环境的虚拟机。具体步骤如下。

下载 Istio 包：

```
$ sudo yum install -y wget
$ wget https://github.com/istio/istio/releases/download/1.0.3/istio-1.0.3-
    linux.tar.gz
```

解压安装：

```
$ tar xf istio-1.0.3-linux.tar.gz
$ sudo mv istio-1.0.3 /usr/local
$ sudo ln -sv /usr/local/istio-1.0.3/ /usr/local/istio
```

添加到 PATH 路径中：

```
$ echo 'export PATH=/usr/local/istio/bin:$PATH' | sudo tee /etc/profile.d/
    istio.sh
```

验证安装：

```
$ source /etc/profile.d/istio.sh
$ istioctl version
Version: 1.0.3
GitRevision: a44d4c8bcb427db16ca4a439adfbd8d9361b8ed3
User: root@0ead81bba27d
Hub: docker.io/istio
GolangVersion: go1.10.4
BuildStatus: Clean
```

2. 最小化部署 Istio

由于实验环境机器资源不足，如果部署全部组件，可能导致实验失败，无法完成实验目标。最小化部署由于没有开启监控数据收集和调用链追踪等功能，可以节省部分资源。因此实验时推荐使用本小节介绍的最小化部署 Istio，如果没有特殊说明，本书中的实验均在此种方式部署的基础上进行的。

1）安装 Helm。

到 Helm 的发布页面 https://github.com/helm/helm/releases，查找当前可用的版本的包地址，由于这个压缩包也在国外服务器上，可能由于网络问题不能顺利下载，可以自己先在网络正常的机器上下载完成后，再上传到实验环境的虚拟机。

下载 Helm 包：

```
$ wget https://storage.googleapis.com/kubernetes-helm/helm-v2.12.2-linux-
    amd64.tar.gz
```

安装并验证 Helm：

```
$ tar xf helm-v2.12.2-linux-amd64.tar.gz
$ sudo mv linux-amd64/ /usr/local/helm-2.12.2
$ sudo ln -sv /usr/local/helm-2.12.2 /usr/local/helm
$ echo 'export PATH=/usr/local/helm:$PATH' | sudo tee /etc/profile.d/helm.sh
$ source /etc/profile.d/helm.sh
$ helm version
Client: &version.Version{SemVer:"v2.12.2", GitCommit:"7d2b0c73d734f6586ed222a5
```

```
    67c5d103fed435be", GitTreeState:"clean"}
Error: could not find tiller
```

2）生成Istio部署文件。

根据配置生成Istio的定制最小化部署文件：

```
$ helm template /usr/local/istio/install/kubernetes/helm/istio  \
--name istio --namespace istio-system \
--set global.hyperkube.hub=registry.cn-shanghai.aliyuncs.com/gcr-k8s \
--set pilot.resources.requests.memory=500Mi \
--set gateways.istio-ingressgateway.enabled=true \
--set gateways.istio-egressgateway.enabled=true \
--set galley.enabled=true \
--set sidecarInjectorWebhook.enabled=true \
--set global.mtls.enabled=false \
--set prometheus.enabled=false \
--set grafana.enabled=false \
--set tracing.enabled=false \
--set global.proxy.envoyStatsd.enabled=false \
--set servicegraph.enabled=false > istio-mini.yaml
```

3）部署Istio。

创建istio-system命名空间：

```
$ kubectl create ns istio-system
```

创建Istio CRD：

```
$ kubectl apply -f /usr/local/istio/install/kubernetes/helm/istio/templates/
    crds.yaml
```

查看Istio CRD：

```
$ kubectl get crd
NAME                                        CREATED AT
...
meshpolicies.authentication.istio.io        2019-01-13T05:37:28Z
metrics.config.istio.io                     2019-01-13T05:37:29Z
noops.config.istio.io                       2019-01-13T05:37:29Z
opas.config.istio.io                        2019-01-13T05:37:29Z
policies.authentication.istio.io            2019-01-13T05:37:28Z
prometheuses.config.istio.io                2019-01-13T05:37:29Z
quotas.config.istio.io                      2019-01-13T05:37:29Z
quotaspecbindings.config.istio.io           2019-01-13T05:37:28Z
quotaspecs.config.istio.io                  2019-01-13T05:37:29Z
...
virtualservices.networking.istio.io         2019-01-13T05:53:28Z
```

部署Istio相关组件：

```
$ kubectl apply -f istio-mini.yaml
```

查看 Istio 组件状态，当组件全部处于 Running 或者 Completed 时再进行之后的实验，由于需要拉取较多镜像，如果网速较慢，可能需要等待很长一段时间。如果没有跟着之前的步骤配置镜像进行加速拉取，会更缓慢，强烈建议配置镜像加速。使用如下命令查看 Istio 组件的部署状态：

```
$ kubectl get deploy -n istio-system
NAME                      DESIRED   CURRENT   UP-TO-DATE   AVAILABLE   AGE
istio-citadel             1         1         1            0           12s
istio-egressgateway       1         1         1            0           13s
istio-galley              1         1         1            0           13s
istio-ingressgateway      1         1         1            0           12s
istio-pilot               1         1         1            0           12s
istio-policy              1         1         1            0           12s
istio-sidecar-injector    1         1         1            0           12s
istio-telemetry           1         1         1            0           12s

$ kubectl get job -n istio-system
NAME                           COMPLETIONS   DURATION   AGE
istio-cleanup-secrets          0/1           37s        37s
istio-security-post-install    0/1           37s        37s

$ kubectl get pod -n istio-system
NAME                                       READY   STATUS      RESTARTS   AGE
istio-citadel-6955bc9cb7-wg4jt             1/1     Running     0          5m32s
istio-cleanup-secrets-7hbxh                0/1     Completed   0          5m35s
istio-egressgateway-7dc5cbbc56-4vx9g       1/1     Running     0          5m33s
istio-galley-545b6b8f5b-zbn8w              1/1     Running     0          5m33s
istio-ingressgateway-7958d776b5-w6ptc      1/1     Running     0          5m32s
istio-pilot-5fb59666cb-75lw5               2/2     Running     0          5m32s
istio-policy-5c689f446f-x9njr              2/2     Running     0          5m32s
istio-security-post-install-xsgkm          0/1     Completed   0          5m35s
istio-sidecar-injector-99b476b7b-zj7b7     1/1     Running     0          5m32s
istio-telemetry-55d68b5dfb-z4pkn           2/2     Running     0          5m32s

$ kubectl get svc -n istio-system
NAME            TYPE        CLUSTER-IP      EXTERNAL-IP   PORT(S)             AGE
...
istio-galley    ClusterIP   10.97.150.161   <none>        443/TCP,9093/TCP    5m56s
istio-ingressgateway   LoadBalancer   10.99.91.255   <pending>   80:31380/
TCP,443:31390/TCP,31400:31400/TCP,15011:32193/TCP,8060:32090/TCP,853:30692/
TCP,15030:30730/TCP,15031:32677/TCP   5m55s
istio-pilot   ClusterIP   10.101.237.88   <none>   15010/TCP,15011/TCP,8080/
TCP,9093/TCP   5m55s
...
```

3. 官方示例部署 Istio

使用官方示例的方式部署 Istio 将部署较多的组件，占用资源多但是功能全，会包含

Prometheus、Grafana、Jaeger、ServiceGraph等组件。本书后面介绍服务监控观测时，需要使用此种部署方式。

1）配置Istio。

复制部署源文件，便于出错时恢复：

```
$ cp /usr/local/istio/install/kubernetes/istio-demo.yaml /usr/local/istio/
install/kubernetes/istio-demo.yaml.ori
```

由于实验使用的虚拟机每台只有2G内存，默认情况下pilot的deployment请求2G内存，为了使实验顺利进行。把13 573行左右关于istio-pilot的内存配置修改成如下内容：

```
containers:
    - name: discovery
        image: "docker.io/istio/pilot:1.0.3"
        imagePullPolicy: IfNotPresent
        args:
        - "discovery"
        ports:
        - containerPort: 8080
        - containerPort: 15010
        ...
        - name: PILOT_PUSH_THROTTLE_COUNT
            value: "100"
        - name: PILOT_TRACE_SAMPLING
            value: "100"
        resources:
            requests:
                cpu: 500m
                memory: 500Mi
```

修改镜像使用国内镜像，加速部署，执行以下命令修改镜像：

```
$ sed -i 's@quay.io/coreos/hyperkube:v1.7.6_coreos.0@registry.cn-shanghai.
aliyuncs.com/gcr-k8s/hyperkube:v1.7.6_coreos.0@g' /usr/local/istio/install/
kubernetes/istio-demo.yaml
```

2）部署Istio。

创建Istio CRD：

```
$ kubectl apply -f /usr/local/istio/install/kubernetes/helm/istio/templates/
crds.yaml
```

查看Istio CRD：

```
$ kubectl get crd
NAME                                                        CREATED AT
...
```

```
meshpolicies.authentication.istio.io          2019-01-13T05:57:28Z
metrics.config.istio.io                       2019-01-13T05:57:29Z
noops.config.istio.io                         2019-01-13T05:57:29Z
opas.config.istio.io                          2019-01-13T05:57:29Z
policies.authentication.istio.io              2019-01-13T05:57:28Z
prometheuses.config.istio.io                  2019-01-13T05:57:29Z
quotas.config.istio.io                        2019-01-13T05:57:29Z
quotaspecbindings.config.istio.io             2019-01-13T05:57:28Z
quotaspecs.config.istio.io                    2019-01-13T05:57:29Z
...
virtualservices.networking.istio.io           2019-01-13T05:57:28Z
```

3）部署Istio相关组件：

```
$ kubectl apply -f /usr/local/istio/install/kubernetes/istio-demo.yaml
```

4）查看Istio组件状态，当组件pod全部处于Running或者Completed状态时再进行之后的实验步骤，由于需要拉取较多镜像，如果网速较慢，可能需要等待较长一段时间。强烈建议配置镜像加速。Istio组件状态如下：

```
$ kubectl get pod -n istio-system
NAME                                      READY   STATUS      RESTARTS   AGE
grafana-546d9997bb-bq5n9                  1/1     Running     0          5m15s
istio-citadel-6955bc9cb7-x65tz            1/1     Running     0          5m13s
istio-cleanup-secrets-khzc5               0/1     Completed   0          5m22s
istio-egressgateway-7dc5cbbc56-w85fr      1/1     Running     0          5m16s
istio-galley-545b6b8f5b-khmck             1/1     Running     0          5m16s
istio-grafana-post-install-vf27h          0/1     Completed   0          5m22s
istio-ingressgateway-7958d776b5-tlnmj     1/1     Running     0          5m16s
istio-pilot-64958c46fc-q46t4              2/2     Running     0          5m14s
istio-policy-5c689f446f-rxbh7             2/2     Running     0          5m14s
istio-security-post-install-6wr8m         0/1     Completed   0          5m22s
istio-sidecar-injector-99b476b7b-m689h    1/1     Running     0          5m13s
istio-telemetry-55d68b5dfb-rrnh8          2/2     Running     0          5m14s
istio-tracing-6445d6dbbf-8vrjf            1/1     Running     0          5m13s
prometheus-65d6f6b6c-gn9rl                1/1     Running     0          5m14s
servicegraph-57c8cbc56f-2bc85             1/1     Running     1          5m13s
$ kubectl get svc -n istio-system
NAME     TYPE     CLUSTER-IP    EXTERNAL-IP     PORT(S)  AGE
grafana      ClusterIP      10.108.59.160      <none>      3000/TCP      5m46s
istio-citadel    ClusterIP    10.109.55.113    <none>    8060/TCP,9093/TCP    5m45s
istio-egressgateway    ClusterIP    10.108.104.70    <none>    80/TCP,443/TCP    5m47s
istio-galley    ClusterIP    10.105.229.22    <none>    443/TCP,9093/TCP    5m47s
istio-ingressgateway    LoadBalancer    10.105.236.32    <pending>
80:31380/TCP,443:31390/TCP,31400:31400/TCP,15011:32075/TCP,8060:32147/
TCP,853:32496/TCP,15030:30157/TCP,15031:30711/TCP    5m46s
istio-pilot    ClusterIP    10.104.117.192    <none>    15010/TCP,15011/TCP,8080/
TCP,9093/TCP    5m46s
istio-policy    ClusterIP    10.99.130.214    <none>    9091/TCP,15004/TCP,9093/
TCP    5m46s
```

```
    istio-sidecar-injector    ClusterIP    10.99.223.75    <none>    443/TCP    5m45s
    istio-telemetry    ClusterIP    10.107.87.105    <none>    9091/TCP,15004/TCP,
9093/TCP,42422/TCP    5m46s
    jaeger-agent    ClusterIP    None    <none>    5775/UDP,6831/UDP,6832/UDP    5m39s
    jaeger-collector    ClusterIP    10.110.145.202    <none>    14267/TCP,14268/TCP    5m39s
    jaeger-query        ClusterIP    10.107.254.90     <none>    16686/TCP    5m39s
    prometheus          ClusterIP    10.108.119.47     <none>    9090/TCP     5m45s
    servicegraph        ClusterIP    10.104.167.142    <none>    8088/TCP     5m45s
    tracing             ClusterIP    10.111.8.183      <none>    80/TCP       5m39s
    zipkin              ClusterIP    10.101.31.65      <none>    9411/TCP     5m39s
```

注意 此时 Istio 已经安装完成，可以对虚拟机环境进行快照保存，方便后续 Istio 实验环境快速创建。快照保存方法可以参考本章后面 5.5 节“注意事项与技巧”。

4. 部署官方 bookinfo 示例

1）开启 default 命名空间的 Istio 自动注入功能：

```
$ kubectl label namespace default istio-injection=enabled
namespace/default labeled
```

2）部署 bookinfo：

```
$ kubectl apply -f /usr/local/istio/samples/bookinfo/platform/kube/bookinfo.yaml
service/details created
deployment.extensions/details-v1 created
service/ratings created
deployment.extensions/ratings-v1 created
service/reviews created
deployment.extensions/reviews-v1 created
deployment.extensions/reviews-v2 created
deployment.extensions/reviews-v3 created
service/productpage created
deployment.extensions/productpage-v1 created
```

3）查看部署状态。

使用如下命令查看部署状态：

```
$ kubectl get pod
NAME                              READY     STATUS     RESTARTS    AGE
details-v1-876bf485f-tpwr8        2/2       Running    0           6m45s
productpage-v1-8d69b45c-dxttl     2/2       Running    0           6m44s
ratings-v1-7c9949d479-nbl58       2/2       Running    0           6m45s
reviews-v1-85b7d84c56-svwp2       2/2       Running    0           6m45s
reviews-v2-cbd94c99b-249rj        2/2       Running    0           6m45s
reviews-v3-748456d47b-bx6v2       2/2       Running    0           6m45s
```

```
$ kubectl get svc
NAME          TYPE        CLUSTER-IP       EXTERNAL-IP   PORT(S)    AGE
details       ClusterIP   10.100.86.70     <none>        9080/TCP   7m
kubernetes    ClusterIP   10.96.0.1        <none>        443/TCP    95m
productpage   ClusterIP   10.98.228.136    <none>        9080/TCP   6m59s
ratings       ClusterIP   10.108.237.160   <none>        9080/TCP   7m
reviews       ClusterIP   10.109.71.86     <none>        9080/TCP   7m
```

当 pod 全部处于 Running 状态时再进行这些步骤，由于需要拉取较多镜像，如果网速较慢，可能需要等待较长一段时间。

4）使用 Gateway 创建访问入口：

```
$ kubectl apply -f /usr/local/istio/samples/bookinfo/networking/bookinfo-gateway.yaml
gateway.networking.istio.io/bookinfo-gateway created
virtualservice.networking.istio.io/bookinfo created
```

5）查看 Gateway：

```
$ kubectl get gateway
NAME               AGE
bookinfo-gateway   18s
```

6）获取访问地址。使用如下的命令获取访问入口地址：

```
$ export INGRESS_PORT=$(kubectl -n istio-system get service istio-ingressgateway -o jsonpath='{.spec.ports[?(@.name=="http2")].nodePort}')
$ export SECURE_INGRESS_PORT=$(kubectl -n istio-system get service istio-ingressgateway -o jsonpath='{.spec.ports[?(@.name=="https")].nodePort}')
$ export INGRESS_HOST=$(kubectl get po -l istio=ingressgateway -n istio-system -o 'jsonpath={.items[0].status.hostIP}')
$ export GATEWAY_URL=$INGRESS_HOST:$INGRESS_PORT
$ echo http://$GATEWAY_URL/productpage
$ curl -I http://$GATEWAY_URL/productpage
HTTP/1.1 200 OK
content-type: text/html; charset=utf-8
content-length: 4415
server: envoy
Date: Sun, 13 Jan 2019 06:24:23 GMT
x-envoy-upstream-service-time: 1865
```

如果没有成功，可以稍等片刻重试，这是由于机器性能不够，导致的 Istio 配置生效会稍慢。根据机器性能，有时可能需要等待更久的时间，此步骤实验失败并不会影响我们后续其他实验。

7）使用浏览器访问。使用浏览器访问 http://11.11.11.111:31380/productpage，多次刷新将可以看到如图 5-1 所示的不同的评价界面。

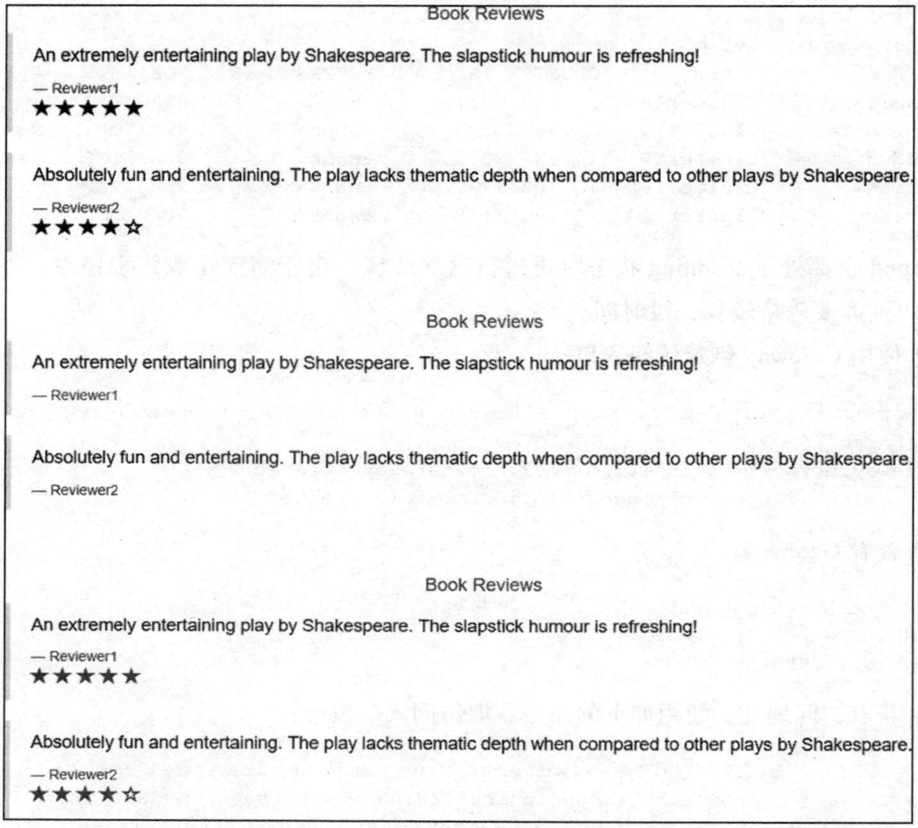

Book Reviews

An extremely entertaining play by Shakespeare. The slapstick humour is refreshing!

— Reviewer1

★★★★★

Absolutely fun and entertaining. The play lacks thematic depth when compared to other plays by Shakespeare.

— Reviewer2

★★★★☆

Book Reviews

An extremely entertaining play by Shakespeare. The slapstick humour is refreshing!

— Reviewer1

Absolutely fun and entertaining. The play lacks thematic depth when compared to other plays by Shakespeare.

— Reviewer2

Book Reviews

An extremely entertaining play by Shakespeare. The slapstick humour is refreshing!

— Reviewer1

★★★★★

Absolutely fun and entertaining. The play lacks thematic depth when compared to other plays by Shakespeare.

— Reviewer2

★★★★☆

图 5-1　部署官方 bookinfo 示例结果

5. 服务使用 Istio 的要求

为了让服务成为服务网格的一部分，部署在 Kubernetes 中的 Service 和 Pod 资源必须要满足如下一些要求。

（1）使用命名端口

必须给 Service 的端口命名，而且命令的格式必须以协议开头，之后可以接连字符和其他字符，例如：http2-foo、http 都是合法的命名方式，而 http2foo 不合法。协议支持 http、http2、grpc、mongo、redis。如果端口命名不符合上述规则或者没有使用命名端口，该端口上的流量将会被视为 TCP 流量，除非 Service 里指明了协议是 UDP。命名端口形式如下：

```
kind: Service
apiVersion: v1
metadata:
  name: service-go
  labels:
```

```
    app: service-go
spec:
  selector:
    app: service-go
  ports:
    - name: http
      port: 80
```

（2）服务关联问题

如果一个 Pod 属于多个 Service，多个 Service 将不能使用同一个端口，并设置多个不同的协议，例如：两个 Service 选择了相同的后端 Pod 实例，但是分别使用了 HTTP 和 TCP 协议，如图 5-2 所示。

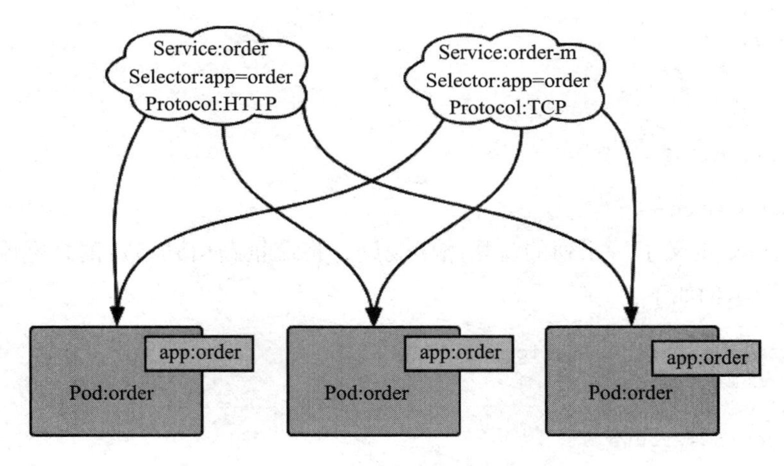

图 5-2 服务关联问题示例

（3）部署使用 app 和 version 标签

使用 Kubernetes 的 Deployment 时，指定明确的 app 和 version 标签是非常推荐的。每个 Deployment 都应该有一个有意义的唯一的 app 标签，并指定一个 version 标签用于指定应用的版本。app 标签用于分布式调用链追踪，app 和 version 标签也会用于 Istio 的度量指标收集。

5.2 常用资源类型

5.2.1 流量控制

流量控制资源有：DestinationRule、VirtualService、Gateway、ServiceEntry、EnvoyFilter，主要用于控制服务的路由，控制集群出入口的流量等功能。下面分别介绍。

（1）DestinationRule

DestinationRule定义了流量路由规则匹配后流量的访问策略。在这些策略中可以定义负载均衡、连接池大小，以及负载均衡池中不健康实例的探测和实例的摘除规则等。示例如下：

```
apiVersion: networking.istio.io/v1alpha3
kind: DestinationRule
metadata:
  name: reviews-destination
spec:
  host: reviews.prod.svc.cluster.local
  subsets:
  - name: v1
    labels:
      version: v1
  - name: v2
    labels:
      version: v2
```

（2）VirtualService

VirtualService定义了一系列的流量路由规则，将流量路由到指定的目标服务或者目标服务的子版本。示例如下：

```
apiVersion: networking.istio.io/v1alpha3
kind: VirtualService
metadata:
  name: reviews-route
spec:
  hosts:
  - reviews.prod.svc.cluster.local
  http:
  - match:
    - uri:
        prefix: "/wpcatalog"
    - uri:
        prefix: "/consumercatalog"
    rewrite:
      uri: "/newcatalog"
    route:
    - destination:
        host: reviews.prod.svc.cluster.local
        subset: v2
  - route:
    - destination:
        host: reviews.prod.svc.cluster.local
        subset: v1
```

（3）Gateway

Gateway 用于配置服务网格流量的边界负载均衡器，负责接收进入服务网格和流出服务网格的 HTTP/TCP 连接。用于定义一系列应该开放的端口及其协议。示例如下：

```
apiVersion: networking.istio.io/v1alpha3
kind: Gateway
metadata:
  name: bookinfo-gateway
spec:
  selector:
    istio: ingressgateway # use istio default controller
  servers:
  - port:
      number: 80
      name: http
      protocol: HTTP
    hosts:
    - "*"
```

（4）ServiceEntry

ServiceEntry 用于添加额外的注册条目到 Istio 内部的注册中心，以此来满足服务网格里自动发现的服务访问或者路由到这些手动指定的服务，这可以把外部服务导入网格内部。示例如下：

```
apiVersion: networking.istio.io/v1alpha3
kind: ServiceEntry
metadata:
  name: external-svc-mongocluster
spec:
  hosts:
  - mymongodb.somedomain # not used
  addresses:
  - 192.192.192.192/24 # VIPs
  ports:
  - number: 27018
    name: mongodb
    protocol: MONGO
  location: MESH_INTERNAL
  resolution: STATIC
  endpoints:
  - address: 2.2.2.2
  - address: 3.3.3.3
```

（5）EnvoyFilter

EnvoyFilter 描述了 Envoy 代理特有的过滤配置，这些配置可用于定制 Envoy 代理，改变 Istio 配置的 Envoy 代理的行为。使用这个特性时要非常的小心，因为错误的配置可能会破坏整个服务网格的稳定性。示例如下：

```
apiVersion: networking.istio.io/v1alpha3
kind: EnvoyFilter
metadata:
  name: httpbin
spec:
  workloadLabels:
    app: httpbin
  filters:
  - listenerMatch:
      portNumber: 8000
      listenerType: SIDECAR_INBOUND
    filterName: envoy.lua
    filterType: HTTP
    filterConfig:
      inlineCode: |
        function envoy_on_request(request_handle)
          request_handle:headers():add("X-Foo", "bar")
        end
        function envoy_on_response(response_handle)
          body_size = response_handle:body():length()
          response_handle:headers():add("X-Response-Body-Size", tostring(body_size))
    end
```

5.2.2 请求配额

请求配额可以对服务请求进行限流配置：

- Quota——指定用于配额限制的数据维度。可以理解为需要从请求中提取的信息及其提取规则，如：源地址、目标地址、目标版本等。示例如下：

```
apiVersion: "config.istio.io/v1alpha2"
kind: quota
metadata:
  name: requestcount
  namespace: istio-system
spec:
  dimensions:
    source: request.headers["x-forwarded-for"] | "unknown"
    destination: destination.labels["app"] | destination.service | "unknown"
    destinationVersion: destination.labels["version"] | "unknown"
```

- Redisquota——基于 Redis 实现的具体 Quota，使用 Quota 定义的规则进行数据维度提取，指定具体的维度来进行配额。
- Memquota——基于内存实现的具体 Quota，使用 Quota 定义的规则进行数据维度提取，指定具体的维度来进行配额。

- Quotaspec——定义指定 Quota 配额标准，例如：每次请求计数加 1，或者每次请求计数加 2，可以方便地用于限流、计费等场景。
- Quotaspecbinding——把 Quotaspec 和具体的 Service 相绑定。
- Rule——把配置的 Instance 对象传递给 Handler 来处理，在配额功能中使用时，负责绑定 Quota 和具体 Quota 实现，把 Quota 对象传递给具体的 Quota 进行处理。

5.2.3　mTLS 认证策略

mTLS 认证策略包括：MeshPolicy、Policy，主要用于服务间的流量加密和认证。

（1）MeshPolicy

MeshPolicy 定义了全局的 mTLS 认证策略，这个资源的定义只能有一个实例。示例如下：

```
apiVersion: authentication.istio.io/v1alpha1
kind: MeshPolicy
metadata:
  name: default
spec:
  peers:
  - mtls: {}
```

（2）Policy

Policy 用于配置命名空间或者服务的 mTLS 认证策略。示例如下：

```
apiVersion: authentication.istio.io/v1alpha1
kind: Policy
metadata:
  name: service-go
spec:
  targets:
  - name: service-go
  peers:
  - mtls: {}
---
apiVersion: networking.istio.io/v1alpha3
kind: DestinationRule
metadata:
  name: service-go
spec:
  host: "service-go.default.svc.cluster.local"
  trafficPolicy:
    tls:
      mode: ISTIO_MUTUAL
```

5.2.4　RBAC 访问权限

RBAC 访问权限包括：ServiceRole、ServiceRoleBinding、RbacConfig，主要用于服务

间的细粒度访问控制。

（1）ServiceRole

ServiceRole用于定义一系列的服务访问控制权限，示例如下：

```
apiVersion: "rbac.istio.io/v1alpha1"
kind: ServiceRole
metadata:
  name: products-viewer
  namespace: default
spec:
  rules:
  - services: ["products.svc.cluster.local"]
    methods: ["GET", "HEAD"]
    constraints:
    - key: "destination.labels[version]"
      value: ["v1", "v2"]
```

（2）ServiceRoleBinding

ServiceRoleBinding给ServiceRole分配一系列的授权对象，示例如下：

```
apiVersion: "rbac.istio.io/v1alpha1"
kind: ServiceRoleBinding
metadata:
  name: test-binding-products
  namespace: default
spec:
  subjects:
  - user: alice@yahoo.com
  - properties:
      source.namespace: "abc"
  roleRef:
    kind: ServiceRole
    name: "products-viewer"
```

（3）RbacConfig

RbacConfig定义全局配置，用来控制Istio RBAC的行为，这个资源的定义只能有一个实例，示例如下：

```
apiVersion: rbac.istio.io/v1alpha1
kind: RbacConfig
metadata:
  name: default
spec:
  mode: 'ON_WITH_INCLUSION'
  inclusion:
    namespaces:
    - default
```

5.3 常用的 kubectl 命令

由于 istioctl 的子命令 create、get、replace、delete 已经全部废弃，不再建议使用，所以对资源的创建、更新、获取、删除，推荐使用 kubectl 来完成。

（1）创建

创建 Istio 相关的新资源，使用参数 --file（-f）指定配置文件，使用形式如下：

```
kubectl create [flags]
```

示例如下：

```
$ kubectl create -f example-routing.yaml
```

（2）更新

更新 Istio 相关的资源，当资源不存在时创建新资源，使用参数 --file(-f) 指定配置文件，使用形式如下：

```
kubectl apply [flags]
```

示例如下：

```
$ kubectl apply -f example-routing.yaml
```

（3）获取

获取 Istio 相关的资源，使用形式如下：

```
kubectl get <type> [<name>] [flags]
```

示例如下：

```
$ kubectl get virtualservices
$ kubectl get destinationrules
$ kubectl get virtualservice bookinfo
```

（4）删除

删除 Istio 相关的资源，可以使用参数 --file（-f）指定配置文件，使用形式如下：

```
kubectl delete <type> <name> [<name2> ... <nameN>] [flags]
```

示例如下：

```
$ kubectl delete -f example-routing.yaml
$ kubectl delete virtualservice bookinfo
```

5.4 常用的 istioctl 命令

本节主要介绍常用的 istioctl 命令。

5.4.1 通用参数说明

- --context：指定 istioctl 使用 kubeconfig 中的哪个 context，默认值为空，表示使用 kubeconfig 里的 default context。
- --istioNamespace：指定 Istio 所在命名空间，默认值为 istio-system，短参数 -i。
- --kubeconfig：指定 kubeconfig 文件的路径，默认值为空，表示使用 ~/.kube/config 文件，短参数 -c。
- --namespace：指定操作的 namespace，默认值为空，表示使用 default 命名空间，短参数 -n。

5.4.2 常用命令

先说明一下，如下的示例中 productpage-v1-8d69b45c-2z8v5 为官方示例 bookinfo 中 productpage 的 Pod 名称，需要通过如下命令先获取 Pod 名称：

```
$ kubectl get pod productpage
```

（1）手动注入 Envoy 代理

手动注入 Envoy 代理到 Pod 中，用于没有开启自动注入的情况，可以使用参数 --file(-f) 指定配置文件，使用形式如下：

```
istioctl kube-inject [flags]
```

示例如下：

```
$ kubectl apply -f <(istioctl kube-inject -f mydeployment.yaml)
$ istioctl kube-inject -f deployment.yaml -o deployment-injected.yaml
$ kubectl get deployment -o yaml | istioctl kube-inject -f - | kubectl apply -f -
```

（2）获取启动时的配置信息

在指定的 Pod 中获取 Envoy 实例启动时的配置信息，使用形式如下：

```
istioctl proxy-config bootstrap <pod-name> [flags]
```

示例如下：

```
$ istioctl proxy-config bootstrap productpage-v1-8d69b45c-2z8v5
```

（3）获取集群配置信息

从指定 Pod 中的 Envoy 实例里读取集群配置信息，使用形式如下：

```
istioctl proxy-config cluster <pod-name> [flags]
```

示例如下：

```
$ istioctl proxy-config cluster productpage-v1-8d69b45c-2z8v5
$ istioctl proxy-config cluster productpage-v1-8d69b45c-2z8v5 --port 9080
$ istioctl proxy-config cluster productpage-v1-8d69b45c-2z8v5 --fqdn details.
default.svc.cluster.local --direction outbound -o json
```

（4）获取监听器信息

从指定 Pod 中的 Envoy 实例里读取监听器信息，使用形式如下：

```
istioctl proxy-config listener <pod-name> [flags]
```

示例如下：

```
$ istioctl proxy-config listener productpage-v1-8d69b45c-2z8v5
$ istioctl proxy-config listener productpage-v1-8d69b45c-2z8v5 --port 9080
$ istioctl proxy-config listener productpage-v1-8d69b45c-2z8v5 --type HTTP
--address 0.0.0.0 -o json
```

（5）获取路由配置信息

从指定 Pod 中的 Envoy 实例里读取路由配置信息，使用形式如下：

```
istioctl proxy-config route <pod-name> [flags]
```

示例如下：

```
$ istioctl proxy-config route productpage-v1-8d69b45c-2z8v5
$ istioctl proxy-config route productpage-v1-8d69b45c-2z8v5 --name 9080
$ istioctl proxy-config route productpage-v1-8d69b45c-2z8v5 --name 9080 -o json
```

（6）获取配置的同步状态

获取 Pilot 到网格中每个 Envoy 的配置同步状态，使用形式如下：

```
istioctl proxy-status [<proxy-name>] [flags]
```

示例如下：

```
$ istioctl proxy-status
$ istioctl proxy-status productpage-v1-8d69b45c-2z8v5.default
```

5.5　注意事项与技巧

1. 清理部署

清理 bookinfo：

```
$ kubectl delete -f /usr/local/istio/samples/bookinfo/platform/kube/bookinfo.yaml
$ kubectl delete -f /usr/local/istio/samples/bookinfo/networking/bookinfo-gateway.
yaml
```

清理 Istio。如果需要继续后续实验，此步骤可以跳过。

使用如下命令清理 Istio 部署：

```
$ kubectl delete -f /usr/local/istio/install/kubernetes/istio-demo.yaml
$ kubectl delete -f /usr/local/istio/install/kubernetes/helm/istio/templates/crds.
yaml
```

2. 使用 Vagrant 保存实验环境

在部署完成 Istio 后，如果直接保存快照，之后使用时再直接恢复快照，这可能会造成 Istio 部署完成的环境无法成功启动。由于快照恢复时从第一台机器 lab1 开始，然后再恢复 lab2、lab3，当 lab1、lab2 恢复完成后，由于 lab3处于关机状态，Kubernetes 检测到 lab3 已经下线，可能会把之前分配到 lab3 节点上的 Pod 再重新分配到 lab1、lab2 上，这可能会导致 lab1、lab2 上的 CPU 内存不够，进而导致 lab1、lab2 节点上的负载过高，导致节点上的其他组件运行出现问题，从而影响整个 Kubernetes 集群的稳定性，即使等到 lab3 恢复完成后，重新加入集群，已经重新分配给 lab1、lab2 节点上的 Pod 也不会自动地再重新迁移到 lab3 节点上，这是由 Kubernetes 的调度特性决定的。所以要参考下面的步骤进行实验环境保存。

（1）暂停虚拟机

虚拟机集群部署完成 Istio 之后，暂停虚拟机（此步骤也可以省略），使用如下命令暂停虚拟机：

```
$ vagrant suspend
==> lab1: Saving VM state and suspending execution...
==> lab2: Saving VM state and suspending execution...
==> lab3: Saving VM state and suspending execution...
```

（2）保存集群快照

虚拟机暂停后，保存集群快照：

```
$ vagrant snapshot save istio-1.0.3
==> lab1: Snapshotting the machine as 'istio-1.0.3'...
==> lab1: Snapshot saved! You can restore the snapshot at any time by
==> lab1: using 'vagrant snapshot restore'. You can delete it using
==> lab1: 'vagrant snapshot delete'.
==> lab2: Snapshotting the machine as 'istio-1.0.3'...
==> lab2: Snapshot saved! You can restore the snapshot at any time by
==> lab2: using 'vagrant snapshot restore'. You can delete it using
==> lab2: 'vagrant snapshot delete'.
==> lab3: Snapshotting the machine as 'istio-1.0.3'...
==> lab3: Snapshot saved! You can restore the snapshot at any time by
==> lab3: using 'vagrant snapshot restore'. You can delete it using
==> lab3: 'vagrant snapshot delete'.
```

（3）暂停当前要恢复的虚拟机集群

暂停当前要恢复的虚拟机集群，防止干扰集群环境的恢复：

```
$ vagrant suspend
==> lab1: Saving VM state and suspending execution...
==> lab2: Saving VM state and suspending execution...
==> lab3: Saving VM state and suspending execution...
```

（4）恢复集群快照

使用如下命令恢复虚拟机集群实验环境：

```
$ vagrant snapshot restore lab2 istio-1.0.3
==> lab2: Discarding saved state of VM...
==> lab2: Restoring the snapshot 'istio-1.0.3'...
==> lab2: Resuming suspended VM...
==> lab2: Booting VM...
==> lab2: Waiting for machine to boot. This may take a few minutes...
    lab2: SSH address: 127.0.0.1:2200
    lab2: SSH username: vagrant
    lab2: SSH auth method: private key
==> lab2: Machine booted and ready!
==> lab2: Running provisioner: shell...
...

$ vagrant snapshot restore lab3 istio-1.0.3
==> lab3: Discarding saved state of VM...
==> lab3: Restoring the snapshot 'istio-1.0.3'...
==> lab3: Resuming suspended VM...
==> lab3: Booting VM...
==> lab3: Waiting for machine to boot. This may take a few minutes...
    lab3: SSH address: 127.0.0.1:2201
    lab3: SSH username: vagrant
    lab3: SSH auth method: private key
==> lab3: Machine booted and ready!
==> lab3: Running provisioner: shell...
...

$ vagrant snapshot restore lab1 istio-1.0.3
==> lab1: Discarding saved state of VM...
==> lab1: Restoring the snapshot 'istio-1.0.3'...
==> lab1: Resuming suspended VM...
==> lab1: Booting VM...
==> lab1: Waiting for machine to boot. This may take a few minutes...
    lab1: SSH address: 127.0.0.1:2222
    lab1: SSH username: vagrant
    lab1: SSH auth method: private key
==> lab1: Machine booted and ready!
==> lab1: Running provisioner: shell...
...
```

代码中把lab1的恢复放在最后，是为了最后恢复启动master节点，因为此时lab2、lab3已经恢复启动完成，Pod也都启动完成，此时lab1恢复启动并不会造成Pod重新分配，所以集群恢复后仍然能正常运行。

（5）暂停后启动集群

当暂停虚拟机集群后，如果要重新启动虚拟机集群时，注意要让lab1机器最后启动，防止发生Pod重新分配，导致集群不能稳定的情况。使用如下命令启动虚拟机集群实验环境：

```
$ vagrant up lab2 lab3 lab1
```

3. Istio注入

（1）开启自动注入的情况下如何实现部分Pod不注入

在指定namespace开启自动注入后，所有部署在该命名空间的Pod默认情况下都会被自动注入Envoy代理。可以通过在Pod的metadata部分添加如下的配置关闭该pod的自动注入功能：

```
template:
  metadata:
    annotations:
      sidecar.istio.io/inject: "false"
```

（2）开启和关闭自动注入功能

Istio的自动注入功能是namespace级别的，当需要在指定namespace开启自动注入功能时，只需要给该namespace打上标签名：istio-injection=enabled即可。当删除该标签时，自动注入功能就会自动关闭，但是对于已经被注入的Pod是不会自动取消注入的，这只对新部署的Pod有效。

开启default命名空间的自动注入功能：

```
$ kubectl label namespace default istio-injection=enabled
```

关闭default命名空间的自动注入功能：

```
$ kubectl label namespace default istio-injection-
```

（3）查看开启自动注入功能的namespace

命令如下：

```
$ kubectl get namespace -L istio-injection
```

（4）如何实现手动注入

当namespace没有开自动注入功能时，可以通过istioctl提供的子命令实现手动注入。

使用方式如下：

```
$ kubectl apply -f <(istioctl kube-inject -f mydeployment.yaml)
```

5.6 本章小结

在 Kubernetes 集群上部署一个 Istio 其实非常简单，只需要两三条命令即可，但是由于实验环境、机器性能问题，本章使用了稍微复杂的部署方式。采用定制化的方式部署 Istio，会减少部署的组件，进而减少 Istio 对虚拟机资源的需求。本章定制化部署 Istio 时开启的那些组件，已经可以满足后续大部分的实验要求，通过降低 Istio 对资源的需求，会使我们的实验更加顺利。

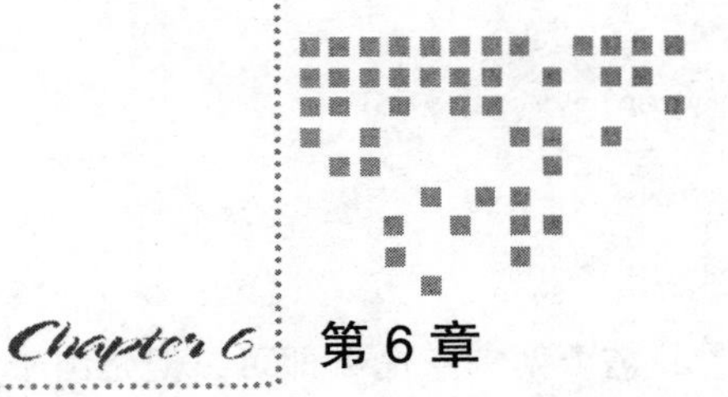

Chapter 6 第 6 章

微服务应用的部署

本章介绍微服务应用架构及其部署，包括服务的整体架构、服务的调用情况、各服务的响应数据结构，然后介绍如何在 Kubernetes 上部署并访问微服务，如何在 Istio 中部署服务，以及如何暴露并访问服务。本章搭建的微服务应用可帮助读者更好地理解后续章节的实验。

6.1 微服务应用架构

在介绍服务部署前，先介绍整个微服务应用的架构，如图 6-1 所示。

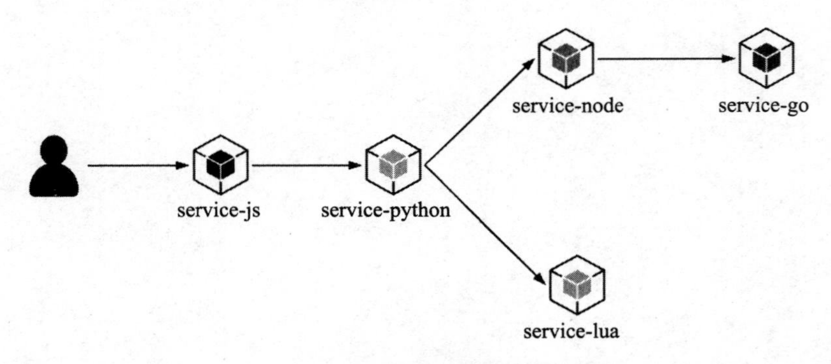

图 6-1 应用的微服务架构

其中，service-js 服务是指由 Vue/React 实现的前端页面，当用户访问前端 Web 页面时，会看到一个静态页面，当用户点击相应的按钮时，前端页面会通过浏览器异步请求后端

service-python 服务提供的 API 接口，service-python 调用后端其他服务完成用户的请求，并把结果合并处理之后发送给前端浏览器，当前端页面收到请求的响应数据时会渲染出新的页面呈现给用户。

service-go 是基础层服务，不调用其他服务，直接返回结果数据。响应结果数据如下：

```
{
    "message": "go v2"
}
```

service-lua 是中间层服务，不调用任何其他服务，直接返回结果数据。响应结果数据如下：

```
{
    "message": "lua v1"
}
```

service-node 是中间层服务，调用底层服务 service-go，整合数据响应。正常响应结果数据如下：

```
{
    "message": "node v1",
    "upstream": [{
        "message": "go v2",
        "response_time": "0.01"
    }]
}
```

service-node 调用底层服务异常时响应数据如下：

```
{
    "message": "node v1",
    "upstream": []
}
```

service-python 服务提供 API 接口给前端调用，调用中间层的 service-node 服务和 service-lua 服务，整合所有服务的数据，发送给前端 API 请求，正常响应数据如下：

```
{
    "message": "python v2",
    "upstream": [{
        "message": "lua v1",
        "response_time": 0.1
    }, {
        "message": "node v2",
        "upstream": [{
            "message": "go v1",
            "response_time": "0.01"
        }],
        "response_time": 0.1
```

```
    }]
}
```

service-python 调用后端服务异常时响应数据如下：

```
{
    "message": "python v1",
    "upstream": []
}
```

service-js 服务在浏览器中异步调用后端 service-python 接口服务，在获取到响应数据后，会通过图表工具 ECharts 库在浏览器中绘制出本次服务调用链上的调用情况，包括调用成功了哪些服务，以及调用成功服务的响应时间。在绘制的服务调用情况图中（见图 6-2），只会显示调用成功的服务，没有调用成功的服务不会展示在图中。后端的每个服务都分别有 v1、v2 两个版本，没有经过路由规则配置的时候，用户请求会以轮询的方式落到每一个版本上。在绘制的服务调用图中，会显示出本次调用的服务的版本，以便在后续的实验中更方便地观察路由配置有没有真正生效。

React 版本所有服务调用成功如图 6-2 所示。

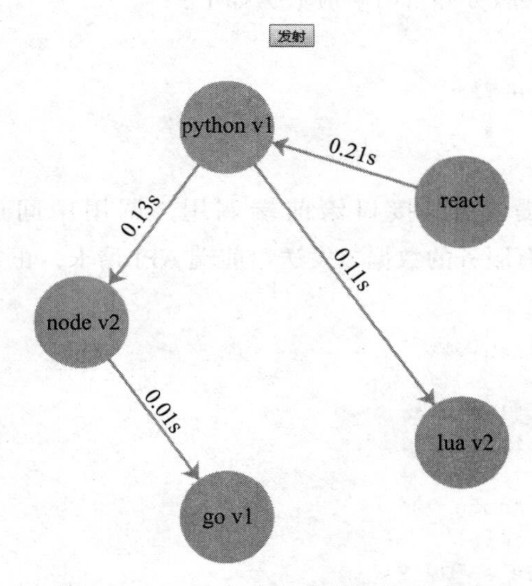

图 6-2　React 版本所有服务调用成功

React 版本部分服务调用失败如图 6-3 所示。

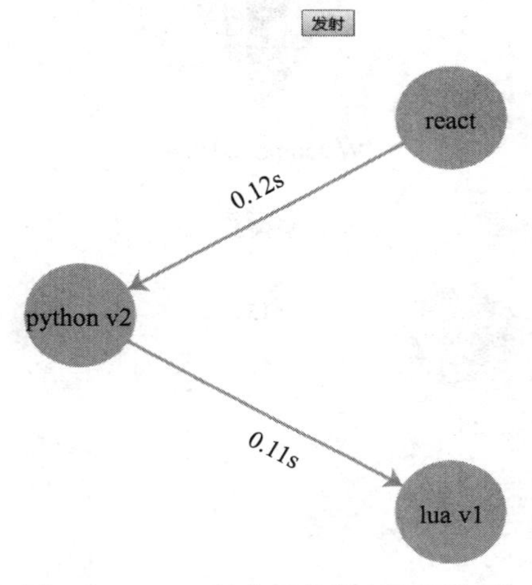

图 6-3　React 版本部分服务调用失败

Vue 版本所有服务调用成功如图 6-4 所示。

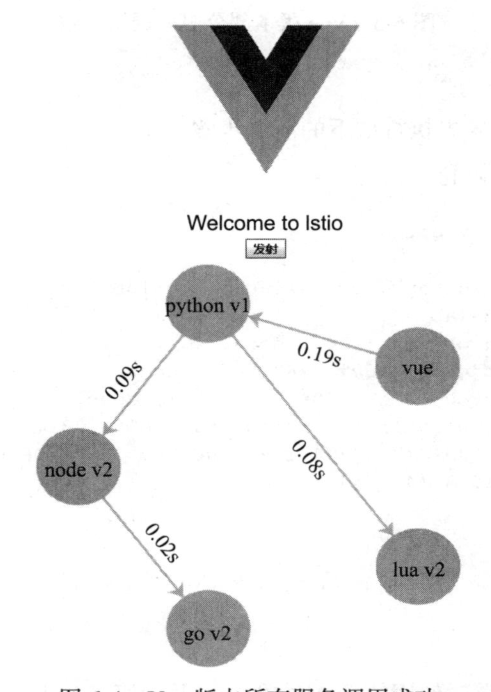

图 6-4　Vue 版本所有服务调用成功

Vue 版本部分服务调用失败如图 6-5 所示。

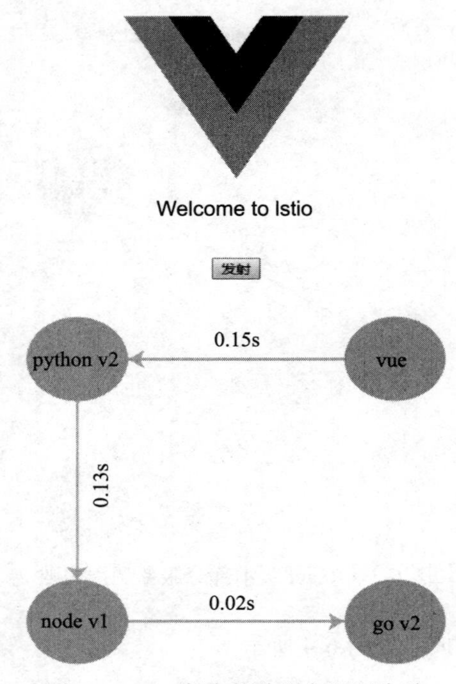

图 6-5　Vue 版本部分服务调用失败

【实验前的准备】

进行本章实验前，需要先执行如下的前置步骤。

下载实验时用到的源码仓库：

```
$ sudo yum install -y git

$ git clone https://github.com/mgxian/istio-lab
Cloning into 'istio-lab'...
remote: Enumerating objects: 247, done.
remote: Counting objects: 100% (247/247), done.
remote: Compressing objects: 100% (173/173), done.
remote: Total 774 (delta 153), reused 164 (delta 73), pack-reused 527
Receiving objects: 100% (774/774), 283.00 KiB | 229.00 KiB/s, done.
Resolving deltas: 100% (447/447), done.

$ cd istio-lab
```

6.2　部署服务

本节主要介绍服务部署时使用的 yaml 文件，以及如何在 Kubernetes 集群中部署查看服务。

1. 服务部署 yaml 文件说明

每个服务部署 yaml 文件的大同小异，这里仅以 service-go 示例说明：

```
 1 kind: Service
 2 apiVersion: v1
 3 metadata:
 4   name: service-go
 5   labels:
 6     app: service-go
 7 spec:
 8   selector:
 9     app: service-go
10   ports:
11     - name: http
12       port: 80
13 ---
14 apiVersion: apps/v1
15 kind: Deployment
16 metadata:
17   name: service-go-v1
18 spec:
19   selector:
20     matchLabels:
21       app: service-go
22       version: v1
23   replicas: 1
24   template:
25     metadata:
26       labels:
27         app: service-go
28         version: v1
29     spec:
30       restartPolicy: Always
31       containers:
32       - name: service-go
33         image: registry.cn-shanghai.aliyuncs.com/istio-lab/service-go:v1
34         ports:
35         - containerPort: 80
36         resources:
37           requests:
38             cpu: 50m
39             memory: 50Mi
40           limits:
41             cpu: 100m
42             memory: 100Mi
43 ---
44 apiVersion: apps/v1
45 kind: Deployment
46 metadata:
47   name: service-go-v2
48 spec:
```

```
49    selector:
50      matchLabels:
51        app: service-go
52        version: v2
53    replicas: 1
54    template:
55      metadata:
56        labels:
57          app: service-go
58          version: v2
59      spec:
60        restartPolicy: Always
61        containers:
62        - name: service-go
63          image: registry.cn-shanghai.aliyuncs.com/istio-lab/service-go:v2
64          ports:
65          - containerPort: 80
66          resources:
67            requests:
68              cpu: 50m
69              memory: 50Mi
70            limits:
71              cpu: 100m
72              memory: 100Mi
```

第1～12行是Kubernetes中的Service定义，第4行指定了服务名为service-go，第8、9行指定选择包含标签app=service-go的Pod，第11、12行指定服务开放在80端口，由于没有指定targetPort，则targetPort默认与port参数一致，即后端Pod开放的端口也应该是80端口。

第14～42行是Kubernetes的Deployment定义，第17行定义了Deployment的名称为service-go-v1。第19到22行定义了Deployment管理的是包含app=service-go，且version=v1标签的Pod。第26到28行的定义表示为启动的Pod添加app=service-go和version=v1的标签。第30行定义了重启策略为Always，此策略表示不论何种情况下，只要Pod状态不正常就持续重启。第32～42行定义了Pod启动一个名称为service-go的容器。第33行定义了service-go容器使用的Docker镜像及版本，第35行定义了容器开放了80端口。第36～42行定义了容器运行时需要的资源情况，requests定义了容器启动时需要的CPU和内存资源，limits定义了容器最多能使用的CPU和内存资源。

第44～72行定义了service-go-v2的Deployment配置，与service-go-v1类似，在此不再赘述。

从上面的Service定义可以知道，Service选择了所有包含了标签名为app、值为service-go的Pod，而这包含service-go-v1和service-go-v2管理的Pod，因此请求流量会轮询地发送给服务的v1版本和v2版本。

如果你是一个Kubernetes的用户，你可能会发现，此次实验编写的Deployment中并没有使用 Kubernetes 的健康查检机制 liveness 和探测机制 readiness。这是由于 Envoy 代理会对服务实例进行健康查检，而且如果你使用了 Istio 的自动 TLS 来加密服务间的通信时，由于 Kubernetes 进行健康检查时没用 TLS，这可能会导致 Kubernetes 健康检查失败，进而导致 Pod 被反复重启，影响服务的稳定性。当然也可以通过其他方式来实现，绕过 Istio 的 TLS 来完成 Kubernetes 的健康检查，在后面的章节中，也会介绍此类方法。

其他服务的 yaml 部署文件类似，只是部署名称和 app 标签值有改动，如 service-node、service-lua 等值会改动。

2. 在 Kubernetes 集群中部署服务

1）部署服务

使用上节介绍的服务部署 yaml 文件进行部署：

```
$ kubectl apply -f service/go/service-go.yaml
service/service-go created
deployment.apps/service-go-v1 created
deployment.apps/service-go-v2 created

$ kubectl apply -f service/node/service-node.yaml
service/service-node created
deployment.apps/service-node-v1 created
deployment.apps/service-node-v2 created

$ kubectl apply -f service/lua/service-lua.yaml
service/service-lua created
deployment.apps/service-lua-v1 created
deployment.apps/service-lua-v2 created

$ kubectl apply -f service/python/service-python.yaml
service/service-python created
deployment.apps/service-python-v1 created
deployment.apps/service-python-v2 created

$ kubectl apply -f service/js/service-js.yaml
service/service-js created
deployment.apps/service-js-v1 created
deployment.apps/service-js-v2 created
```

2）查看服务部署状态：

```
$ kubectl get deploy
NAME                 DESIRED   CURRENT   UP-TO-DATE   AVAILABLE   AGE
service-go-v1        1         1         1            1           3h28m
service-go-v2        1         1         1            1           3h28m
service-js-v1        1         1         1            1           3h28m
service-js-v2        1         1         1            1           3h28m
service-lua-v1       1         1         1            1           3h28m
service-lua-v2       1         1         1            1           3h28m
```

```
service-node-v1      1       1       1       1       3h28m
service-node-v2      1       1       1       1       3h28m
service-python-v1    1       1       1       1       3h28m
service-python-v2    1       1       1       1       3h28m

$ kubectl get pod
NAME                                  READY   STATUS    RESTARTS   AGE
service-go-v1-7cc5c6f574-2dn6n        1/1     Running   0          3h29m
service-go-v2-7656dcc478-2dgc6        1/1     Running   0          3h29m
service-js-v1-55756d577-49cl5         1/1     Running   0          3h29m
service-js-v2-86bdfc86d9-pbq2d        1/1     Running   0          3h29m
service-lua-v1-5c9bcb7778-hsbr6       1/1     Running   0          3h29m
service-lua-v2-75cb5cdf8-75tm4        1/1     Running   0          3h29m
service-node-v1-d44b9bf7b-6hrzd       1/1     Running   0          3h29m
service-node-v2-86545d9796-4ttsf      1/1     Running   0          3h29m
service-python-v1-79fc5849fd-vtzsf    1/1     Running   0          3h29m
service-python-v2-7b6864b96b-9qk8c    1/1     Running   0          3h29m

$ kubectl get svc
NAME             TYPE        CLUSTER-IP       EXTERNAL-IP   PORT(S)   AGE
kubernetes       ClusterIP   10.96.0.1        <none>        443/TCP   4h47m
service-go       ClusterIP   10.108.207.28    <none>        80/TCP    3h29m
service-js       ClusterIP   10.97.251.7      <none>        80/TCP    3h29m
service-lua      ClusterIP   10.103.165.139   <none>        80/TCP    3h29m
service-node     ClusterIP   10.102.21.34     <none>        80/TCP    3h29m
service-python   ClusterIP   10.103.119.8     <none>        80/TCP    3h29m
```

6.3 访问服务

对于服务的访问，我们先在Kubernetes集群内进行访问测试，有两种访问方式如下所示：

- ❑ 通过Cluster IP访问。
- ❑ 通过集群内的DNS访问。

当如上的两种服务访问方式正常后，我们可以通过NodePort暴露给外部，然后使用浏览器进行访问测试。

1. 在kubenetes集群内访问测试

（1）通过Cluster IP访问

```
$ kubectl get svc
NAME             TYPE        CLUSTER-IP       EXTERNAL-IP   PORT(S)   AGE
kubernetes       ClusterIP   10.96.0.1        <none>        443/TCP   6h6m
service-go       ClusterIP   10.108.207.28    <none>        80/TCP    4h48m
service-js       ClusterIP   10.97.251.7      <none>        80/TCP    4h48m
service-lua      ClusterIP   10.103.165.139   <none>        80/TCP    4h48m
service-node     ClusterIP   10.102.21.34     <none>        80/TCP    4h48m
service-python   ClusterIP   10.103.119.8     <none>        80/TCP    4h48m
```

```
$ curl http://10.108.207.28/env
{"message":"go v2"}

$ curl http://10.103.165.139/env
{"message":"lua v1"}

$ curl http://10.102.21.34/env
{"message":"node v1","upstream":[{"message":"go v1","response_time":"0.01"}]}

$ curl http://10.103.119.8/env
{"message":"python v1","upstream":[{"message":"lua v2","response_
time":0.12},{"message":"node v2","response_time":0.12,"upstream":[{"message":"go
v2","response_time":"0.01"}]}]}

$ curl -I http://10.97.251.7/
HTTP/1.1 200 OK
last-modified: Sun, 16 Dec 2018 07:49:03 GMT
content-length: 548
content-disposition: inline; filename="index.html"
accept-ranges: bytes
content-type: text/html; charset=utf-8
vary: Accept-Encoding
date: Sun, 13 Jan 2019 10:27:13 GMT
Connection: keep-alive
```

（2）通过集群内的 DNS 访问

1）创建用于访问的测试容器：

```
$ kubectl apply -f kubernetes/dns-test.yaml
```

2）服务解析测试：

```
$ kubectl exec dns-test -c dns-test -- nslookup service-go
Server:    10.96.0.10
Address 1: 10.96.0.10 kube-dns.kube-system.svc.cluster.local

Name:      service-go
Address 1: 10.108.207.28 service-go.default.svc.cluster.local

$ kubectl exec dns-test -c dns-test -- nslookup service-node
Server:    10.96.0.10
Address 1: 10.96.0.10 kube-dns.kube-system.svc.cluster.local

Name:      service-node
Address 1: 10.102.21.34 service-node.default.svc.cluster.local

$ kubectl exec dns-test -c dns-test -- nslookup service-lua
Server:    10.96.0.10
Address 1: 10.96.0.10 kube-dns.kube-system.svc.cluster.local
```

```
Name:      service-lua
Address 1: 10.103.165.139 service-lua.default.svc.cluster.local

$ kubectl exec dns-test -c dns-test -- nslookup service-python
Server:    10.96.0.10
Address 1: 10.96.0.10 kube-dns.kube-system.svc.cluster.local

Name:      service-python
Address 1: 10.103.119.8 service-python.default.svc.cluster.local

$ kubectl exec dns-test -c dns-test -- nslookup service-js
Server:    10.96.0.10
Address 1: 10.96.0.10 kube-dns.kube-system.svc.cluster.local

Name:      service-js
Address 1: 10.97.251.7 service-js.default.svc.cluster.local
```

3）服务访问测试：

```
$ kubectl exec dns-test -c dns-test -- curl -s http://service-go/env
{"message":"go v2"}

$ kubectl exec dns-test -c dns-test -- curl -s http://service-node/env
{"message":"node v1","upstream":[{"message":"go v2","response_time":"0.01"}]}

$ kubectl exec dns-test -c dns-test -- curl -s http://service-lua/env
{"message":"lua v1"}

$ kubectl exec dns-test -c dns-test -- curl -s http://service-python/env
{"message":"python v2","upstream":[{"message":"lua v2","response_
time":0.05},{"message":"node v2","upstream":[{"message":"go v1","response_
time":"0.00"}],"response_time":0.05}]}

$ kubectl exec dns-test -c dns-test -- curl -s -I http://service-js/
HTTP/1.1 200 OK
last-modified: Sun, 16 Dec 2018 07:49:13 GMT
content-length: 505
content-disposition: inline; filename="index.html"
accept-ranges: bytes
content-type: text/html; charset=utf-8
vary: Accept-Encoding
date: Sun, 13 Jan 2019 10:31:23 GMT
Connection: keep-alive
```

2. 在浏览器中访问测试

1）使用 NodePort 暴露服务。使用 NodePort 暴露 sevice-js 和 service-python 服务的 v1 版本：

```
$ kubectl expose deployment service-js-v1 --type=NodePort --name=service-js-
nodeport --port=80 --target-port=80
service/service-js-nodeport exposed
```

```
$ kubectl expose deployment service-python-v1 --type=NodePort --name=service-python-nodeport --port=80 --target-port=80
service/service-python-nodeport exposed
```

2）在浏览器中访问 sevice-js 和 service-python 服务。

获取 service-js 外部访问地址：

```
$ SERVICE_JS_PORT=$(kubectl get service service-js-nodeport -o jsonpath='{.spec.ports[0].nodePort}')
$ HOST=$(kubectl get node lab2 -o 'jsonpath={.status.addresses[0].address}')
$ GATEWAY_URL=$HOST:$SERVICE_JS_PORT
$ echo http://$GATEWAY_URL/
$ curl -I http://$GATEWAY_URL/
```

获取 service-python 外部访问地址：

```
$ SERVICE_PYTHON_PORT=$(kubectl get service service-python-nodeport -o jsonpath='{.spec.ports[0].nodePort}')
$ HOST=$(kubectl get node lab2 -o 'jsonpath={.status.addresses[0].address}')
$ GATEWAY_URL=$HOST:$SERVICE_PYTHON_PORT
$ echo http://$GATEWAY_URL/env
$ curl http://$GATEWAY_URL/env
```

使用上面步骤获取的 URL 地址在浏览器中访问。

访问 service-js 地址，结果如图 6-6 所示。

图 6-6　在浏览器中访问测试

访问 service-python 服务地址，结果如下：

```
{"message":"python v1","upstream":[{"message":"lua v1","response_time":0.15},{"message":"node v1","response_time":0.25,"upstream":[{"message":"go v2","response_time":"0.01"}]}]}
```

【清理】

删除访问测试服务时创建的资源：

```
$ kubectl delete service service-js-nodeport service-python-nodeport
```

```
service "service-js-nodeport" deleted
service "service-python-nodeport" deleted

$ kubectl delete -f kubernetes/dns-test.yaml
pod "dns-test" deleted

$ kubectl delete -f service/go/service-go.yaml
$ kubectl delete -f service/node/service-node.yaml
$ kubectl delete -f service/lua/service-lua.yaml
$ kubectl delete -f service/python/service-python.yaml
$ kubectl delete -f service/js/service-js.yaml
```

6.4 在 Istio 中部署微服务

在前面的实验中，服务部署完成之后，可以单独访问 service-js 服务的前端静态页面，也可以单独访问 service-python 实现的 API 接口，但是仍然不能实现两者相互配合调用，展现整个服务调用链情况。本节就带你体验一下 Istio 的简单路由功能，并了解整个微服务的调用流程，对于本节使用的 Istio 路由配置文件，不必太过关注细节，这些功能会在后续章节中详细说明。

1）在 default 命令空间开启自动注入功能：

```
$ kubectl label namespace default istio-injection=enabled
```

2）部署服务：

```
$ kubectl apply -f service/go/service-go.yaml
$ kubectl apply -f service/node/service-node.yaml
$ kubectl apply -f service/lua/service-lua.yaml
$ kubectl apply -f service/python/service-python.yaml
$ kubectl apply -f service/js/service-js.yaml
```

3）查看服务部署状态。

使用如下的命令查看服务的部署状态：

```
$ kubectl get pod
NAME                                   READY     STATUS    RESTARTS   AGE
service-go-v1-7cc5c6f574-sjqmk         2/2       Running   0          39s
service-go-v2-7656dcc478-mfxsk         2/2       Running   0          39s
service-js-v1-55756d577-nzwmb          2/2       Running   0          21s
service-js-v2-86bdfc86d9-w6t9s         2/2       Running   0          21s
service-lua-v1-5c9bcb7778-ccxxq        2/2       Running   0          37s
service-lua-v2-75cb5cdf8-zbl2r         2/2       Running   0          37s
service-node-v1-d44b9bf7b-lpkqn        2/2       Running   0          37s
service-node-v2-86545d9796-5ct7q       2/2       Running   0          37s
service-python-v1-79fc5849fd-hl4vn     2/2       Running   0          36s
service-python-v2-7b6864b96b-4xjlp     2/2       Running   0          36s
```

与之前部署不同，现在每个服务启动的 Pod 有两个容器，其中一个容器是被注入的

Envoy 代理，当所有服务的 Pod 处于 Running 状态时，再进行后续的实验步骤。

4）应用路由规则，将 service-js 服务的流量全部导入 v1 版本：

```
$ kubectl apply -f istio/route/gateway-js-v1.yaml
gateway.networking.istio.io/istio-lab-gateway created
destinationrule.networking.istio.io/service-js created
virtualservice.networking.istio.io/istio-lab created

$ kubectl get gateway
NAME                AGE
istio-lab-gateway   21s

$ kubectl get virtualservice
NAME                AGE
istio-lab           35s

$ kubectl get destinationrule
NAME                AGE
service-js          51s
```

5）在浏览器中访问测试。

此时通过浏览器访问地址 http://11.11.11.112:31380/，并点击“发射”按钮，会看到如图 6-7 所示的服务调用图。

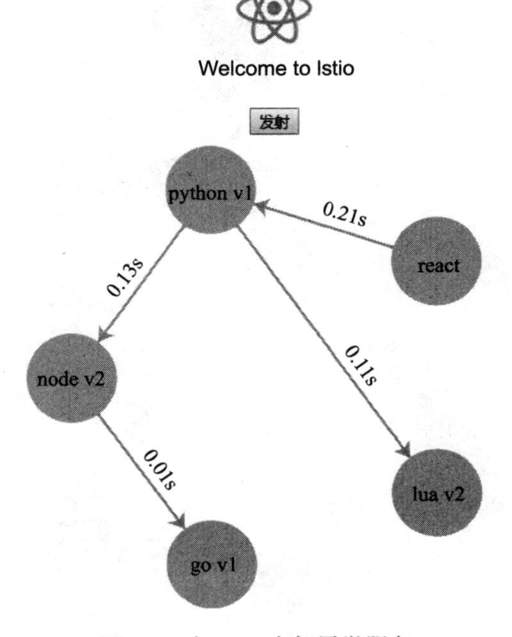

图 6-7　在 Istio 中部署微服务

6）应用路由规则，将流量全部导入 service-js 的 v2 版本：

```
$ kubectl delete -f istio/route/gateway-js-v1.yaml
gateway.networking.istio.io "istio-lab-gateway" deleted
destinationrule.networking.istio.io "service-js" deleted
virtualservice.networking.istio.io "istio-lab" deleted

$ kubectl apply -f istio/route/gateway-js-v2.yaml
gateway.networking.istio.io/istio-lab-gateway created
destinationrule.networking.istio.io/service-js created
virtualservice.networking.istio.io/istio-lab created

$ kubectl get gateway
NAME                AGE
istio-lab-gateway   16s

$ kubectl get virtualservice
NAME              AGE
istio-lab         31s

$ kubectl get destinationrule
NAME              AGE
service-js        47s
```

7）在浏览器中访问测试。

此时通过刷新浏览器，点击“发射”按钮，应该看到 Vue 版本的服务调用图，如图 6-8 所示。

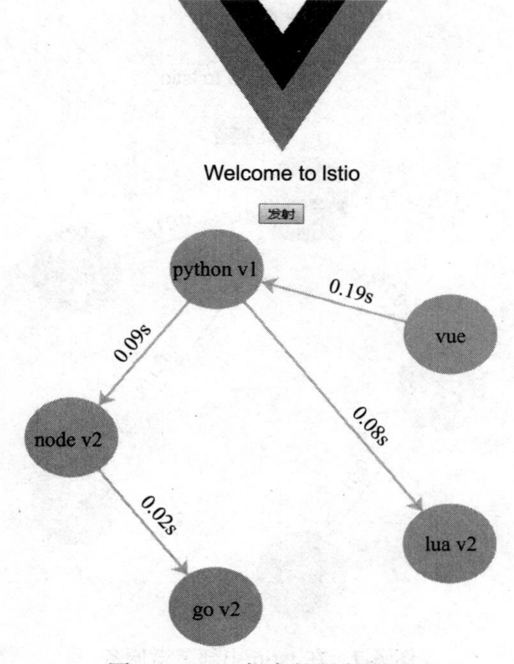

图 6-8　Vue 版本的服务调用

8）清理：

```
$ kubectl label namespace default istio-injection-
$ kubectl delete -f istio/route/gateway-js-v2.yaml
$ kubectl delete -f service/go/service-go.yaml
$ kubectl delete -f service/node/service-node.yaml
$ kubectl delete -f service/lua/service-lua.yaml
$ kubectl delete -f service/python/service-python.yaml
$ kubectl delete -f service/js/service-js.yaml
```

实验时可能会由于虚拟机性能不足，出现创建了路由但是没有达到实现的效果，这是由于路由虽然已经创建成功，但可能由于机器性能问题，导致路由规则并没能生效，此时可能需要多等一段时间或者多次删除重建。如果还是无法达到实验效果，可以参考第 2 章 2.1.3 节“问题及解决方案”。

6.5　本章小结

本章介绍的微服务应用是后续实验的基础。使用图的方式把服务间的调用链详情展现出来，可以更好地演示在 Istio 中配置的路由规则。本章中使用的 Istio 路由规则在后续章节会再次用到，到时候会逐一介绍。

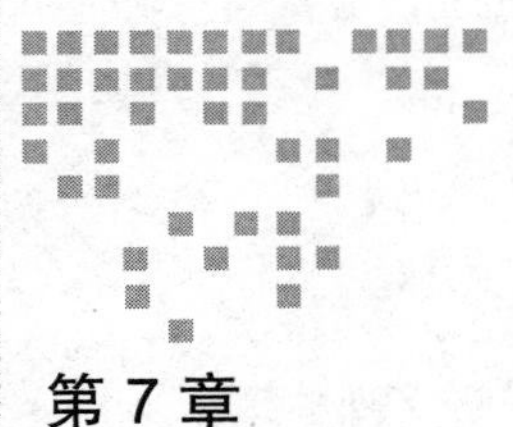

Chapter 7 第7章

让服务流量控制更简单

本章主要介绍 Istio 服务路由相关的功能，来展示 Istio 如何让服务流量控制更简单。主要内容包括如何管理集群的入口流量、出口流量，如何将请求路由到指定的服务版本，如何根据服务版本的权重来拆分流量，如何根据请求信息来路由服务流量，如何引入线上真实的流量测试即将上线的新版本服务，如何借助 Istio 实现 A/B 测试、灰度发布，以及如何将两者结合使用。

7.1 整体介绍

流量路由管理是 Istio 最重要的功能。我们可以通过路由规则轻松地把请求流量导入不同的服务版本中，可以很容易地实现 A/B 测试、灰度发布这类需要复杂技术才能实现的功能。Istio 的流量路由管理如图 7-1 所示。

Envoy 代理会根据路由规则把请求转发到相应服务实例的 Pod 上，默认情况下 Enovy 代理会把流量以轮询的方式转到后端服务实例的 Pod 上。通过设置路由规则，我们可以把对服务的请求流量按百分比转发到不同的版本上。如图 7-1 的上半部分所示，把 Service A 请求 Service B 的流量进行拆分，95% 的流量路由到生产环境正在使用的版本，5% 的流量路由到新的实验版本。

Envoy 代理还可以提取请求的信息，根据请求的相关信息来拆分流量，如图 7-1 的下半部分所示，我们可以根据服务的请求头 User-agent 把来自 Android 和 iPhone 用户请求转发到不同的服务版本，把 iPhone 用户流量导入到实验版本中的。

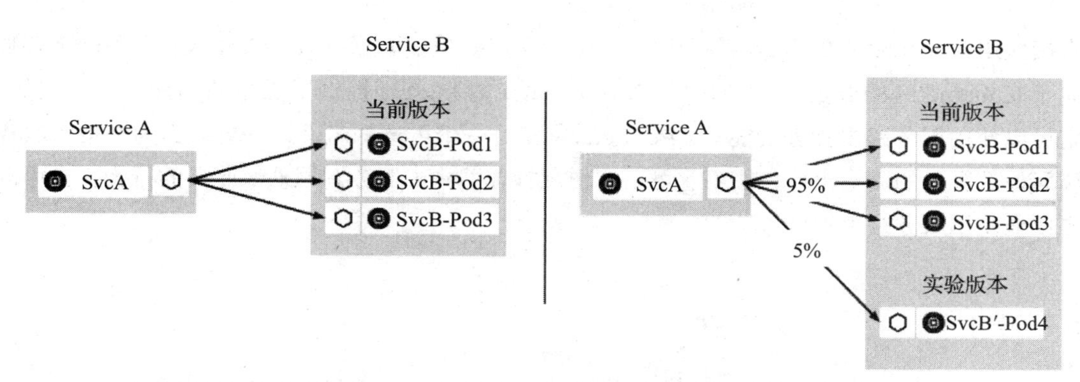

流量与基础设施伸缩解耦——路由到某个版本的流量比例与支持该版本的实例数无关。

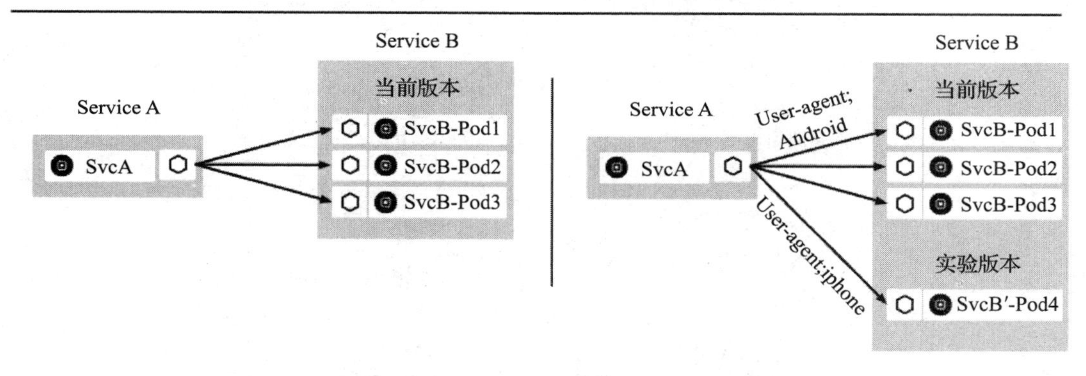

根据请求内容来操作流量——请求的内容可以用于决定该请求的目的地址。

图 7-1　Istio 流量路由管理（图片来源：Istio 官方网站）

Istio 的流量路由管理流程如图 7-2 所示。

通过规则配置 API，用户可配置路由规则，Pilot 将用户配置的路由规则转换为 Envoy 代理需要的配置格式，并分发给 Envoy 代理。此处用户配置的规则是：对于 http://serviceb.example.com 的访问，99% 的流量转发到 v1.5 版本，1% 的流量转发到 v2.0-alpha 版本。这是为了把线上流量引入测试待上线的版本。

用户通过 Pilot 提供的路由规则管理 API 来创建、更新、删除路由规则，当路由规则发生改变时，Pilot 会把路由规则下发到每一个 Envoy 代理上，由于 Envoy 会拦截服务间的所有流量，Envoy 代理根据这些路由规则来决定服务的请求应该发送到哪一个后端服务实例 Pod 上，而这些对服务调用方来说是完全透明的，服务调用方不需要做出任何改变。

请求进入服务网格后的数据流向如图 7-3 所示。

Ingress gateway 是外部访问网格内服务的流量入口，所有外部请求流量都会经过这里。Ingress gateway 类似于 Kubernetes 的 Ingress Controller 的实现，但是 Ingress gateway 并不

直接使用Kubernetes的Ingress规则，在Istio中需要单独配置Ingress gateway。正由于有独立于Kubernetes的Ingress，所以它比Kubernetes的Ingress提供了更多的功能特性，比如，它可以使用服务路由管理功能。Egress gateway是一个可选的组件，它可以作为网格内服务访问外部服务的流量出口，使用它可以方便地管理网格内服务调用外部服务的行为，也可以方便地使用服务路由管理等功能。

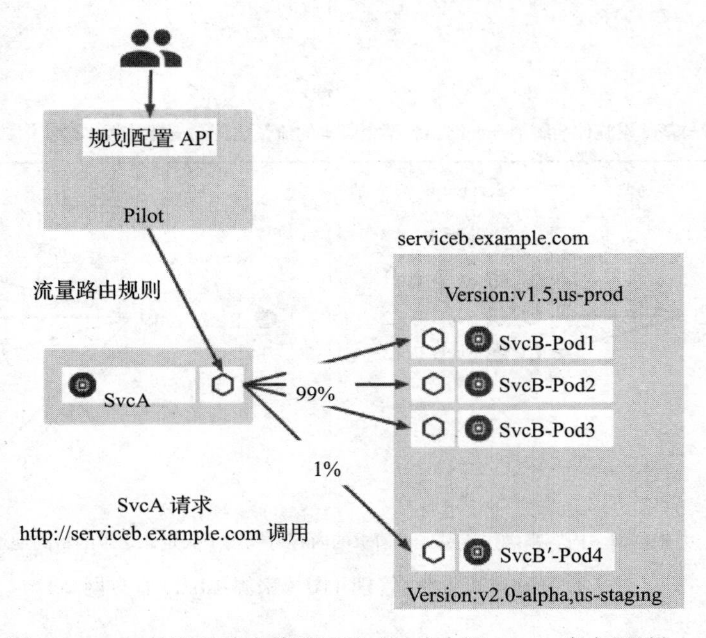

图7-2　Istio的流量路由管理流程（图片来源：Istio官方网站）

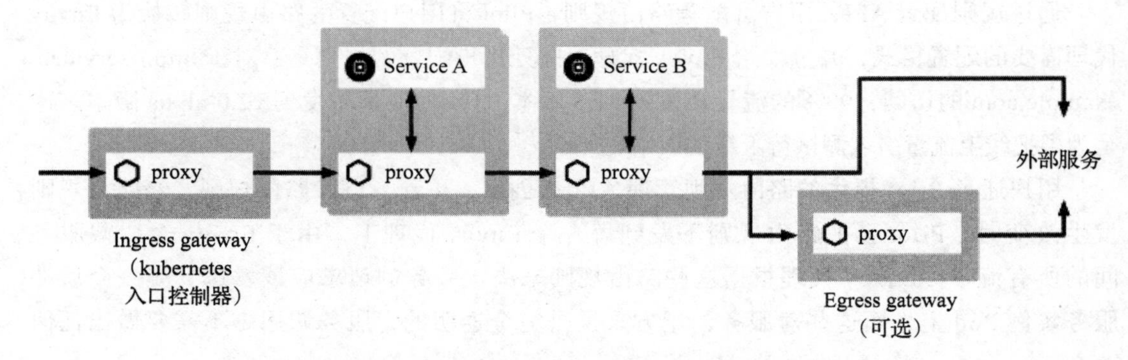

图7-3　服务网格中的数据流向（图片来源：Istio官方网站）

外部请求通过Ingress gateway进入网格内部，然后被路由到网格内部的具体服务上，服务之间相互调用配合来完成用户的请求，最后由Ingress gateway把请求的响应数据返回给外部调用方。当网格中的服务需要访问外部服务时，可以直接通过跟随服务一起部署的 Envoy 代理来访问外部服务，也可以通过统一部署的 Egress gateway 来访问外部服务。

【实验前的准备】

进行本章实验前，需要先执行如下的前置步骤。

1）下载实验时用到的源码仓库：

```
$ sudo yum install -y git

$ git clone https://github.com/mgxian/istio-lab
Cloning into 'istio-lab'...
remote: Enumerating objects: 247, done.
remote: Counting objects: 100% (247/247), done.
remote: Compressing objects: 100% (173/173), done.
remote: Total 774 (delta 153), reused 164 (delta 73), pack-reused 527
Receiving objects: 100% (774/774), 283.00 KiB | 229.00 KiB/s, done.
Resolving deltas: 100% (447/447), done.

$ cd istio-lab
```

2）开启 default 命名空间的自动注入功能：

```
$ kubectl label namespace default istio-injection=enabled
namespace/default labeled
```

3）部署用于测试的服务：

```
$ kubectl apply -f service/go/service-go.yaml
$ kubectl apply -f service/node/service-node.yaml
$ kubectl apply -f service/lua/service-lua.yaml
$ kubectl apply -f service/python/service-python.yaml
$ kubectl apply -f service/js/service-js.yaml

$ kubectl get pod
NAME                                   READY   STATUS    RESTARTS   AGE
service-go-v1-7cc5c6f574-tndgb         2/2     Running   0          3m11s
service-go-v2-7656dcc478-xtqwz         2/2     Running   0          3m11s
service-js-v1-55756d577-jn4w9          2/2     Running   0          3m4s
service-js-v2-86bdfc86d9-5gbxm         2/2     Running   0          3m4s
service-lua-v1-5c9bcb7778-tj87t        2/2     Running   0          3m8s
service-lua-v2-75cb5cdf8-nldtq         2/2     Running   0          3m8s
service-node-v1-d44b9bf7b-fh257        2/2     Running   0          3m10s
service-node-v2-86545d9796-scztl       2/2     Running   0          3m10s
service-python-v1-79fc5849fd-f8bxp     2/2     Running   0          3m6s
service-python-v2-7b6864b96b-5drn4     2/2     Running   0          3m6s
```

7.2 管理集群的入口流量

主动进入服务网格的流量称为入口流量，这是获取服务的入口，所有外部的请求流量都会最先经过这里，这里是外部流量与内部服务通信的入口。在Istio中这个入口称为Ingress gateway，它是一个边界负载均衡器，负责接收外部的请求流量，并按相应的规则转发到网格内部服务上。Istio使用一个叫Gateway的资源对象来管理服务网格的入口。

为service-js和service-python服务创建Gateway的使用示例如下：

```
 1 apiVersion: networking.istio.io/v1alpha3
 2 kind: Gateway
 3 metadata:
 4   name: istio-lab-gateway
 5 spec:
 6   selector:
 7     istio: ingressgateway # use Istio default gateway implementation
 8   servers:
 9   - port:
10       number: 80
11       name: http
12       protocol: HTTP
13     hosts:
14     - "*"
15 ---
16 apiVersion: networking.istio.io/v1alpha3
17 kind: DestinationRule
18 metadata:
19   name: service-js
20 spec:
21   host: service-js
22   subsets:
23   - name: v1
24     labels:
25       version: v1
26   - name: v2
27     labels:
28       version: v2
29 ---
30 apiVersion: networking.istio.io/v1alpha3
31 kind: VirtualService
32 metadata:
33   name: istio-lab
34 spec:
35   hosts:
36   - "*"
37   gateways:
38   - istio-lab-gateway
```

```
39   http:
40   - match:
41     - uri:
42         prefix: /env
43     route:
44     - destination:
45         host: service-python
46   - route:
47     - destination:
48         host: service-js
49         subset: v1
```

第 1 ～ 14 行定义了名为 istio-lab-gateway 的 Gateway，监听 80 端口并接收 HTTP 协议的请求，由于 hosts 使用 *，所以任何域名的请求都会被接受。第 6 ～ 7 行选择了包含标签名为 Istio、值为 ingressgateway 的 Ingress gateway，这是 Istio 默认实现的 Ingress gateway，本书介绍的两种部署方式都包含了该组件。

第 16 ～ 28 行定义了名为 service-js 的 DestinationRule，指定了 service-js 有两个版本的服务，根据 version 标签的值来区分服务的实例版本。

第 30 ～ 49 行定义了名为 istio-lab 的 VirtualService，指定了与名为 istio-lab-gateway 的 Gateway 绑定，它接受任何域名的请求流量。当请求的 URI 以 /env 为前缀时，请求转发到 service-python 服务上，其他请求都转发到 service-js 服务的 v1 版本上。字段 host 的值完整地址为 service-js.default.svc.cluster.local，当规则配置与要请求的服务在同一个命名空间时，可以只简写成 Service 的名称。由于我们没有在 metadata 的定义中设置 namespace 字段来指定命名空间，默认情况下，在 default 命名空间中创建规则。

> 注意　上述 Gateway 中，我们配置监听 80 端口，其实是监听容器的 80 端口，如果你使用的是云平台的 LoadBalancer，那么会直接监听在 LoadBalancer 的 80 端口。由于我们的实验是在本地虚拟机环境进行，只能使用 NodePort 方式，NodePort 监听的 31380 端口对应于容器中 80 端口，也就是我们在 Gateway 中指定的 80 端口。

7.3　把请求路由到服务的指定版本

当服务有多个版本，而我们只希望把流量全部转发到某个特定版本，这时候可以使用如下的方式配置路由规则：

```
1 apiVersion: networking.istio.io/v1alpha3
2 kind: DestinationRule
3 metadata:
4   name: service-lua
```

```
 5 spec:
 6   host: service-lua
 7   subsets:
 8   - name: v1
 9     labels:
10       version: v1
11   - name: v2
12     labels:
13       version: v2
14 ---
15 apiVersion: networking.istio.io/v1alpha3
16 kind: VirtualService
17 metadata:
18   name: service-lua
19 spec:
20   hosts:
21   - service-lua
22   http:
23   - route:
24     - destination:
25         host: service-lua
26         subset: v1
```

第1～13行定义了名为service-lua的DestinationRule，version标签用于区分service-lua服务的实例版本，host字段表示要访问的后端的服务名称，一般为Kubernetes中定义的Service名称。字段host的值完整地址为service-lua.default.svc.cluster.local，当路由规则配置与要请求的服务在同一个命名空间时，可以只简写成Service的名称。

第15～26行定义名为service-lua的VirtualService，表示对service-lua服务的所有请求转发到v1版本，此处的v1为DestinationRule中定义的名为v1的subset，hosts字段表示接收请求的域名，一般为Kubernetes中定义的Service名称。host字段的值与对应DestinationRule中的host字段值保持一致。

【实验】

1）创建Gateway和service-js的路由规则：

```
$ kubectl apply -f istio/route/gateway-js-v1.yaml
```

2）创建service-lua的指定版本路由规则：

```
$ kubectl apply -f istio/route/virtual-service-lua-v1.yaml
```

3）浏览器访问。

此时打开浏览器访问http：//11.11.11.112：31380/，并多次点击“发射”，你会看到service-lua服务只会请求到v1版本。

4）清理：

```
$ kubectl delete -f istio/route/gateway-js-v1.yaml
$ kubectl delete -f istio/route/virtual-service-lua-v1.yaml
```

7.4 根据服务版本权重拆分流量

默认情况下，当服务有多个版本时，请求流量会以轮询方式发送给多个版本，我们可以把流量按百分比分配给多个服务版本，来实现服务的逐渐过渡升级。我们可以使用如下的方式配置路由规则：

```
apiVersion: networking.istio.io/v1alpha3
kind: DestinationRule
metadata:
  name: service-node
spec:
  host: service-node
  subsets:
  - name: v1
    labels:
      version: v1
  - name: v2
    labels:
      version: v2
---
apiVersion: networking.istio.io/v1alpha3
kind: VirtualService
metadata:
  name: service-node
spec:
  hosts:
  - service-node
  http:
  - route:
    - destination:
        host: service-node
        subset: v1
      weight: 20
    - destination:
        host: service-node
        subset: v2
      weight: 80
```

通过上面的配置，我们把 service-node 服务的 v2 版本权重设置为 80，而把 v1 版本的权重设置为 20。这样设置权重比值之后，对 service-node 服务的请求，会有 80% 的概率被路由到 v2 版本，只有 20% 的概率被路由到 v1 版本。

【实验】

1）创建 Gateway 和 service-js 的路由规则：

```
$ kubectl apply -f istio/route/gateway-js-v1.yaml
```

2）创建 service-node 的路由规则：

```
$ kubectl apply -f istio/route/virtual-service-node-v1-v2.yaml
```

3）浏览器访问。

此时打开浏览器访问 http：//11.11.11.112：31380/，并多次点击“发射”按钮，你会看到 service-node 服务请求大概率会路由到 v2 版本，只有少部分请求会落到 v1 版本上。

4）清理：

```
$ kubectl delete -f istio/route/gateway-js-v1.yaml
$ kubectl delete -f istio/route/virtual-service-node-v1-v2.yaml
```

7.5 根据请求信息路由到服务的不同版本

有时候我们希望根据请求的信息（如请求头、请求地址、请求协议等）来决定请求的路由规则[⊖]。根据请求的 URI 以及请求头来转发请求的配置规则示例如下：

```
 1 ...
 2 apiVersion: networking.istio.io/v1alpha3
 3 kind: VirtualService
 4 metadata:
 5   name: istio-lab
 6 spec:
 7   hosts:
 8   - "*"
 9   gateways:
10   - istio-lab-gateway
11   http:
12   - match:
13     - uri:
14         prefix: /env
15       headers:
16         app-client:
17           regex: "react"
18     route:
19     - destination:
20         host: service-python
21         subset: v1
22   - match:
```

⊖ 具体配置项可以参考官方文档，链接地址 https://istio.io/docs/reference/config/istio.networking.v1alpha3/#HTTPMatchRequest。

```
23      - uri:
24          prefix: /env
25      route:
26      - destination:
27          host: service-python
28          subset: v2
29    - route:
30      - destination:
31          host: service-js
32          subset: v1
```

第 12 ～ 28 行定义了两个 match 路由规则，第一个 match 规则表示当服务请求的 URI 是以 /env 开头，并且包含 app-client 值为 react 的请求头时，请求转发到服务 service-python 的 v1 版本上。第二个 match 规则表示当服务请求的 URI 是以 /env 开头时，请求转发到服务 service-python 的 v2 版本上。由于请求路由匹配规则是从上到下，当匹配到第一个规则时就进入请求转发阶段，所以这两个 match 规则的顺序不能相互调换。

第 15 行使用的 headers 下定义的请求头字段必须使用小写字母，并且只能使用连字符连接，比如：x-request-id、app-client。

第 29 ～ 32 行定义了默认的路由规则，当请求不满足上述的两个 match路由规则时，把请求转发到默认路由。

【实验】

1）创建 Gateway 和 service-js 的路由规则：

```
$ kubectl apply -f istio/route/gateway-js-react-v1-v2.yaml
```

2）浏览器访问。

此时打开浏览器访问 http：//11.11.11.112：31380/，并多次点击“发射”按钮，你会看到 service-python 服务的请求会一直落到 v1 版本。这是由于使用 React 框架实现的 service-js 服务的 v1 版本请求 service-python 服务时都携带了字段名为 app-client、值为 react 的请求头，所以根据上述的路由规则，对于 service-python 服务的请求都会落到 v1 版本。

3）清理：

```
$ kubectl delete -f istio/route/gateway-js-react-v1-v2.yaml
```

7.6　流量镜像

有时候当我们需要将服务的新版本上线时，虽然我们已经在线下经过测试，但是并没有经过生产环境的真实流量验证，对服务的性能和是否有 bug 并没有太大的把握，我们希望通过线上的真实流量来验证一下新版本服务，经过生产环境流量验证无误之后，再把生产环境的流量切换到新版本的服务上。在此种场景下，我们就可以使用 Istio 提供的**流量镜像**功能，实时复制线上真实的请求流量到我们的新版本服务上，验证新版本的服务。下面以 service-go 服务为例：

```
 1 apiVersion: networking.istio.io/v1alpha3
 2 kind: DestinationRule
 3 metadata:
 4   name: service-go
 5 spec:
 6   host: service-go
 7   subsets:
 8   - name: v1
 9     labels:
10       version: v1
11   - name: v2
12     labels:
13       version: v2
14 ---
15 apiVersion: networking.istio.io/v1alpha3
16 kind: VirtualService
17 metadata:
18   name: service-go
19 spec:
20   hosts:
21   - service-go
22   http:
23   - route:
24     - destination:
25         host: service-go
26         subset: v1
27     mirror:
28       host: service-go
29       subset: v2
```

第 24 ～ 26 行定义了默认路由，表明把所有请求 service-go 服务的流量都路由到 v1 版本。

第 27 ～ 29 行的定义表明，把所有请求 service-go 服务的流量都镜像路由到 v2 版本上。

【实验】

1）创建测试 Pod：

```
$ kubectl apply -f kubernetes/dns-test.yaml
```

2）创建 service-go 服务的路由规则：

```
$ kubectl apply -f istio/route/virtual-service-go-v1-mirror-v2.yaml
```

3）打开一个新的终端查看 service-go-v2 实例的日志信息：

```
$ POD=$(kubectl get pod | grep service-go-v2 | awk '{print $1}')
$ kubectl logs -f $POD service-go
```

4）测试访问 service-go 服务：

```
$ kubectl exec dns-test -c dns-test -- curl -s http://service-go/env
{"message":"go v1"}
```

多次执行上述命令，访问 service-go 服务。你只会访问到 service-go 服务的 v1 版本，但是在 service-go 服务的 v2 版本的 Pod 上，你可以查看到如下所示的请求日志，这说明对 service-go 服务的 v1 版本的请求被镜像到了 v2 版本中：

```
[GIN] 2019/01/13 - 12:44:16 | 200 | 397.207µs | 10.244.1.18 | GET /env
[GIN] 2019/01/13 - 12:44:36 | 200 | 28.28µs   | 10.244.1.18 | GET /env
[GIN] 2019/01/13 - 12:44:38 | 200 | 98.168µs  | 10.244.1.18 | GET /env
[GIN] 2019/01/13 - 12:45:25 | 200 | 47.049µs  | 10.244.1.18 | GET /env
[GIN] 2019/01/13 - 12:45:26 | 200 | 32.157µs  | 10.244.1.18 | GET /env
[GIN] 2019/01/13 - 12:45:27 | 200 | 33.921µs  | 10.244.1.18 | GET /env
```

5）清理：

```
$ kubectl delete -f kubernetes/dns-test.yaml
$ kubectl delete -f istio/route/virtual-service-go-v1-mirror-v2.yaml
```

7.7　管理集群的出口流量

默认情况下，服务网格中的服务无法访问外部的服务。要想访问外部服务，可以创建 ServiceEntry 对象，把外部服务注册到服务网格的注册中心，这样，网格中的服务就可以访问外部服务了。同时，可以配合 VirtualService 更好地控制网格内服务访问外部服务的行为。本节介绍通过 ServiceEntry 来导入外部服务，让服务网格中的服务也可以访问外部的服务的方法[⊖]。也可以使用 Egress gateway 配合 ServiceEntry 统一管理集群的出口流量；此外，还可以通过修改 Istio 的部署配置文件，直接放开所有对外部服务的访问流量。

1. 使用 ServiceEntry 导入网格外部服务

（1）配置访问 HTTP 协议的外部服务

导入 httpbin.org 服务示例：

```
1 apiVersion: networking.istio.io/v1alpha3
2 kind: ServiceEntry
3 metadata:
4   name: httpbin-ext
5 spec:
6   hosts:
```

⊖ 详细文档可参考官方文档 https://istio.io/docs/reference/config/istio.networking.v1alpha3/#ServiceEntry。

```
 7    - httpbin.org
 8    ports:
 9    - number: 80
10      name: http
11      protocol: HTTP
12    resolution: DNS
13    location: MESH_EXTERNAL
14 ---
15 apiVersion: networking.istio.io/v1alpha3
16 kind: VirtualService
17 metadata:
18   name: httpbin-ext
19 spec:
20   hosts:
21     - httpbin.org
22   http:
23   - timeout: 3s
24     route:
25       - destination:
26         host: httpbin.org
```

第 1 ～ 13 行定义了名为 httpbin-ext 的 ServiceEntry。

第 6 ～ 7 行定义的 hosts 字段指定了要访问的外部服务的 DNS 域名，可以使用泛域名匹配，如 *.httpbin.org。如果是非 HTTP 协议的流量，会使用 ports 和 addresses 字段的定义来区分目标地址。

第 8 ～ 9 行定义了要访问的外部服务的端口和协议。

第 12 行定义的 resolution 字段定义了外部服务的解析方式，取值包括 DNS、STATIC、NONE，一般情况下使用 DNS 方式，设置为 STATIC 时需要在 endpoints 中配置 IP 地址，当应用程序自己负责解析时，使用 NONE 方式。

第 13 行定义的 location 字段表示服务所处的位置，取值可以为 MESH_EXTERNAL 和 MESH_INTERNAL 两种，MESH_EXTERNAL 一般用于导入外部服务，并不作为网格的一部分，通常服务以 API 的方式提供使用。MESH_INTERNAL 一般用于把网格外的服务作为网络中的一部分，这个配置会影响 Istio 对服务的 mTLS 认证及策略的实施等。当网格内的服务与网格外的服务通信时，mTLS 认证会被关闭，策略的实施也会由客户端实施，而不是由服务端实施。

第 15 ～ 26 行定义了同名的 VirtualService，进一步配合 ServiceEntry 实现更好地控制网格内服务访问外部服务的行为。hosts 字段中的值 httpbin.org 表示网格内服务访问时使用的域名，route 里的 host 字段应设置为在 ServiceEntry 定义中 hosts 字段指定的域名。

（2）配置访问 HTTPS 协议的外部服务

导入 baidu.com 服务示例：

```
apiVersion: networking.istio.io/v1alpha3
kind: ServiceEntry
metadata:
  name: baidu-ext
spec:
  hosts:
  - baidu.com
  ports:
  - number: 443
    name: https
    protocol: HTTPS
  resolution: DNS
  location: MESH_EXTERNAL
```

与上述导入 HTTP 协议的外部服务定义基本一致，只是协议字段变成 HTTPS 协议，并且减少了对应的 VirtualService 的定义。HTTPS 协议的服务也可以配置 VirtualService 进行更细致的流量管理，与 HTTP 协议的使用方法基本一致。

2. 使用 Egress gateway 统一管理访问外部服务的流量

配置 Egress gateway 示例：

```
 1 apiVersion: networking.istio.io/v1alpha3
 2 kind: Gateway
 3 metadata:
 4   name: istio-egressgateway
 5 spec:
 6   selector:
 7     istio: egressgateway
 8   servers:
 9   - port:
10       number: 80
11       name: http
12       protocol: HTTP
13     hosts:
14     - "*"
15   - port:
16       number: 443
17       name: tls
18       protocol: TLS
19     tls:
20       mode: PASSTHROUGH
21     hosts:
22     - "*"
```

第 6 ～ 7 行选择 Istio 部署时默认创建的 egressgateway 作为网格访问外部服务的流量出口。

第 9 ～ 14 行定义了一个支持 HTTP 协议的网关出口。

第15～22行定义了一个支持HTTPS协议的网关出口。tls字段中的mode模式指定为PASSTHROUGH，表示不进行TLS终止，也就是说，服务通过Egress gateway直接访问HTTPS协议的外部服务。

（1）配置通过Egress gateway访问外部HTTP协议服务

导入httpbin.org服务示例：

```
 1 apiVersion: networking.istio.io/v1alpha3
 2 kind: ServiceEntry
 3 metadata:
 4   name: external-svc-httpbin
 5 spec:
 6   hosts:
 7   - httpbin.org
 8   location: MESH_EXTERNAL
 9   ports:
10   - number: 80
11     name: http
12     protocol: HTTP
13   resolution: DNS
14 ---
15 apiVersion: networking.istio.io/v1alpha3
16 kind: VirtualService
17 metadata:
18   name: gateway-routing-http-httpbin
19 spec:
20   hosts:
21   - httpbin.org
22   gateways:
23   - mesh
24   - istio-egressgateway
25   http:
26   - match:
27     - port: 80
28       gateways:
29       - mesh
30     route:
31     - destination:
32         host: istio-egressgateway.istio-system.svc.cluster.local
33   - match:
34     - port: 80
35       gateways:
36       - istio-egressgateway
37     route:
38     - destination:
39         host: httpbin.org
```

第26～32行的路由规则定义表明，对于网格内部请求httpbin.org的流量，全部转发到istio-egressgateway外部网关服务上。

第 33 ～ 39 行的路由规则定义表明，对于不是网格内部访问 httpbin.org 的流量，也就是说，对于从 istio-egressgateway 外部网关中请求 httpbin.org 的流量，全部转发到 ServiceEntry 中定义的 httpbin.org 地址上。

（2）配置通过 Egress gateway 访问外部 HTTPS 协议服务

导入 baidu.com 服务示例：

```
apiVersion: networking.istio.io/v1alpha3
kind: ServiceEntry
metadata:
  name: external-svc-baidu
spec:
  hosts:
  - www.baidu.com
  location: MESH_EXTERNAL
  ports:
  - number: 443
    name: tls
    protocol: TLS
  resolution: DNS
---
apiVersion: networking.istio.io/v1alpha3
kind: VirtualService
metadata:
  name: gateway-routing-https-baidu
spec:
  hosts:
  - www.baidu.com
  gateways:
  - mesh
  - istio-egressgateway
  tls:
  - match:
    - port: 443
      gateways:
      - mesh
      sni_hosts:
      - www.baidu.com
    route:
    - destination:
        host: istio-egressgateway.istio-system.svc.cluster.local
  - match:
    - port: 443
      gateways:
      - istio-egressgateway
      sni_hosts:
      - www.baidu.com
    route:
    - destination:
        host: www.baidu.com
```

路由规则的配置方式与配置HTTP协议的路由规则基本一致，只是换成了tls的相关配置。

3. 网格内服务直接调用外部所有服务

在Istio的部署配置中，可以指定服务网格只拦截指定IP段的流量，这样，当服务访问网格内的服务时，流量经过Envoy代理；而访问网格外部的服务时，Envoy代理将不再拦截流量，完全由调用方自己来处理流量。如果使用这种方式来开放网格内服务调用外部服务，你将不再能享受到服务网格带来的服务路由相关的功能，比如重试、超时、熔断等。所以不到万不得已的时候，请不要使用这种方式。下面展示了本书实验部署的Istio集群开启此功能的步骤[⊖]。

1）修改Istio部署文件。

找到/usr/local/istio/install/kubernetes/istio-demo.yaml文件的10503行左右的配置行或者istio-mini.yaml文件的326行，修改成如下的形式：

```
- "[[ annotation .ObjectMeta `traffic.sidecar.istio.io/includeOutboundIPRanges`
"10.244.0.0/16,10.96.0.0/12"  ]]"
```

10.244.0.0/16是kube-controller-manager中配置的cluster-cidr的参数值，也就是Pod的IP地址段。10.96.0.0/12是kube-apiserver中配置的service-cluster-ip-range的参数值，也就是Service的IP地址段。

如上配置表明，网格只拦截10.244.0.0/16和10.96.0.0/12网段的流量，其他网段的流量全部放行，由应用自己处理。

2）应用配置文件：

```
$ kubectl apply -f /usr/local/istio/install/kubernetes/istio-demo.yaml
```

3）部署用于测试的Pod，新配置的规则只对新创建的Pod生效：

```
$ kubectl apply -f kubernetes/dns-test.yaml
```

4）访问服务：

```
$ kubectl exec dns-test -c dns-test -- curl -s http://httpbin.org/user-agent
{
  "user-agent": "curl/7.35.0"
}

$ kubectl exec dns-test -c dns-test -- curl -s http://httpbin.org/ip
{
```

⊖ 具体配置方式可参考官方文档https://istio.io/docs/tasks/traffic-management/egress/#calling-external-services-directly。

```
  "origin": "116.226.24.128"
}

$ kubectl exec dns-test -c dns-test -- curl -s -I https://www.baidu.com/
HTTP/1.1 200 OK
Accept-Ranges: bytes
Cache-Control: private, no-cache, no-store, proxy-revalidate, no-transform
Connection: Keep-Alive
Content-Length: 277
Content-Type: text/html
Date: Mon, 14 Jan 2019 15:01:31 GMT
Etag: "575e1f8a-115"
Last-Modified: Mon, 13 Jun 2016 02:50:50 GMT
Pragma: no-cache
Server: bfe/1.0.8.18
```

5）清理：

```
$ kubectl delete -f kubernetes/dns-test.yaml
```

【实验一】

以 ServiceEntry 方式访问外部服务。

1）创建测试 Pod：

```
$ kubectl create -f kubernetes/dns-test.yaml
```

2）测试访问：

```
$ kubectl exec dns-test -c dns-test -- curl -I -s http://httpbin.org/user-agent
HTTP/1.1 404 Not Found
date: Mon, 14 Jan 2019 14:35:51 GMT
server: envoy
transfer-encoding: chunked
$ kubectl exec dns-test -c dns-test -- curl -s -I https://www.baidu.com/
command terminated with exit code 35
```

3）创建 httpbin 和 baidu 的 ServiceEntry：

```
$ kubectl apply -f istio/route/service-entry-httpbin.yaml

$ kubectl apply -f istio/route/service-entry-baidu.yaml
```

4）测试访问：

```
$ kubectl exec dns-test -c dns-test -- curl -s http://httpbin.org/user-agent
{
  "user-agent": "curl/7.35.0"
}
```

```
$ kubectl exec dns-test -c dns-test -- curl -s http://httpbin.org/ip
{
  "origin": "116.226.24.128"
}

$ kubectl exec dns-test -c dns-test -- curl -s -I https://www.baidu.com/
HTTP/1.1 200 OK
Accept-Ranges: bytes
Cache-Control: private, no-cache, no-store, proxy-revalidate, no-transform
Connection: Keep-Alive
Content-Length: 277
Content-Type: text/html
Date: Mon, 14 Jan 2019 14:40:52 GMT
Etag: "575e1f8a-115"
Last-Modified: Mon, 13 Jun 2016 02:50:50 GMT
Pragma: no-cache
Server: bfe/1.0.8.18
```

5）清理：

```
$ kubectl delete -f istio/route/service-entry-baidu.yaml
$ kubectl delete -f istio/route/service-entry-httpbin.yaml
$ kubectl delete -f kubernetes/dns-test.yaml
```

【实验二】

通过 Egress gateway 方式访问外部服务。

1）创建测试 Pod：

```
$ kubectl create -f kubernetes/dns-test.yaml
```

2）测试访问：

```
$ kubectl exec dns-test -c dns-test -- curl -I -s http://httpbin.org/user-
agent
HTTP/1.1 404 Not Found
date: Mon, 14 Jan 2019 14:55:51 GMT
server: envoy
transfer-encoding: chunked
$ kubectl exec dns-test -c dns-test -- curl -s -I https://www.baidu.com/
command terminated with exit code 35
```

3）创建 Egress gateway：

```
$ kubectl create -f istio/route/gateway-egress.yaml
```

4）创建 httpbin 和 baidu 的路由规则：

```
$ kubectl create -f istio/route/service-entry-egress-httpbin.yaml

$ kubectl create -f istio/route/service-entry-egress-baidu.yaml
```

5）测试访问：

```
$ kubectl exec dns-test -c dns-test -- curl -s http://httpbin.org/user-agent
{
  "user-agent": "curl/7.35.0"
}

$ kubectl exec dns-test -c dns-test -- curl -s http://httpbin.org/ip
{
  "origin": "10.244.1.43, 116.226.24.128"
}

$ kubectl exec dns-test -c dns-test -- curl -s -I https://www.baidu.com/
HTTP/1.1 200 OK
Accept-Ranges: bytes
Cache-Control: private, no-cache, no-store, proxy-revalidate, no-transform
Connection: Keep-Alive
Content-Length: 277
Content-Type: text/html
Date: Mon, 14 Jan 2019 14:49:14 GMT
Etag: "575e1f8a-115"
Last-Modified: Mon, 13 Jun 2016 02:50:50 GMT
Pragma: no-cache
Server: bfe/1.0.8.18
```

6）清理：

```
$ kubectl delete -f istio/route/service-entry-egress-baidu.yaml
$ kubectl delete -f istio/route/service-entry-egress-httpbin.yaml
$ kubectl delete -f istio/route/gateway-egress.yaml
$ kubectl delete -f kubernetes/dns-test.yaml
```

【总结】

默认情况下，网格内服务是不能访问任何外部服务的，当前可以使用如下三种方式来满足网格内服务调用外部服务的需求：

1）使用 ServiceEntry，让每一个 Envoy 代理都可以访问特定外部服务。

2）使用 Egress gateway 管理出口流量，统一管理对外部服务的调用出口。

3）配置 Istio 直接开放对任何外部服务的调用，调用外部服务的流量由网格内服务自行管理。

第一种方式适用于集群中每台服务器都能和外部互通的网络环境。第二种方式适用于集群中存在边界服务器，或者集群中只有部分节点能与外部网络互通，所有外部流量都经过此服务器的情况。当然使用第二种方式也可以只是为了使用 Egress gateway 对网格的出口流量进行统一管理。第三种方式是最直接、暴力的方式，也是最不推荐的方式，因为使用这种方式的流量不经过网格，所有流量全部由网格内的服务自行处理，此时出口流量就不能享受到网格带来的路由管理等功能了，而且这也会降低服务网格的安全性。

7.8 实现服务 A/B 测试

通过 Istio 提供的服务路由功能，我们可以根据请求的信息轻松实现 A/B 测试。本节演示根据用户使用的浏览器来展示不同版本的 Web 界面。当用户使用 Firefox 浏览器访问时，会得到 Vue 版本实现的前端界面；当用户使用其他浏览器访问时，会得到 React 版本实现的前端界面。配置示例如下：

```
 1 apiVersion: networking.istio.io/v1alpha3
 2 kind: Gateway
 3 metadata:
 4   name: istio-lab-gateway
 5 spec:
 6   selector:
 7     istio: ingressgateway # use Istio default gateway implementation
 8   servers:
 9   - port:
10       number: 80
11       name: http
12       protocol: HTTP
13     hosts:
14     - "*"
15 ---
16 apiVersion: networking.istio.io/v1alpha3
17 kind: DestinationRule
18 metadata:
19   name: service-js
20 spec:
21   host: service-js
22   subsets:
23   - name: v1
24     labels:
25       version: v1
26   - name: v2
27     labels:
28       version: v2
29 ---
30 apiVersion: networking.istio.io/v1alpha3
31 kind: VirtualService
32 metadata:
33   name: istio-lab
34 spec:
35   hosts:
```

```
36    - "*"
37    gateways:
38    - istio-lab-gateway
39    http:
40    - match:
41      - uri:
42          prefix: /env
43      route:
44      - destination:
45          host: service-python
46    - match:
47      - uri:
48          prefix: /
49        headers:
50          user-agent:
51            regex: ".*?(Firefox).*?"
52      route:
53      - destination:
54          host: service-js
55          subset: v2
56    - route:
57      - destination:
58          host: service-js
59          subset: v1
```

第 46 ～ 55 行定义的路由规则表明，当用户的请求携带的 user-agent 请求头正则匹配 Firefox 时，转发请求到 sevice-js 服务的 v2 版本。

第 56 ～ 59 行定义了默认的路由规则，当上面的条件都不符合时，把请求转发到 sevice-js 服务的 v1 版本。

通过上面的路由规则就会实现，当用户使用 Chrome 浏览器访问时，会得到 React 版本实现的前端界面；当用户使用 Firefox 浏览器访问时，会得到 Vue 版本实现的前端界面。这样就轻松实现用户界面的 A/B 版本测试。

【实验】

1）创建 A/B 测试路由规则：

```
$ kubectl apply -f istio/route/gateway-js-user-agent-v1-v2.yaml
```

2）浏览器访问。

通过浏览器访问地址 http：//11.11.11.112：31380/，并多次点击“发射”按钮。

使用 Chrome 浏览器会得到如图 7-4 所示的界面。

使用 Firefox 浏览器会得到如图 7-5 所示的界面。

3）清理：

```
$ kubectl delete -f istio/route/gateway-js-user-agent-v1-v2.yaml
```

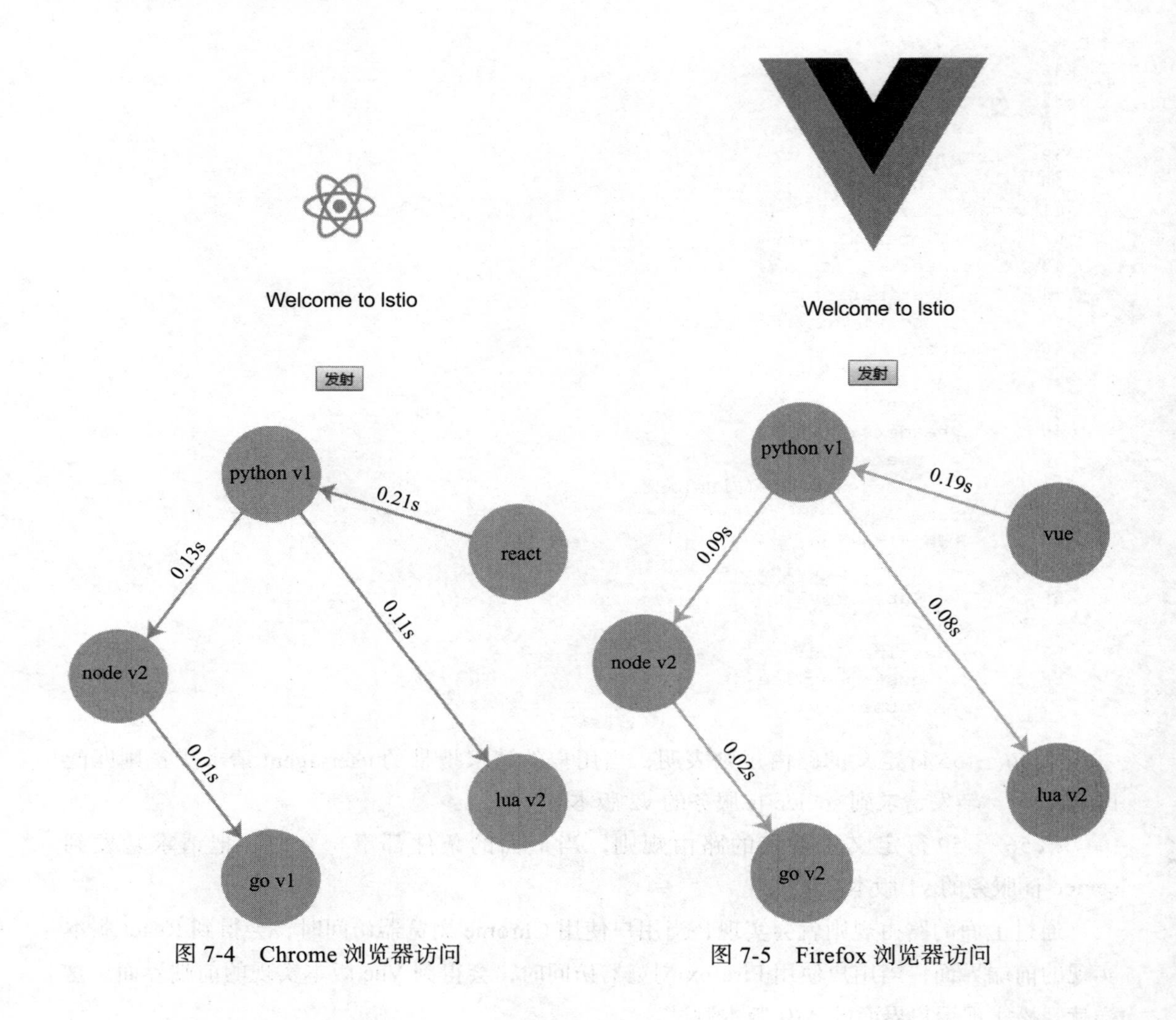

图 7-4　Chrome 浏览器访问　　图 7-5　Firefox 浏览器访问

7.9　实现服务灰度发布

灰度发布又叫金丝雀发布，通过 Istio 提供的服务路由功能可以轻松实现灰度发布。当我们的服务需要上线新版本时，可以先只把部分流量引入到新版本，等到确认新版本没有问题时，再把全部流量切换到新版本上。通过调节服务版本路由的权重，就可以达到上述目的。当然，为了不浪费资源，对服务的新版本我们只需要部署很少的实例，然后确认服务没有问题后，再增加新版本的服务实例，减少旧版本的服务实例。

对 service-go 服务实行灰度发布，使用的路由规则配置文件如下：

```
apiVersion: networking.istio.io/v1alpha3
kind: DestinationRule
```

```
metadata:
  name: service-go
spec:
  host: service-go
  subsets:
  - name: v1
    labels:
      version: v1
  - name: v2
    labels:
      version: v2
---
apiVersion: networking.istio.io/v1alpha3
kind: VirtualService
metadata:
  name: service-go
spec:
  hosts:
  - service-go
  http:
  - route:
    - destination:
        host: service-go
        subset: v1
      weight: 70
    - destination:
        host: service-go
        subset: v2
      weight: 30
```

【实验】

1）删除 service-go 服务的 v2 版本的 Pod：

```
$ kubectl delete deploy service-go-v2
$ kubectl get pod -l app=service-go
NAME                              READY     STATUS      RESTARTS    AGE
service-go-v1-7cc5c6f574-2k6wv    2/2       Running     0           4m38s
```

2）为了模拟真实环境，我们提前把 service-go 服务扩容到 5 个实例，也就是 5 个 Pod，使用如下方式增加实例个数：

```
$ kubectl scale --replicas=5 deployment/service-go-v1

$ kubectl get pod -l app=service-go
service-go-v1-7cc5c6f574-9g52l    2/2       Running     0          29s
service-go-v1-7cc5c6f574-kkmmh    2/2       Running     0          29s
service-go-v1-7cc5c6f574-kqwg5    2/2       Running     0          7m24s
service-go-v1-7cc5c6f574-pl4dt    2/2       Running     0          29s
service-go-v1-7cc5c6f574-snd7b    2/2       Running     0          29s
```

3）创建用于测试的Pod：

```
$ kubectl apply -f kubernetes/dns-test.yaml
```

4）模拟生产环境，流量全部转发到服务service-go的v1版本：

```
$ kubectl apply -f istio/route/virtual-service-go-v1.yaml
```

5）访问service-go服务，此时所有的请求只会落到v1版本上：

```
$ kubectl exec dns-test -c dns-test -- curl -s http://service-go/env
{"message":"go v1"}
```

6）部署新版本服务，此时只创建两个v2版本的service-go服务实例用来准备接收请求：

```
$ kubectl apply -f service/go/service-go-v2.yaml
$ kubectl scale --replicas=2 deployment/service-go-v2

$ kubectl get pod -l app=service-go
NAME                               READY     STATUS     RESTARTS    AGE
service-go-v1-7cc5c6f574-9g52l     2/2       Running    0           6m40s
service-go-v1-7cc5c6f574-kkmmh     2/2       Running    0           6m40s
service-go-v1-7cc5c6f574-kqwg5     2/2       Running    0           13m
service-go-v1-7cc5c6f574-pl4dt     2/2       Running    0           6m40s
service-go-v1-7cc5c6f574-snd7b     2/2       Running    0           6m40s
service-go-v2-7656dcc478-7xn4h     2/2       Running    0           19s
service-go-v2-7656dcc478-qmkr8     2/2       Running    0           8s
```

7）给service-go服务v2版本分配30%流量：

```
$ kubectl apply -f istio/route/virtual-service-go-canary.yaml
```

8）访问service-go服务，此时请求会分别落到v1、v2两个版本上：

```
$ kubectl exec dns-test -c dns-test -- curl -s http://service-go/env
{"message":"go v1"}

$ kubectl exec dns-test -c dns-test -- curl -s http://service-go/env
{"message":"go v2"}
```

9）观察服务的性能、响应时间、错误率等数据，当服务达到要求时，增加service-go服务的v2版本实例数：

```
$ kubectl scale --replicas=3 deployment/service-go-v2
$ kubectl get pod -l app=service-go
NAME                               READY     STATUS     RESTARTS    AGE
service-go-v1-7cc5c6f574-9g52l     2/2       Running    0           15m
```

```
service-go-v1-7cc5c6f574-kkmmh      2/2     Running   0          15m
service-go-v1-7cc5c6f574-kqwg5      2/2     Running   0          22m
service-go-v1-7cc5c6f574-pl4dt      2/2     Running   0          15m
service-go-v1-7cc5c6f574-snd7b      2/2     Running   0          15m
service-go-v2-7656dcc478-78jt7      2/2     Running   0          116s
service-go-v2-7656dcc478-7xn4h      2/2     Running   0          9m
service-go-v2-7656dcc478-qmkr8      2/2     Running   0          8m49s
```

10）给 service-go 服务 v2 版本分配 50% 流量：

```
$ kubectl apply -f istio/route/virtual-service-go-canary-50.yaml
```

11）减少 v1 版本的实例数，增加 service-go 服务的 v2 版本实例数：

```
$ kubectl scale --replicas=2 deployment/service-go-v1
$ kubectl scale --replicas=4 deployment/service-go-v2
$ kubectl get pod -l app=service-go
NAME                                READY   STATUS    RESTARTS   AGE
service-go-v1-7cc5c6f574-kkmmh      2/2     Running   0          17m
service-go-v1-7cc5c6f574-kqwg5      2/2     Running   0          24m
service-go-v2-7656dcc478-78jt7      2/2     Running   0          4m26s
service-go-v2-7656dcc478-7xn4h      2/2     Running   0          11m
service-go-v2-7656dcc478-qmkr8      2/2     Running   0          11m
service-go-v2-7656dcc478-t2t57      2/2     Running   0          60s
```

12）给 service-go 服务 v2 版本分配 70% 流量：

```
$ kubectl apply -f istio/route/virtual-service-go-canary-70.yaml
```

13）增加 service-go 服务的 v2 版本实例数：

```
$ kubectl scale --replicas=5 deployment/service-go-v2
$ kubectl get pod -l app=service-go
NAME                                READY   STATUS    RESTARTS   AGE
service-go-v1-7cc5c6f574-kkmmh      2/2     Running   0          19m
service-go-v1-7cc5c6f574-kqwg5      2/2     Running   0          26m
service-go-v2-7656dcc478-4nn2q      2/2     Running   0          19s
service-go-v2-7656dcc478-78jt7      2/2     Running   0          5m53s
service-go-v2-7656dcc478-7xn4h      2/2     Running   0          12m
service-go-v2-7656dcc478-qmkr8      2/2     Running   0          12m
service-go-v2-7656dcc478-t2t57      2/2     Running   0          2m27s
```

14）把流量全部引入 service-go 服务的 v2 版本：

```
$ kubectl apply -f istio/route/virtual-service-go-v2.yaml
```

15）访问 service-go 服务，此时所有的请求只会落到 v2 版本上：

```
$ kubectl exec dns-test -c dns-test -- curl -s http://service-go/env
{"message":"go v2"}
```

16）减少或者删除 service-go 服务 v1 版本的实例，完成服务的灰度发布：

```
$ kubectl scale --replicas=0 deployment/service-go-v1
$ kubectl get pod -l app=service-go
NAME                             READY     STATUS    RESTARTS   AGE
service-go-v2-7656dcc478-4nn2q   2/2       Running   0          2m35s
service-go-v2-7656dcc478-78jt7   2/2       Running   0          8m9s
service-go-v2-7656dcc478-7xn4h   2/2       Running   0          15m
service-go-v2-7656dcc478-qmkr8   2/2       Running   0          15m
service-go-v2-7656dcc478-t2t57   2/2       Running   0          4m43s

$ kubectl delete deployment service-go-v1
```

17）清理。

恢复环境到实验的初始状态：

```
$ kubectl delete -f kubernetes/dns-test.yaml
$ kubectl delete -f istio/route/virtual-service-go-v2.yaml
$ kubectl scale --replicas=1 deployment/service-go-v2
$ kubectl apply -f service/go/service-go.yaml
$ kubectl get pod -l app=service-go
NAME                             READY     STATUS    RESTARTS   AGE
service-go-v1-7cc5c6f574-2282d   2/2       Running   0          16s
service-go-v2-7656dcc478-7xn4h   2/2       Running   0          22m
```

7.10 灰度发布与 A/B 测试结合

上节介绍了简单的灰度发布实验，在真实的生产环境中，可能需要根据用户信息、应用版本等信息进行灰度发布，比如我们可能只想在安卓手机用户中挑选部分用户进行灰度测试，这时候我们就需要把灰度发布和 A/B 测试进行结合，来达到目的。路由配置示例如下：

```
 1 apiVersion: networking.istio.io/v1alpha3
 2 kind: Gateway
 3 metadata:
 4   name: istio-lab-gateway
 5 spec:
 6   selector:
 7     istio: ingressgateway # use Istio default gateway implementation
 8   servers:
 9   - port:
10       number: 80
11       name: http
12       protocol: HTTP
13     hosts:
```

```
14      - "*"
15 ---
16 apiVersion: networking.istio.io/v1alpha3
17 kind: DestinationRule
18 metadata:
19   name: service-js
20 spec:
21   host: service-js
22   subsets:
23   - name: v1
24     labels:
25       version: v1
26   - name: v2
27     labels:
28       version: v2
29 ---
30 apiVersion: networking.istio.io/v1alpha3
31 kind: DestinationRule
32 metadata:
33   name: service-python
34 spec:
35   host: service-python
36   subsets:
37   - name: v1
38     labels:
39       version: v1
40   - name: v2
41     labels:
42       version: v2
43 ---
44 apiVersion: networking.istio.io/v1alpha3
45 kind: VirtualService
46 metadata:
47   name: istio-lab
48 spec:
49   hosts:
50   - "*"
51   gateways:
52   - istio-lab-gateway
53   http:
54   - match:
55     - uri:
56         prefix: /env
57       headers:
58         user-agent:
59           regex: ".*?(Firefox).*?"
60     route:
61     - destination:
```

```
62          host: service-python
63          subset: v1
64        weight: 70
65      - destination:
66          host: service-python
67          subset: v2
68        weight: 30
69    - match:
70      - uri:
71          prefix: /env
72      route:
73      - destination:
74          host: service-python
75          subset: v1
76    - route:
77      - destination:
78          host: service-js
79          subset: v1
```

第 54 ～ 68 行的路由规则定义表示，对于使用 Firefox 浏览器访问 /env 的用户请求，会有 30% 概率的请求转发到 service-python 的 v2 版本，70% 的请求转发到 v1 版本。

第 69 ～ 75 行的路由规则定义表示，对于使用其他浏览器访问 /env 的用户请求只会转发到 service-python 的 v1 版本。

第 76 ～ 79 行定义了默认了路由规则，当没有规则匹配时，使用此默认路由。当访问除了 /env 的链接时，请求转发到 service-js 的 v1 版本。

通过如上的路由规则配置，就可以实现对于使用 Firefox 浏览器的用户进行灰度上线 service-python 服务 v2 版本的需求。

【实验】

1）创建路由规则：

```
$ kubectl apply -f istio/route/virtual-service-python-ab-canary.yaml
```

2）浏览器访问。

此时打开浏览器访问 http：//11.11.11.112：31380/，并多次点击“发射”按钮，如果你使用的是 Firefox 浏览器，就会看到大部请求都会落到 service-python 服务的 v1 版本，只有部分请求落到 service-python 服务的 v2 版本。如果使用其他浏览器访问，就会看到所有请求只会落到 service-python 服务的 v1 版本。

3）清理：

```
$ kubectl delete -f istio/route/virtual-service-python-ab-canary.yaml
```

注意事项　实验中可能会遇到一些问题，如路由创建不生效、应用路由规则时出现超时错误、自动注入失败等，这些问题都可以通过重启电脑即可解决。另外，实验时有可能第一次访问服务时响应时间过长，也有可能出现中间部分请求响应时间过长，这些都是由于实验时的机器性能不足导致的，直接忽略，进行多次访问即可。

7.11　本章小结

借助 Istio 提供的路由功能，我们只需要创建简单的路由规则配置文件，就可以轻松实现服务路由管理，而完全不需要对应用做任何改变。我们可以使用 Gateway 和 ServiceEntry 更方便地管理集群的入口与出口流量，这也可以增强集群的安全性。另外，我们还可以很容易地实现复杂的 A/B 测试和灰度发布功能。

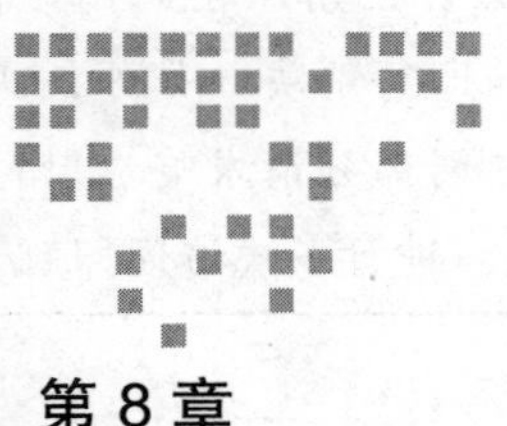

Chapter 8 第 8 章

让服务更具弹性

本章介绍如何借助 Istio 提供的功能，让我们的服务更具弹性。主要包括配置服务的负载均衡策略、连接池、健康检测、熔断、超时、重试、限流等。通过上述这些配置，可以让我们服务在遇到故障时更具弹性。

8.1 整体介绍

服务具备弹性是指系统能够优雅地处理故障并从故障中恢复。尽管我们在进行服务设计时会尽量做高可用方案，但是服务出现故障或者服务间通信的网络出现故障的情况是无法避免的，当故障出现时，我们应该尽量保证服务的可用性，不要让故障变得更加严重，影响整个应用的稳定性。

Istio 提供了许多开箱即用的提升服务弹性的功能，包括负载均衡、连接池、健康检测、熔断、超时、重试、限流，这些功能都是服务治理的必备功能。通过组合使用这些功能，可以让服务更具弹性。

- **负载均衡**。当服务有多个实例时，通过负载均衡器来分发请求，负载均衡器一般会提供轮询、随机、最少连接、哈希等算法。
- **连接池**。当服务调用上游服务时，可以提前创建好到上游服务的连接，当请求到来时，通过已经创建好的连接直接发送请求给上游服务，减少连接的创建时间，从而降低请求的整体耗时。我们把这些提前创建好的连接集合称为连接池。
- **健康检测**。当上游服务部分实例出现故障时，健康检测机制能自动检测到上游服务的不可用实例，从而避免把请求发送给上游不可用的服务实例。

- ❑ **限流**。当服务并发请求激增，流量增大，如果没有使用任何限流措施，这很有可能导致我们的服务无法承受如此之多的请求，进而导致服务崩溃，还可能会影响整个应用的稳定性。限流功能是当请求数过多时，直接丢掉过多的流量，防止服务被压垮，保证服务稳定。
- ❑ **超时**。当上游服务响应时间过长，很有可能会增加请求的整体耗时，影响服务调用方。通过设置服务调用的超时时间，当服务没有在规定的时间内返回数据，就直接取消此次请求，这样可以防止服务提供方拖垮服务调用方，防止请求的总耗时时间过长。
- ❑ **重试**。网络出现抖动，或者被调用的服务出现瞬时故障，这些问题都是偶发瞬时的，只需要再重新调用一次后端服务，可能请求就会成功。这种场景下就需要服务的重试功能，防止由于被调用方的服务偶发瞬时故障，导致出现服务调用不可用的情况，从而影响应用的整体稳定性与可用性。
- ❑ **熔断**。后端服务调用偶尔出现失败是正常情况，但是当对后端服务的调用出现大比例的调用失败，此时可能由于后端服务已经无法承受当前压力，如果我们还是继续调用后端服务，不仅不能得到响应，还有可能会把后端服务整个压垮。所以当服务出现大比例调用失败时，应停止调用后端服务，经过短暂时间间隔后，再尝试让部分请求调用后端服务，如果服务返回正常，我们就可以让更多的请求调用后端服务，直到服务恢复为正常情况。如果仍然出现无法响应的情况，我们将再次停止调用服务，这种处理后端服务调用的机制就称为熔断。

虽然一个服务出现故障的概率可能很低，但是当服务数量变多，低概率的故障事件就会必然发生。当服务数量持续增长，服务间的调用关系复杂，如果不做服务故障处理，提升服务的弹性，就很有可能因为一个服务的故障导致其他服务也出现故障，进而导致服务出现级联故障，甚至可能导致整个应用出现不可用的现象。这会严重影响我们的应用拆分为微服务后的服务可用性指标，使整个应用的用户体验变差，导致用户流失，进而影响业务发展。如果经常出现这种故障，这对业务来说无疑是毁灭性的打击。

【实验前的准备】

进行本章实验前，需要先执行如下的前置步骤。

1）下载实验时用到的源码仓库：

```
$ sudo yum install -y git

$ git clone https://github.com/mgxian/istio-lab
Cloning into 'istio-lab'...
remote: Enumerating objects: 252, done.
```

```
remote: Counting objects: 100% (252/252), done.
remote: Compressing objects: 100% (177/177), done.
remote: Total 779 (delta 157), reused 166 (delta 74), pack-reused 527
Receiving objects: 100% (779/779), 283.37 KiB | 243.00 KiB/s, done.
Resolving deltas: 100% (451/451), done.

$ cd istio-lab
```

2）开启 default 命名空间的自动注入功能：

```
$ kubectl label namespace default istio-injection=enabled
namespace/default labeled
```

3）部署用于测试的服务：

```
$ kubectl apply -f service/go/service-go.yaml
$ kubectl get pod
NAME                              READY     STATUS     RESTARTS    AGE
service-go-v1-7cc5c6f574-lrp2h    2/2       Running    0           76s
service-go-v2-7656dcc478-svn5c    2/2       Running    0           76s
```

8.2 负载均衡

Istio 通过设置 DestinationRule 来指定服务的负载均衡策略，Istio 提供了两种常用的负载均衡策略：简单负载均衡（simple）和一致性哈希负载均衡（consistentHash）。可以为服务设置默认的负载均衡策略，也可以在单独的服务子集中设置负载均衡策略，服务子集中的设置会覆盖服务设置的默认负载均衡策略。我们还可以设置端口级别的负载均衡策略（portLevelSettings），可以为服务设置默认的端口级别的负载均衡策略。当然，我们也可以在单独的服务子集上设置端口级别的负载均衡策略。

1. 简单负载均衡

简单负载均衡策略提供了如下 4 种负载均衡算法：

- 轮询（ROUND_ROBIN）：把请求依次转发给后端健康实例，这是默认算法。
- 最少连接（LEAST_CONN）：把请求转发给活跃请求最少的后端健康实例，此处的活跃请求数是 Istio 自己维护的，是 Istio 调用后端实例且正在等待返回响应的请求数。由于服务实例可能还有其他客户端在调用，没有经过 Istio 统计，所以 Istio 维护的活跃请求数并不是此时实例真正的活跃请求数。由于 Envoy 与服务部署在一个 Pod 中，并拦截所有流量，因此一般情况下，可以把 Istio 维护的活跃请求数看成是服务实例的真正活跃请求数。
- 随机（RANDOM）：把请求随机转发给后端健康实例。
- 直连（PASSTHROUGH）：将连接转发到调用方请求的原始 IP 地址，而不进行任何形式的负载平衡，这是高级用法，一般情况下不会使用。

使用示例：

```
 1 apiVersion: networking.istio.io/v1alpha3
 2 kind: DestinationRule
 3 metadata:
 4   name: service-go
 5 spec:
 6   host: service-go
 7   trafficPolicy:
 8     loadBalancer:
 9       simple: ROUND_ROBIN
10   subsets:
11   - name: v1
12     labels:
13       version: v1
14   - name: v2
15     labels:
16       version: v2
17     trafficPolicy:
18       loadBalancer:
19         simple: LEAST_CONN
20       portLevelSettings:
21       - port:
22           number: 80
23         loadBalancer:
24           simple: RANDOM
```

第 7 ～ 9 行定义了默认的负载均衡策略为轮询方式。

第 17 ～ 19 行定义了对于名称为 v2 的实例子集负载均衡策略为最少连接方式。

第 20 ～ 24 行定义了端口级别的负载均衡策略，指定 80 端口的负载均衡策略为随机方式。

2. 一致性哈希负载均衡

一致性哈希负载均衡策略只适用于使用 HTTP 类协议（HTTP 1.1/HTTPS/HTTP2）的请求，可以基于请求头、Cookie 或者来源 IP 做会话保持，让同一用户的请求一直转发到后端同一实例，当实例出现故障时会选择新的实例。当添加删除新实例时，会有部分用户的会话保持失效。

使用示例：

```
1 apiVersion: networking.istio.io/v1alpha3
2 kind: DestinationRule
3 metadata:
4   name: service-go
5 spec:
```

```
 6   host: service-go
 7   trafficPolicy:
 8     loadBalancer:
 9       consistentHash:
10         httpHeaderName: x-lb-test
11   subsets:
12   - name: v1
13     labels:
14       version: v1
15   - name: v2
16     labels:
17       version: v2
```

第7～10行定义了默认的负载均衡策略为一致性哈希，基于x-lb-test请求头进行一致性哈希负载均衡。

【实验】

1）创建测试Pod：

```
$ kubectl apply -f kubernetes/dns-test.yaml
```

2）创建service-go服务的路由规则：

```
$ kubectl apply -f istio/route/virtual-service-go.yaml
```

3）访问service-go服务：

```
    $ kubectl exec dns-test -c dns-test -- curl -s -H "X-lb-test: 1" http://
service-go/env
    {"message":"go v1"}

    $ kubectl exec dns-test -c dns-test -- curl -s -H "X-lb-test: 1" http://
service-go/env
    {"message":"go v2"}
```

在没有创建一致性哈希负载均衡规则时访问service-go服务，同样的请求头，会落到service-go服务的v1和v2两个版本。

4）创建一致性哈希负载均衡规则：

```
$ kubectl apply -f istio/resilience/destination-rule-go-lb-hash.yaml
```

5）访问service-go服务：

```
    $ kubectl exec dns-test -c dns-test -- curl -s -H "X-lb-test: 1" http://
service-go/env
    {"message":"go v2"}

    $ kubectl exec dns-test -c dns-test -- curl -s -H "X-lb-test: 2" http://
service-go/env
```

```
{"message":"go v2"}

$ kubectl exec dns-test -c dns-test -- curl -s -H "X-lb-test: 3" http://
service-go/env
{"message":"go v1"}
```

在创建一致性哈希负载均衡规则后访问 service-go 服务，同样的请求头，只会落到 service-go 服务的同一个版本上。

6）清理：

```
$ kubectl delete -f kubernetes/dns-test.yaml
$ kubectl delete -f istio/route/virtual-service-go.yaml
$ kubectl delete -f istio/resilience/destination-rule-go-lb-hash.yaml
```

总结：从以上的实验结果可以看出，实验部署了两个版本的 service-go 服务实例，每个版本一个 Pod，默认情况下访问 service-go 服务会轮询地转发到后端 Pod 上，因此多次访问会看到两个版本的响应结果；当配置了一致性哈希负载均衡规则以后，以固定的 X-lb-test 请求头值请求时，多次访问只能获取到同一个版本的服务实例响应信息。

实验结果可能与上面的结果并不完全一致，但是结论是一致的，当请求携带同样的 X-lb-test 请求头访问 service-go 服务时，只会得到同一个版本的服务响应结果。

8.3　连接池

Istio 通过设置 DestinationRule 来指定服务的连接池配置，Istio 提供了两类常用协议的连接池配置：TCP 连接池和 HTTP 连接池。与负载均衡策略设置类似，连接池的配置也支持服务默认级别的配置、服务子集的配置以及端口级别的连接池配置。TCP 连接池和 HTTP 连接池可以一同配合使用。

1. TCP 连接池

TCP 连接池对 TCP 和 HTTP 类协议均提供支持，示例如下：

```
1 apiVersion: networking.istio.io/v1alpha3
2 kind: DestinationRule
3 metadata:
4   name: service-go
5 spec:
6   host: service-go
7   trafficPolicy:
8     connectionPool:
9       tcp:
```

```
10          maxConnections: 10
11          connectTimeout: 30ms
12    subsets:
13    - name: v1
14      labels:
15        version: v1
16    - name: v2
17      labels:
18        version: v2
```

第 8 ～ 11 行设置 TCP 连接池中的最大连接数为 10，连接超时时间为 30 毫秒，当连接池中连接不够用时，服务调用会返回 503 响应码。

2. HTTP 连接池

HTTP 连接池对 HTTP 类协议和 gRPC 协议均提供支持，示例如下：

```
 1 apiVersion: networking.istio.io/v1alpha3
 2 kind: DestinationRule
 3 metadata:
 4   name: service-go
 5 spec:
 6   host: service-go
 7   trafficPolicy:
 8     connectionPool:
 9       http:
10         http2MaxRequests: 10
11         http1MaxPendingRequests: 5
12         maxRequestsPerConnection: 2
13         maxRetries: 3
14   subsets:
15   - name: v1
16     labels:
17       version: v1
18   - name: v2
19     labels:
20       version: v2
```

第 8 ～ 13 行设置 HTTP 连接池的后端实例的最大并发请求数为 10，每个目标的最大待处理请求数为 5，连接池中每个连接最多处理 2 个请求后就关闭，并根据需要重新创建连接池中的连接，请求在服务后端实例集群中失败后的最大重试次数为 3。几个参数说明如下：

- http2MaxRequests 表示对后端的最大并发请求数，默认值 1024。
- http1MaxPendingRequests 表示每个目标的最大待处理请求数，这里的目标指的是 VirtualService 路由规则中配置的 destination，当连接池中连接不够用时请求就处于待处理状态。默认值 1024。
- maxRequestsPerConnection 表示每个连接最多处理多少个请求后关闭，设置为 1 时表示关闭 keep alive 特性，每次请求都创建一个新的请求。

- maxRetries 表示请求后端失败后重试其他后端实例的总次数，默认值 3。

【实验】

1）启动用于并发测试的 Pod：

```
$ kubectl apply -f kubernetes/fortio.yaml
```

2）创建 service-go 服务的路由规则：

```
$ kubectl apply -f istio/route/virtual-service-go.yaml
```

3）创建连接池规则：

```
$ kubectl apply -f istio/resilience/destination-rule-go-pool-http.yaml
```

4）并发访问服务：

```
$ kubectl exec fortio -c fortio /usr/local/bin/fortio -- load -curl http://
service-go/env
HTTP/1.1 200 OK
content-type: application/json; charset=utf-8
date: Wed, 16 Jan 2019 10:12:35 GMT
content-length: 19
x-envoy-upstream-service-time: 4
server: envoy

{"message":"go v2"}

# 10 并发
$ kubectl exec fortio -c fortio /usr/local/bin/fortio -- load -c 10 -qps 0 -n
100 -loglevel Error http://service-go/env
09:40:38 I logger.go:97> Log level is now 4 Error (was 2 Info)
Fortio 1.0.1 running at 0 queries per second, 2->2 procs, for 100 calls:
http://service-go/env
Aggregated Function Time : count 100 avg 0.01652562 +/- 0.013 min 0.002576677
max 0.064653438 sum 1.65256199
# target 50% 0.0119375
# target 75% 0.018
# target 90% 0.035
# target 99% 0.06
# target 99.9% 0.0641881
Sockets used: 15 (for perfect keepalive, would be 10)
Code 200 : 95 (95.0 %)
Code 503 : 5 (5.0 %)
All done 100 calls (plus 0 warmup) 16.526 ms avg, 563.4 qps

# 20 并发
$ kubectl exec fortio -c fortio /usr/local/bin/fortio -- load -c 20 -qps 0 -n
200 -loglevel Error http://service-go/env
```

```
09:41:32 I logger.go:97> Log level is now 4 Error (was 2 Info)
Fortio 1.0.1 running at 0 queries per second, 2->2 procs, for 200 calls:
http://service-go/env
Aggregated Function Time : count 200 avg 0.023987068 +/- 0.01622 min
0.001995258 max 0.067905383 sum 4.79741353
# target 50% 0.0194286
# target 75% 0.0357692
# target 90% 0.05
# target 99% 0.0626351
# target 99.9% 0.0673784
Sockets used: 43 (for perfect keepalive, would be 20)
Code 200 : 177 (88.5 %)
Code 503 : 23 (11.5 %)
All done 200 calls (plus 0 warmup) 23.987 ms avg, 711.9 qps

# 30 并发
$ kubectl exec fortio -c fortio /usr/local/bin/fortio -- load -c 30 -qps 0 -n
300 -loglevel Error http://service-go/env
09:42:05 I logger.go:97> Log level is now 4 Error (was 2 Info)
Fortio 1.0.1 running at 0 queries per second, 2->2 procs, for 300 calls:
http://service-go/env
Aggregated Function Time : count 300 avg 0.034233818 +/- 0.02268 min
0.002354402 max 0.114700368 sum 10.2701455
# target 50% 0.0285417
# target 75% 0.0446667
# target 90% 0.0686957
# target 99% 0.1
# target 99.9% 0.11323
Sockets used: 137 (for perfect keepalive, would be 30)
Code 200 : 192 (64.0 %)
Code 503 : 108 (36.0 %)
All done 300 calls (plus 0 warmup) 34.234 ms avg, 702.1 qps
```

从压测结果可以看出，当并发逐渐增大时，服务不可用的响应码（503）所占比例逐渐升高，说明我们配置的 HTTP 连接池参数已经生效。

5）清理：

```
$ kubectl delete -f kubernetes/fortio.yaml
$ kubectl delete -f istio/route/virtual-service-go.yaml
$ kubectl delete -f istio/resilience/destination-rule-go-pool-http.yaml
```

8.4 健康检测

Istio 通过设置 DestinationRule 来指定服务实例健康检测的配置，可以设置服务实例健康检测计算的时间间隔、实例移出负载均衡池的条件、实例移出负载均衡池时间的基础值，以及服务实例移出负载均衡池的最大比例值。设置移出负载均衡池的最大比例，可以防止异常情况移出负载均衡池过多的服务实例，导致服务实例不够，剩余实例承受流量过大，压跨整个服务。与负载均衡策略设置类似，服务实例健康检测的配置也支持服务默认级别的配置、服务子集的配置、端口级别的配置。

对于 HTTP 协议的服务，当后端实例返回 5xx 的响应码时，代表后端实例出现错误，后端实例的错误计数会增加。对于 TCP 协议的服务，连接超时、连接失败会被认为后端实例出现错误，后端实例的错误计数会增加。

配置示例：

```
 1 apiVersion: networking.istio.io/v1alpha3
 2 kind: DestinationRule
 3 metadata:
 4   name: service-go
 5 spec:
 6   host: service-go
 7   trafficPolicy:
 8     outlierDetection:
 9       consecutiveErrors: 3
10       interval: 10s
11       baseEjectionTime: 30s
12       maxEjectionPercent: 10
13   subsets:
14   - name: v1
15     labels:
16       version: v1
17   - name: v2
18     labels:
19       version: v2
```

第 8 ～ 12 行为后端实例健康检测配置的定义，配置最大实例移出比例（maxEjectionPercent）不超过 10%，基础移出时间（baseEjectionTime）为 30 秒，当实例恢复健康加入集群中后，再次出现故障被移出时，移出时间会根据此值增加，每隔 10 秒（interval）检测一次后端实例是否应该被移出负载均衡池，当在 10 秒内出现 3 次错误时，实例会被移出负载均衡池。其中参数说明如下：

- consecutiveErrors 表示服务实例移出负载均衡池之前的最大出错次数阈值，当负载均衡池中的实例在指定的时间间隔内出错次数到达该值时，会被移出负载均衡池。
- interval 表示两次后端实例健康检查时间间隔，默认值为 10 秒。
- baseEjectionTime 表示移出负载均衡池的基础时间，默认值为 30 秒。
- maxEjectionPercent 表示后端实例最大移出百分比，默认值为 10。

8.5 熔断

Istio 通过结合连接池和实例健康检测机制实现熔断功能。当后端实例出现故障时就移出负载均衡池，当负载均衡池中无可用健康实例时，服务请求会立即得到服务不可用的响应码，此时服务就处于熔断状态了。当服务实例被移出的时间结束后，服务实例会被再次添加到负载均衡池中，等待下一轮的服务健康检测。

配置示例：

```
 1 apiVersion: networking.istio.io/v1alpha3
 2 kind: DestinationRule
 3 metadata:
 4   name: service-go
 5 spec:
 6   host: service-go
 7   trafficPolicy:
 8     connectionPool:
 9       tcp:
10         maxConnections: 10
11       http:
12         http2MaxRequests: 10
13         maxRequestsPerConnection: 10
14     outlierDetection:
15       consecutiveErrors: 3
16       interval: 3s
17       baseEjectionTime: 3m
18       maxEjectionPercent: 100
19   subsets:
20   - name: v1
21     labels:
22       version: v1
23   - name: v2
24     labels:
25       version: v2
```

第8～13行定义了连接池配置，并发请求设置为10。

第14～18行定义了后端实例健康检测配置，允许全部实例移出连接池。

【实验】

1）启动用于并发测试的Pod：

```
$ kubectl apply -f kubernetes/fortio.yaml
```

2）创建service-go服务的路由规则：

```
$ kubectl apply -f istio/route/virtual-service-go.yaml
```

3）创建熔断规则：

```
$ kubectl apply -f istio/resilience/destination-rule-go-cb.yaml
```

4）访问：

```
$ kubectl exec fortio -c fortio /usr/local/bin/fortio -- load -curl http://service-go/env
HTTP/1.1 200 OK
content-type: application/json; charset=utf-8
date: Wed, 16 Jan 2019 10:22:35 GMT
content-length: 19
x-envoy-upstream-service-time: 3
server: envoy

{"message":"go v2"}

# 20 并发
$ kubectl exec fortio -c fortio /usr/local/bin/fortio -- load -c 20 -qps 0 -n 200 -loglevel Error http://service-go/env
10:25:21 I logger.go:97> Log level is now 4 Error (was 2 Info)
Fortio 1.0.1 running at 0 queries per second, 2->2 procs, for 200 calls: http://service-go/env
Aggregated Function Time : count 200 avg 0.023687933 +/- 0.01781 min 0.002302379 max 0.082312522 sum 4.73758658
# target 50% 0.0175385
# target 75% 0.029375
# target 90% 0.0533333
# target 99% 0.0766667
# target 99.9% 0.08185
Sockets used: 22 (for perfect keepalive, would be 20)
Code 200 : 198 (99.0 %)
Code 503 : 2 (1.0 %)
All done 200 calls (plus 0 warmup) 23.688 ms avg, 631.3 qps

# 30 并发
$ kubectl exec fortio -c fortio /usr/local/bin/fortio -- load -c 30 -qps 0 -n 300 -loglevel Error http://service-go/env
10:26:49 I logger.go:97> Log level is now 4 Error (was 2 Info)
Fortio 1.0.1 running at 0 queries per second, 2->2 procs, for 300 calls: http://service-go/env
Aggregated Function Time : count 300 avg 0.055940327 +/- 0.04215 min 0.001836339 max 0.207798702 sum 16.782098
# target 50% 0.0394737
# target 75% 0.0776471
# target 90% 0.123333
# target 99% 0.18
# target 99.9% 0.205459
Sockets used: 94 (for perfect keepalive, would be 30)
Code 200 : 236 (78.7 %)
Code 503 : 64 (21.3 %)
All done 300 calls (plus 0 warmup) 55.940 ms avg, 486.3 qps
```

```
# 40 并发
$ kubectl exec fortio -c fortio /usr/local/bin/fortio -- load -c 40 -qps 0 -n
400 -loglevel Error http://service-go/env
10:27:17 I logger.go:97> Log level is now 4 Error (was 2 Info)
Fortio 1.0.1 running at 0 queries per second, 2->2 procs, for 400 calls:
http://service-go/env
Aggregated Function Time : count 400 avg 0.034048003 +/- 0.02541 min
0.001808212 max 0.144268023 sum 13.6192011
# target 50% 0.028587
# target 75% 0.0415789
# target 90% 0.0588889
# target 99% 0.132
# target 99.9% 0.143414
Sockets used: 203 (for perfect keepalive, would be 40)
Code 200 : 225 (56.2 %)
Code 503 : 175 (43.8 %)
All done 400 calls (plus 0 warmup) 34.048 ms avg, 951.0 qps

# 查看 istio-proxy 状态
$ kubectl exec fortio  -c istio-proxy  -- curl -s localhost:15000/stats | grep
service-go | grep pending
cluster.outbound|80|v1|service-go.default.svc.cluster.local.upstream_rq_
pending_active: 0
cluster.outbound|80|v1|service-go.default.svc.cluster.local.upstream_rq_
pending_failure_eject: 0
cluster.outbound|80|v1|service-go.default.svc.cluster.local.upstream_rq_
pending_overflow: 0
cluster.outbound|80|v1|service-go.default.svc.cluster.local.upstream_rq_
pending_total: 0
cluster.outbound|80|v2|service-go.default.svc.cluster.local.upstream_rq_
pending_active: 0
cluster.outbound|80|v2|service-go.default.svc.cluster.local.upstream_rq_
pending_failure_eject: 0
cluster.outbound|80|v2|service-go.default.svc.cluster.local.upstream_rq_
pending_overflow: 0
cluster.outbound|80|v2|service-go.default.svc.cluster.local.upstream_rq_
pending_total: 0
cluster.outbound|80||service-go.default.svc.cluster.local.upstream_rq_pending_
active: 0
cluster.outbound|80||service-go.default.svc.cluster.local.upstream_rq_pending_
failure_eject: 0
cluster.outbound|80||service-go.default.svc.cluster.local.upstream_rq_pending_
overflow: 551
cluster.outbound|80||service-go.default.svc.cluster.local.upstream_rq_pending_
total: 1282
```

实验部署了两个版本的service-go服务实例，每个版本一个Pod，每个Pod的并发数为10，所以总的最大并发数就为20。从压测结果可以看出，当并发逐渐增大时，服务不可用的响应码（503）所占比例逐渐升高。但是从结果看，熔断器并不是非常准确地拦截了高于设置并发值的请求，Istio 允许有部分请求遗漏。

5）清理：

```
$ kubectl delete -f kubernetes/fortio.yaml
$ kubectl delete -f istio/route/virtual-service-go.yaml
$ kubectl delete -f istio/resilience/destination-rule-go-cb.yaml
```

8.6　超时

Istio 通过设置 VirtualService 中的 timeout 字段来指定服务的调用超时时间。

使用示例如下：

```
 1 apiVersion: networking.istio.io/v1alpha3
 2 kind: VirtualService
 3 metadata:
 4   name: service-node
 5 spec:
 6   hosts:
 7   - service-node
 8   http:
 9   - route:
10     - destination:
11         host: service-node
12     timeout: 500ms
```

第 12 行指定服务的调用时间不能超过 500 毫秒，当调用 service-node 服务时，如果超过 500 毫秒请求还没有完成，就直接给调用方返回超时错误。

【实验】

1）部署 service-node 服务：

```
$ kubectl apply -f service/node/service-node.yaml
$ kubectl get pod
NAME                                  READY     STATUS      RESTARTS     AGE
service-go-v1-7cc5c6f574-lrp2h        2/2       Running     0            4m
service-go-v2-7656dcc478-svn5c        2/2       Running     0            4m
service-node-v1-d44b9bf7b-ppn26       2/2       Running     0            24s
service-node-v2-86545d9796-rgmb7      2/2       Running     0            24s
```

2）启动用于并发测试的 Pod：

```
$ kubectl apply -f kubernetes/fortio.yaml
```

3）创建 service-node 服务的超时规则：

```
$ kubectl apply -f istio/resilience/virtual-service-node-timeout.yaml
```

4）访问 service-node 服务：

```
$ kubectl exec fortio -c fortio /usr/local/bin/fortio -- load -curl http://service-node/env
HTTP/1.1 200 OK
content-type: application/json; charset=utf-8
content-length: 77
date: Wed, 16 Jan 2019 10:33:57 GMT
x-envoy-upstream-service-time: 18
server: envoy

{"message":"node v1","upstream":[{"message":"go v1","response_time":"0.01"}]}

# 10 并发
$ kubectl exec fortio -c fortio /usr/local/bin/fortio -- load -c 10 -qps 0 -n 100 -loglevel Error http://service-node/env
11:08:24 I logger.go:97> Log level is now 4 Error (was 2 Info)
Fortio 1.0.1 running at 0 queries per second, 2->2 procs, for 100 calls: http://service-node/env
Aggregated Function Time : count 100 avg 0.19270902 +/- 0.1403 min 0.009657651 max 0.506141264 sum 19.2709017
# target 50% 0.173333
# target 75% 0.3
# target 90% 0.421429
# target 99% 0.505118
# target 99.9% 0.506039
Sockets used: 15 (for perfect keepalive, would be 10)
Code 200 : 94 (94.0 %)
Code 504 : 6 (6.0 %)
All done 100 calls (plus 0 warmup) 192.709 ms avg, 45.4 qps

# 20 并发
$ kubectl exec fortio -c fortio /usr/local/bin/fortio -- load -c 20 -qps 0 -n 200 -loglevel Error http://service-node/env
11:08:47 I logger.go:97> Log level is now 4 Error (was 2 Info)
Fortio 1.0.1 running at 0 queries per second, 2->2 procs, for 200 calls: http://service-node/env
Aggregated Function Time : count 200 avg 0.44961158 +/- 0.122 min 0.006904922 max 0.524347684 sum 89.9223153
# target 50% 0.50864
# target 75% 0.516494
# target 90% 0.521206
# target 99% 0.524034
# target 99.9% 0.524316
Sockets used: 163 (for perfect keepalive, would be 20)
Code 200 : 46 (23.0 %)
Code 504 : 154 (77.0 %)
All done 200 calls (plus 0 warmup) 449.612 ms avg, 39.2 qps
```

当并发逐渐增大时，service-node 服务的响应时间逐渐增大，服务请求响应超时的响应码（504）所占比例逐渐升高。这说明我们配置的服务超时时间已经生效。

5）清理：

```
$ kubectl delete -f kubernetes/fortio.yaml
$ kubectl delete -f service/node/service-node.yaml
$ kubectl delete -f istio/resilience/virtual-service-node-timeout.yaml
```

8.7　重试

Istio 通过设置 VirtualService 中的 retries 字段来指定服务的重试机制。

使用示例如下：

```
 1 apiVersion: networking.istio.io/v1alpha3
 2 kind: VirtualService
 3 metadata:
 4   name: service-node
 5 spec:
 6   hosts:
 7   - service-node
 8   http:
 9   - route:
10     - destination:
11         host: service-node
12     retries:
13       attempts: 3
14       perTryTimeout: 2s
```

第 12 ～ 14 行定义了重试规则，当调用 service-node 服务时，如果服务出错，就需要进行重试，最多可以重试 3 次，每次调用超时为 2 秒，每次重试的时间间隔由 Istio 决定，重试时间间隔一般会大于 25 毫秒。

【实验】

1）创建并发测试的 Pod：

```
$ kubectl apply -f kubernetes/fortio.yaml
```

2）部署 httpbin 服务：

```
$ kubectl apply -f kubernetes/httpbin.yaml

$ kubectl get pod -l app=httpbin
NAME                          READY     STATUS      RESTARTS     AGE
httpbin-b67975b8f-vmbtv       2/2       Running     0            49s
```

3）创建 httpbin 服务路由规则：

```
$ kubectl apply -f istio/route/virtual-service-httpbin.yaml
```

4）访问 httpbin 服务：

```
$ kubectl exec fortio -c fortio /usr/local/bin/fortio -- load -curl http://
httpbin:8000/status/200
    HTTP/1.1 200 OK
    server: envoy
    date: Wed, 16 Jan 2019 14:03:00 GMT
    content-type: text/html; charset=utf-8
    access-control-allow-origin: *
    access-control-allow-credentials: true
    content-length: 0
    x-envoy-upstream-service-time: 33

    $ kubectl exec fortio -c fortio /usr/local/bin/fortio -- load -c 10 -qps 0 -n
100 -loglevel Error http://httpbin:8000/status/200%2C200%2C200%2C200%2C500
    14:18:37 I logger.go:97> Log level is now 4 Error (was 2 Info)
    Fortio 1.0.1 running at 0 queries per second, 2->2 procs, for 100 calls:
http://httpbin:8000/status/200%2C200%2C200%2C200%2C500
    Aggregated Function Time : count 100 avg 0.24802899 +/- 0.06426 min 0.016759858
max 0.390472066 sum 24.8028985
    # target 50% 0.252941
    # target 75% 0.289706
    # target 90% 0.326667
    # target 99% 0.376981
    # target 99.9% 0.389123
    Sockets used: 30 (for perfect keepalive, would be 10)
    Code 200 : 78 (78.0 %)
    Code 500 : 22 (22.0 %)
    All done 100 calls (plus 0 warmup) 248.029 ms avg, 38.5 qps
```

5）创建 httpbin 服务重试路由规则：

```
$ kubectl apply -f istio/resilience/virtual-service-httpbin-retry.yaml
```

6）访问 httpbin 服务：

```
$ kubectl exec fortio -c fortio /usr/local/bin/fortio -- load -curl http://
httpbin:8000/status/200
    HTTP/1.1 200 OK
    server: envoy
    date: Wed, 16 Jan 2019 14:19:14 GMT
    content-type: text/html; charset=utf-8
    access-control-allow-origin: *
    access-control-allow-credentials: true
    content-length: 0
    x-envoy-upstream-service-time: 5

    $ kubectl exec fortio -c fortio /usr/local/bin/fortio -- load -c 10 -qps 0 -n
100 -loglevel Error http://httpbin:8000/status/200%2C200%2C200%2C200%2C500
    14:19:32 I logger.go:97> Log level is now 4 Error (was 2 Info)
```

```
    Fortio 1.0.1 running at 0 queries per second, 2->2 procs, for 100 calls:
http://httpbin:8000/status/200%2C200%2C200%2C200%2C500
    Aggregated Function Time : count 100 avg 0.23708609 +/- 0.1323 min 0.017537636
max 0.793965189 sum 23.7086086
    # target 50% 0.226471
    # target 75% 0.275
    # target 90% 0.383333
    # target 99% 0.7
    # target 99.9% 0.784569
    Sockets used: 13 (for perfect keepalive, would be 10)
    Code 200 : 97 (97.0 %)
    Code 500 : 3 (3.0 %)
    All done 100 calls (plus 0 warmup) 237.086 ms avg, 35.5 qps
```

从上面的测试结果可以看出，当没有开启服务重试时，服务有大概 1/4 的请求失败，当开启服务重试之后，服务只有极少数的请求失败。

7）清理：

```
$ kubectl delete -f kubernetes/fortio.yaml
$ kubectl delete -f kubernetes/httpbin.yaml
$ kubectl delete -f istio/resilience/virtual-service-httpbin-retry.yaml
```

8.8　限流

Istio 提供两种限流的实现：基于内存限流和基于 Redis 的限流。基于内存的限流方式适用于只部署一个 Mixer 的集群，而且由于使用内存存储的数据来限流，Mixer 重启后限流数据会丢失。因此在生产环境，建议使用基于 Redis 的限流方式，这种方式可存储限流数据。在 Istio 中，服务被限流的请求会得到 429（Too Many Requests）响应码。

限流的配置分为客户端和 Mixer 端两个部分。

客户端配置：

- QuotaSpec 定义了 Quota 实例和对应的每次请求消耗的配额数。
- QuotaSpecBinding 将 QuotaSpec 与一个或多个服务相关联绑定，只有被关联绑定的服务限流才会生效。

Mixer 端配置：

- quota 实例定义了 Mixer 如何区别度量一个请求的限流配额，用来描述请求数据收集的维度。
- memquota/redisquota 适配器定义了 memquota/redisquota 的配置，根据 quota 实例定义的请求数据收集维度来区分并定义一个或多个限流配额。
- rule 规则定义了 quota 实例应该何时分发给 memquota/redisquota 适配器处理。

1. 基于内存的限流

使用示例如下：

```
 1 apiVersion: "config.istio.io/v1alpha2"
 2 kind: quota
 3 metadata:
 4   name: requestcount
 5   namespace: istio-system
 6 spec:
 7   dimensions:
 8     source: request.headers["x-forwarded-for"] | "unknown"
 9     destination: destination.labels["app"] | destination.service.name | "unknown"
10     destinationVersion: destination.labels["version"] | "unknown"
11 ---
12 apiVersion: "config.istio.io/v1alpha2"
13 kind: memquota
14 metadata:
15   name: handler
16   namespace: istio-system
17 spec:
18   quotas:
19   - name: requestcount.quota.istio-system
20     maxAmount: 500
21     validDuration: 1s
22     overrides:
23     - dimensions:
24         destination: service-go
25       maxAmount: 50
26       validDuration: 1s
27     - dimensions:
28         destination: service-node
29         source: "10.28.11.20"
30       maxAmount: 50
31       validDuration: 1s
32     - dimensions:
33         destination: service-node
34       maxAmount: 20
35       validDuration: 1s
36     - dimensions:
37         destination: service-python
38       maxAmount: 2
39       validDuration: 5s
40 ---
41 apiVersion: config.istio.io/v1alpha2
42 kind: rule
43 metadata:
44   name: quota
45   namespace: istio-system
```

```
46 spec:
47   actions:
48   - handler: handler.memquota
49     instances:
50     - requestcount.quota
51 ---
52 apiVersion: config.istio.io/v1alpha2
53 kind: QuotaSpec
54 metadata:
55   name: request-count
56   namespace: istio-system
57 spec:
58   rules:
59   - quotas:
60     - charge: 1
61       quota: requestcount
62 ---
63 apiVersion: config.istio.io/v1alpha2
64 kind: QuotaSpecBinding
65 metadata:
66   name: request-count
67   namespace: istio-system
68 spec:
69   quotaSpecs:
70   - name: request-count
71     namespace: istio-system
72   services:
73   - name: service-go
74     namespace: default
75   - name: service-node
76     namespace: default
77   - name: service-python
78     namespace: default
```

第 1 ～ 10 行定义了名为 requestcount 的 quota 实例，获取请求的 source、destination、destinationVersion 值供 memquota 适配器来区分请求的限流配额。取值规则如下：

- source 获取请求的 x-forwarded-for 请求头的值作为 source 的取值，不存在时，source 取值 "unknown"。
- destination 获取请求的目标服务标签中的 app 标签的值，不存在时，取目标服务的 service.name 字段值，否则 destination 取值 "unknown"。
- destinationVersion 获取请求目标服务标签中的 version 标签的值，不存在时，destinationVersion 取值 "unknown"。

第 12 ～ 39 行定义了名为 handler 的 memquota 适配器。19 行中的 name 字段值为上面定义的 quota 实例名称。第 20 行定义了默认的限流配额为 500。第 21 行定义默认的限流计算周期为1s，即默认情况下每秒最高500个请求。第23～39行为具体的限流配置。第23～26行定义了当destination是service-go时，每秒不能高于50个请求。第27～31行定义了当destination为service-node且source为"10.28.11.20"时，每秒不能高于50个请求。第32～35行定义了当destination为service-node时，每秒不能高于20个请求。第36～39行定义了当destination为service-python时，每5秒内不能高于2个请求。

第41～50行定义了名为quota的rule规则，由于没有指定条件，会把所有相关联的服务请求都分发给memquota适配器处理。

第52～61行定义了名为request-count的QuotaSpec，指定了名为requestcount的quota实例每次消耗一个配额。

第63～78行定义了名为request-count的QuotaSpecBinding，把default命名空间的service-go、service-node、service-python服务与名为request-count的QuotaSpec关联起来。

在memquota适配器配置的所有限流规则中，执行限流时会从第一条限流规则开始匹配。当遇到第一条匹配的规则后，后面的规则不再匹配。如果没有匹配到任何具体的规则，则使用默认的规则。所以第27～31行定义的限流规则不能与第32～35行定义的限流规则交换位置，如果交换位置就会导致第27～31行定义的限流规则永远不会被匹配到，所以配置限流规则的时候，越具体的匹配规则应该放在越靠前的位置，否则可能会出现达不到预期的限流效果。

quota实例具体可以使用获取哪些值用于区分请求，可以参考官方文档[⊖]。

2. 基于Redis的限流

使用示例如下：

```
 1 apiVersion: "config.istio.io/v1alpha2"
 2 kind: quota
 3 metadata:
 4   name: requestcount
 5   namespace: istio-system
 6 spec:
 7   dimensions:
 8     source: request.headers["x-forwarded-for"] | "unknown"
 9     destination: destination.labels["app"] | destination.workload.name | "unknown"
10     destinationVersion: destination.labels["version"] | "unknown"
11 ---
12 apiVersion: "config.istio.io/v1alpha2"
13 kind: redisquota
14 metadata:
15   name: handler
16   namespace: istio-system
```

⊖ 官方文档链接是 https://istio.io/docs/reference/config/policy-and-telemetry/attribute-vocabulary/。

```
17 spec:
18   redisServerUrl: redis-ratelimit.istio-system:6379
19   connectionPoolSize: 10
20   quotas:
21   - name: requestcount.quota.istio-system
22     maxAmount: 500
23     validDuration: 1s
24     bucketDuration: 500ms
25     rateLimitAlgorithm: ROLLING_WINDOW
26     overrides:
27     - dimensions:
28         destination: service-go
29       maxAmount: 50
30     - dimensions:
31         destination: service-node
32         source: "10.28.11.20"
33       maxAmount: 50
34     - dimensions:
35         destination: service-node
36       maxAmount: 20
37     - dimensions:
38         destination: service-python
39       maxAmount: 2
40 ---
41 apiVersion: config.istio.io/v1alpha2
42 kind: rule
43 metadata:
44   name: quota
45   namespace: istio-system
46 spec:
47   actions:
48   - handler: handler.redisquota
49     instances:
50     - requestcount.quota
51 ---
52 apiVersion: config.istio.io/v1alpha2
53 kind: QuotaSpec
54 metadata:
55   name: request-count
56   namespace: istio-system
57 spec:
58   rules:
59   - quotas:
60     - charge: 1
61       quota: requestcount
62 ---
63 apiVersion: config.istio.io/v1alpha2
64 kind: QuotaSpecBinding
65 metadata:
66   name: request-count
```

```
67   namespace: istio-system
68 spec:
69   quotaSpecs:
70   - name: request-count
71     namespace: istio-system
72   services:
73   - name: service-go
74     namespace: default
75   - name: service-node
76     namespace: default
77   - name: service-python
78     namespace: default
```

第 12 ~ 39 行定义了名为 handler 的 redisquota 适配器，第 18 行定义了 Redis 的连接地址，19 行定义了 Redis 的连接池大小。

第 22 行定义了默认配额为 500，23 行定义了默认限流周期为 1 秒，即默认情况下每秒最高 500 个请求。

第 25 行定义了使用的限流算法有两种：FIXED_WINDOW 和 ROLLING_WINDOW，其中，FIXED_WINDOW 为默认的算法：

- FIXED_WINDOW 算法可以设置请求速率峰值高达 2 倍。
- ROLLING_WINDOW 算法可以提高精确度，这也会额外消耗 Redis 的资源。

第 27 ~ 39 行定义了具体的限流规则，与 memquota 不同，这里不允许再单独为限流规则设置限流周期，只能使用默认的限流周期。

其余部分的配置与 memquota 的限流配置一样。下面举例说明限流方法。

（1）基于条件的限流

如下配置表示只对 cookie 中不存在 user 的请求做限流：

```
apiVersion: config.istio.io/v1alpha2
kind: rule
metadata:
  name: quota
  namespace: istio-system
spec:
  match: match(request.headers["cookie"], "user=*") == false
  actions:
  - handler: handler.memquota
    instances:
    - requestcount.quota
```

（2）对所有服务限流

如下的配置表示对所有服务进行限流。

```
apiVersion: config.istio.io/v1alpha2
```

```
kind: QuotaSpecBinding
metadata:
  name: request-count
  namespace: istio-system
spec:
  quotaSpecs:
  - name: request-count
    namespace: istio-system
  services:
    - service: '*'
```

【实验】

本次实验使用基于内存的 memquota 适配器来进行服务限流测试。如果使用基于 Redis 的 redisquota 适配器进行实验，可能会由于实验环境机器性能问题，导致 Mixer 访问 Redis 出现错误，进而导致请求速率还没有到达设置值时就出现被限流的情况，影响实验结果的准确性。

1）部署其他服务：

```
$ kubectl apply -f service/node/service-node.yaml
$ kubectl apply -f service/lua/service-lua.yaml
$ kubectl apply -f service/python/service-python.yaml
$ kubectl get pod
NAME                                   READY     STATUS     RESTARTS    AGE
service-go-v1-7cc5c6f574-488rs         2/2       Running    0           15m
service-go-v2-7656dcc478-bfq5x         2/2       Running    0           15m
service-lua-v1-5c9bcb7778-d7qwp        2/2       Running    0           3m12s
service-lua-v2-75cb5cdf8-g9vht         2/2       Running    0           3m12s
service-node-v1-d44b9bf7b-z7vbr        2/2       Running    0           3m11s
service-node-v2-86545d9796-rgtxw       2/2       Running    0           3m10s
service-python-v1-79fc5849fd-xgfkn     2/2       Running    0           3m9s
service-python-v2-7b6864b96b-5w6cj     2/2       Running    0           3m15s
```

2）启动用于并发测试的 Pod：

```
$ kubectl apply -f kubernetes/fortio.yaml
```

3）创建限流规则：

```
$ kubectl apply -f istio/resilience/quota-mem-ratelimit.yaml
```

4）访问 service-go 服务，测试限流是否生效：

```
$ kubectl exec fortio -c fortio /usr/local/bin/fortio -- load -curl http://
service-go/env
HTTP/1.1 200 OK
content-type: application/json; charset=utf-8
date: Wed, 16 Jan 2019 15:33:02 GMT
content-length: 19
```

```
x-envoy-upstream-service-time: 226
server: envoy

{"message":"go v1"}

# 30 qps
$ kubectl exec fortio -c fortio /usr/local/bin/fortio -- load -qps 30 -n 300
-loglevel Error http://service-go/env
15:33:36 I logger.go:97> Log level is now 4 Error (was 2 Info)
Fortio 1.0.1 running at 30 queries per second, 2->2 procs, for 300 calls:
http://service-go/env
Aggregated Function Time : count 300 avg 0.0086544419 +/- 0.005944 min
0.002929143 max 0.065596074 sum 2.59633258
# target 50% 0.007375
# target 75% 0.00938095
# target 90% 0.0115
# target 99% 0.0325
# target 99.9% 0.0647567
Sockets used: 4 (for perfect keepalive, would be 4)
Code 200 : 300 (100.0 %)
All done 300 calls (plus 0 warmup) 8.654 ms avg, 30.0 qps

# 50 qps
$ kubectl exec fortio -c fortio /usr/local/bin/fortio -- load -qps 50 -n 500
-loglevel Error http://service-go/env
15:34:17 I logger.go:97> Log level is now 4 Error (was 2 Info)
Fortio 1.0.1 running at 50 queries per second, 2->2 procs, for 500 calls:
http://service-go/env
Aggregated Function Time : count 500 avg 0.0086848862 +/- 0.005076 min
0.00307391 max 0.05419281 sum 4.34244311
# target 50% 0.0075
# target 75% 0.00959459
# target 90% 0.0132857
# target 99% 0.03
# target 99.9% 0.0531446
Sockets used: 4 (for perfect keepalive, would be 4)
Code 200 : 500 (100.0 %)
All done 500 calls (plus 0 warmup) 8.685 ms avg, 50.0 qps

# 60 qps
$ kubectl exec fortio -c fortio /usr/local/bin/fortio -- load -qps 60 -n 600
-loglevel Error http://service-go/env
15:35:28 I logger.go:97> Log level is now 4 Error (was 2 Info)
Fortio 1.0.1 running at 60 queries per second, 2->2 procs, for 600 calls:
http://service-go/env
Aggregated Function Time : count 600 avg 0.0090870522 +/- 0.008314 min
0.002537502 max 0.169680378 sum 5.45223134
# target 50% 0.00748529
# target 75% 0.0101538
```

```
# target 90% 0.0153548
# target 99% 0.029375
# target 99.9% 0.163872
Sockets used: 23 (for perfect keepalive, would be 4)
Code 200 : 580 (96.7 %)
Code 429 : 20 (3.3 %)
All done 600 calls (plus 0 warmup) 9.087 ms avg, 59.9 qps
```

5）访问 service-node 服务，测试限流是否生效：

```
$ kubectl exec fortio -c fortio /usr/local/bin/fortio -- load -curl http://
service-node/env
HTTP/1.1 200 OK
content-type: application/json; charset=utf-8
content-length: 77
date: Wed, 16 Jan 2019 15:36:13 GMT
x-envoy-upstream-service-time: 1187
server: envoy

{"message":"node v2","upstream":[{"message":"go v1","response_time":"0.51"}]}

# 20 qps
$ kubectl exec fortio -c fortio /usr/local/bin/fortio -- load -qps 20 -n 200
-loglevel Error http://service-node/env
15:37:51 I logger.go:97> Log level is now 4 Error (was 2 Info)
Fortio 1.0.1 running at 20 queries per second, 2->2 procs, for 200 calls:
http://service-node/env
Aggregated Sleep Time : count 196 avg -0.21285915 +/- 1.055 min
-4.8433788589999995 max 0.190438028 sum -41.7203939
# range, mid point, percentile, count
>= -4.84338 <= -0.001 , -2.42219 , 18.37, 36
> 0.003 <= 0.004 , 0.0035 , 20.41, 4
> 0.011 <= 0.013 , 0.012 , 20.92, 1
> 0.015 <= 0.017 , 0.016 , 21.43, 1
> 0.069 <= 0.079 , 0.074 , 21.94, 1
> 0.089 <= 0.099 , 0.094 , 24.49, 5
> 0.099 <= 0.119 , 0.109 , 28.57, 8
> 0.119 <= 0.139 , 0.129 , 33.67, 10
> 0.139 <= 0.159 , 0.149 , 38.27, 9
> 0.159 <= 0.179 , 0.169 , 68.37, 59
> 0.179 <= 0.190438 , 0.184719 , 100.00, 62
# target 50% 0.166797
WARNING 18.37% of sleep were falling behind
Aggregated Function Time : count 200 avg 0.07655831 +/- 0.3601 min 0.007514854
max 5.046878744 sum 15.311662
# target 50% 0.0258696
# target 75% 0.045
# target 90% 0.104
# target 99% 0.55
# target 99.9% 5.0375
```

```
Sockets used: 4 (for perfect keepalive, would be 4)
Code 200 : 200 (100.0 %)
All done 200 calls (plus 0 warmup) 76.558 ms avg, 18.1 qps

# 30 qps
$ kubectl exec fortio -c fortio /usr/local/bin/fortio -- load -qps 30 -n 300
-loglevel Error http://service-node/env
15:38:36 I logger.go:97> Log level is now 4 Error (was 2 Info)
Fortio 1.0.1 running at 30 queries per second, 2->2 procs, for 300 calls:
http://service-node/env
Aggregated Sleep Time : count 296 avg 0.035638851 +/- 0.1206 min -0.420611573
max 0.132597685 sum 10.5491
# range, mid point, percentile, count
>= -0.420612 <= -0.001 , -0.210806 , 24.66, 73
> -0.001 <= 0 , -0.0005 , 25.00, 1
...
# target 50% 0.0934
WARNING 24.66% of sleep were falling behind
Aggregated Function Time : count 300 avg 0.06131494 +/- 0.08193 min 0.001977589
max 0.42055696 sum 18.3944819
# target 50% 0.03
# target 75% 0.0628571
# target 90% 0.175
# target 99% 0.4
# target 99.9% 0.418501
Sockets used: 55 (for perfect keepalive, would be 4)
Code 200 : 249 (83.0 %)
Code 429 : 51 (17.0 %)
All done 300 calls (plus 0 warmup) 61.315 ms avg, 29.9 qps

# 30 qps
$ kubectl exec fortio -c fortio /usr/local/bin/fortio -- load -qps 30 -n 300
-loglevel Error -H "x-forwarded-for: 10.28.11.20" http://service-node/env
15:40:34 I logger.go:97> Log level is now 4 Error (was 2 Info)
Fortio 1.0.1 running at 30 queries per second, 2->2 procs, for 300 calls:
http://service-node/env
Aggregated Sleep Time : count 296 avg -1.4901022 +/- 1.952 min -6.08576837 max
0.123485559 sum -441.070241
# range, mid point, percentile, count
>= -6.08577 <= -0.001 , -3.04338 , 69.59, 206
...
# target 50% -1.72254
WARNING 69.59% of sleep were falling behind
Aggregated Function Time : count 300 avg 0.1177745 +/- 0.4236 min 0.008494289
max 5.14910151 sum 35.332351
# target 50% 0.0346875
# target 75% 0.0985714
# target 90% 0.25
# target 99% 0.55
# target 99.9% 5.12674
```

```
Sockets used: 4 (for perfect keepalive, would be 4)
Code 200 : 300 (100.0 %)
All done 300 calls (plus 0 warmup) 117.775 ms avg, 24.7 qps

# 50 qps
$ kubectl exec fortio -c fortio /usr/local/bin/fortio -- load -qps 50 -n 500
-loglevel Error -H "x-forwarded-for: 10.28.11.20" http://service-node/env
15:45:31 I logger.go:97> Log level is now 4 Error (was 2 Info)
Fortio 1.0.1 running at 50 queries per second, 2->2 procs, for 500 calls:
http://service-node/env
Aggregated Sleep Time : count 496 avg 0.0015264793 +/- 0.1077 min -0.382731569
max 0.078526418 sum 0.757133711
# range, mid point, percentile, count
>= -0.382732 <= -0.001 , -0.191866 , 25.40, 126
> -0.001 <= 0 , -0.0005 , 25.60, 1
...
> 0.069 <= 0.0785264 , 0.0737632 , 100.00, 34
# target 50% 0.0566056
WARNING 25.40% of sleep were falling behind
Aggregated Function Time : count 500 avg 0.039103632 +/- 0.05723 min
0.001972061 max 0.450959277 sum 19.5518159
# target 50% 0.0175385
# target 75% 0.0323529
# target 90% 0.0975
# target 99% 0.3
# target 99.9% 0.450719
Sockets used: 7 (for perfect keepalive, would be 4)
Code 200 : 497 (99.4 %)
Code 429 : 3 (0.6 %)
All done 500 calls (plus 0 warmup) 39.104 ms avg, 48.4 qps

# 60 qps
$ kubectl exec fortio -c fortio /usr/local/bin/fortio -- load -qps 60 -n 600
-loglevel Error -H "x-forwarded-for: 10.28.11.20" http://service-node/env
15:50:24 I logger.go:97> Log level is now 4 Error (was 2 Info)
Fortio 1.0.1 running at 60 queries per second, 2->2 procs, for 600 calls:
http://service-node/env
Aggregated Sleep Time : count 596 avg -0.081667759 +/- 0.1592 min -0.626635518
max 0.064876123 sum -48.6739846
# range, mid point, percentile, count
>= -0.626636 <= -0.001 , -0.313818 , 51.01, 304
> 0 <= 0.001 , 0.0005 , 51.34, 2
...
> 0.059 <= 0.0648761 , 0.0619381 , 100.00, 14
# target 50% -0.0133888
WARNING 51.01% of sleep were falling behind
Aggregated Function Time : count 600 avg 0.04532505 +/- 0.04985 min 0.001904423
max 0.304644243 sum 27.1950299
# target 50% 0.0208163
```

```
# target 75% 0.07
# target 90% 0.1025
# target 99% 0.233333
# target 99.9% 0.303251
Sockets used: 19 (for perfect keepalive, would be 4)
Code 200 : 585 (97.5 %)
Code 429 : 15 (2.5 %)
All done 600 calls (plus 0 warmup) 45.325 ms avg, 59.9 qps
```

6）访问 service-python 服务，测试限流是否生效：

```
$ kubectl exec fortio -c fortio /usr/local/bin/fortio -- load -curl http://
service-python/env
HTTP/1.1 200 OK
content-type: application/json
content-length: 178
server: envoy
date: Wed, 16 Jan 2019 15:47:30 GMT
x-envoy-upstream-service-time: 366

{"message":"python v2","upstream":[{"message":"lua v2","response_
time":0.19},{"message":"node v2","response_time":0.18,"upstream":[{"message":"go
v1","response_time":"0.02"}]}]}


$ kubectl exec fortio -c fortio /usr/local/bin/fortio -- load -qps 1 -n 10
-loglevel Error http://service-python/env
15:48:02 I logger.go:97> Log level is now 4 Error (was 2 Info)
Fortio 1.0.1 running at 1 queries per second, 2->2 procs, for 10 calls:
http://service-python/env
Aggregated Function Time : count 10 avg 0.45553668 +/- 0.5547 min 0.003725253
max 1.4107851249999999 sum 4.55536678
# target 50% 0.18
# target 75% 1.06846
# target 90% 1.27386
# target 99% 1.39709
# target 99.9% 1.40942
Sockets used: 6 (for perfect keepalive, would be 4)
Code 200 : 5 (50.0 %)
Code 429 : 5 (50.0 %)
All done 10 calls (plus 0 warmup) 455.537 ms avg, 0.6 qps
```

从上面的实验结果，可以得出如下的结论：

- 对于 service-go 服务，当 qps 低于 50 时，请求几乎全部正常通过，当 qps 大于 50 时，会有部分请求得到 429 的响应码，这说明我们针对 service-go 服务配置的限流规则已经生效。
- 对于 service-node 服务，普通调用时，当 qps 大于 20 时，就会出现部分请求得到

429 响应码。但是当添加 "x-forwarded-for: 10.28.11.20" 请求头时，只有 qps 大于 50 时，才会出现部分请求得到 429 响应码，这说明我们针对 service-node 服务配置的两条限流规则都已经生效。

- 对于 service-python 服务，我们限定每 5 秒只允许 2 次请求的限制，当以每秒 1qps 请求时，10 个请求只有 3 个请求通过，其他请求均得到 429 响应码。这说明我们针对 service-python 服务配置的限流规则也已经生效。

注意　Istio 通过 quota 实现限流，但是限流控制并不是非常精确，可能会存在部分误差，使用时需要注意。

7）清理：

```
$ kubectl delete -f kubernetes/fortio.yaml
$ kubectl delete -f istio/resilience/quota-mem-ratelimit.yaml
$ kubectl delete -f service/node/service-node.yaml
$ kubectl delete -f service/lua/service-lua.yaml
$ kubectl delete -f service/python/service-python.yaml
```

8.9　本章小结

借助 Istio 提供的负载均衡、连接池、熔断和限流等机制，我们可以把服务设置得更具弹性，在遇到故障时能更好地应对，以及快速从故障中恢复，而这些功能几乎不需要人的参与。

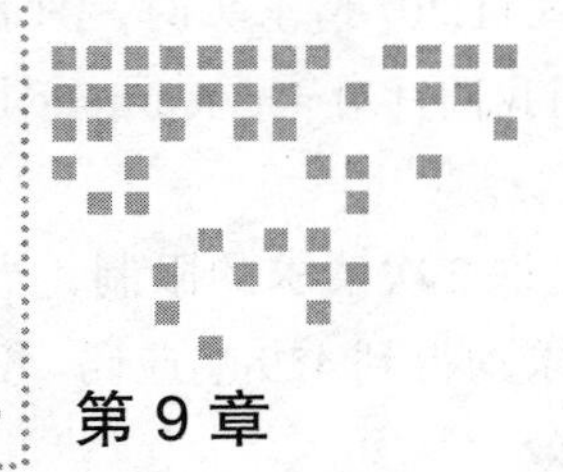

Chapter 9 第 9 章

让服务故障检测更容易

本章介绍如何检测服务的故障。Istio 提供了延时和错误注入功能，使我们可以方便地给服务增加延时，模拟服务过载的故障；我们也可以给服务注入故障，模拟服务不可用的场景；我们还可以将两个功能结合起来，更好地检测服务在遇到故障时的稳定性。

9.1 整体介绍

为了提高服务的稳定性，我们可以通过让服务调用失败或者增加服务的响应时间等方式，来主动模拟服务故障的场景，这种给服务注入故障来测试服务稳定性的行为称为故障注入（Fault Injection），也称为混沌工程（Chaos Engineering）。我们可以随机地给服务注入故障，观察服务的表现，检测服务的稳定性；故障的范围可大可小，小到给部分服务增加错误或者时延，大到给整个机房断电，来检测整个应用的稳定性。

虽然 Istio 的 Envoy 代理已经提供了许多的服务故障恢复机制，但是端到端地检测整个应用的故障恢复能力仍然很有必要。配置不当的失败恢复策略（例如，在整个服务调用链中设置了不合适的服务调用超时）可能会导致应用的核心服务持续不可用，给用户带来糟糕的体验。

Istio 选择了支持协议相关的故障注入，而不是杀死 Pod 或者延时、破坏 TCP 层的包，因为不管是软件的故障还是硬件的故障，对于应用层来说，故障的表现都是一样的。我们可以选择更有意义的故障来注入应用层，以此来检测并提高应用的弹性。例如：对于 HTTP 协议来说，HTTP 错误响应码注入是个不错的选择。

在Istio中，可以给满足特定条件的请求注入故障，也可以进一步地控制故障出现的比例。有两种可以选择的故障：服务延时和服务请求出错，服务延时是响应时间相关的故障，用来模拟增加网络的延迟或者一个过载的服务。服务请求出错是崩溃故障，用来模拟服务失败的故障，服务请求出错一般以HTTP错误响应码或者TCP连接故障居多。

【实验前的准备】

进行本章实验前，需要先执行如下的前置步骤。

1）下载实验时用到的源码仓库：

```
$ sudo yum install -y git

$ git clone https://github.com/mgxian/istio-lab
Cloning into 'istio-lab'...
remote: Enumerating objects: 247, done.
remote: Counting objects: 100% (247/247), done.
remote: Compressing objects: 100% (173/173), done.
remote: Total 774 (delta 153), reused 164 (delta 73), pack-reused 527
Receiving objects: 100% (774/774), 283.00 KiB | 229.00 KiB/s, done.
Resolving deltas: 100% (447/447), done.

$ cd istio-lab
```

2）开启default命名空间的自动注入功能：

```
$ kubectl label namespace default istio-injection=enabled
namespace/default labeled
```

3）部署用于测试的服务：

```
$ kubectl apply -f service/go/service-go.yaml
$ kubectl apply -f service/node/service-node.yaml
$ kubectl apply -f service/lua/service-lua.yaml
$ kubectl apply -f service/python/service-python.yaml
$ kubectl apply -f service/js/service-js.yaml

$ kubectl get pod
NAME                                   READY     STATUS     RESTARTS    AGE
service-go-v1-7cc5c6f574-svn5c         2/2       Running    0           3m38s
service-go-v2-7656dcc478-lrp2h         2/2       Running    0           3m38s
service-js-v1-55756d577-674p5          2/2       Running    0           3m33s
service-js-v2-86bdfc86d9-5v9xw         2/2       Running    0           3m33s
service-lua-v1-5c9bcb7778-7g9d5        2/2       Running    0           3m36s
service-lua-v2-75cb5cdf8-t79dh         2/2       Running    0           3m36s
service-node-v1-d44b9bf7b-ppn26        2/2       Running    0           3m37s
service-node-v2-86545d9796-rgmb7       2/2       Running    0           3m37s
service-python-v1-79fc5849fd-5bx5w     2/2       Running    0           3m35s
service-python-v2-7b6864b96b-7jnt6     2/2       Running    0           3m34s
```

9.2 给服务增加时延

给服务调用增加时延，模拟网络出现故障或者服务过载时响应变慢的情况。

使用示例如下：

```
 1 apiVersion: networking.istio.io/v1alpha3
 2 kind: DestinationRule
 3 metadata:
 4   name: service-go
 5 spec:
 6   host: service-go
 7   subsets:
 8   - name: v1
 9     labels:
10       version: v1
11   - name: v2
12     labels:
13       version: v2
14 ---
15 apiVersion: networking.istio.io/v1alpha3
16 kind: VirtualService
17 metadata:
18   name: service-go
19 spec:
20   hosts:
21   - service-go
22   http:
23   - route:
24     - destination:
25         host: service-go
26         subset: v1
27     fault:
28       delay:
29         percent: 30
30         fixedDelay: 5s
```

第27～30行定义了服务故障注入规则，delay表示给要路由的服务注入时延，percent表示请求注入时延的比例，默认值为0；fixedDelay表示注入时延的时间。代码中定义的故障注入表明，要在调用service-go服务v1版本的所有请求中随机抽取30%的请求注入5秒的时延。故障注入规则是跟路由规则一起配置的，可以根据路由规则指定多个故障注入规则。

【实验】

1）创建基础路由规则，创建Gateway应用访问入口，service-js服务路由规则：

```
$ kubectl apply -f istio/route/gateway-js-v1.yaml
```

2）创建 service-go 服务的时延故障注入规则：

```
$ kubectl apply -f istio/fault/virtual-service-go-delay.yaml
```

3）浏览器访问。

浏览器访问地址为 http：//11.11.11.111：31380/，多次点击“发射”按钮，查看服务调用耗时，可以看到有部分请求出现 5 秒左右的调用时延，如图 9-1 所示。

Welcome to Istio

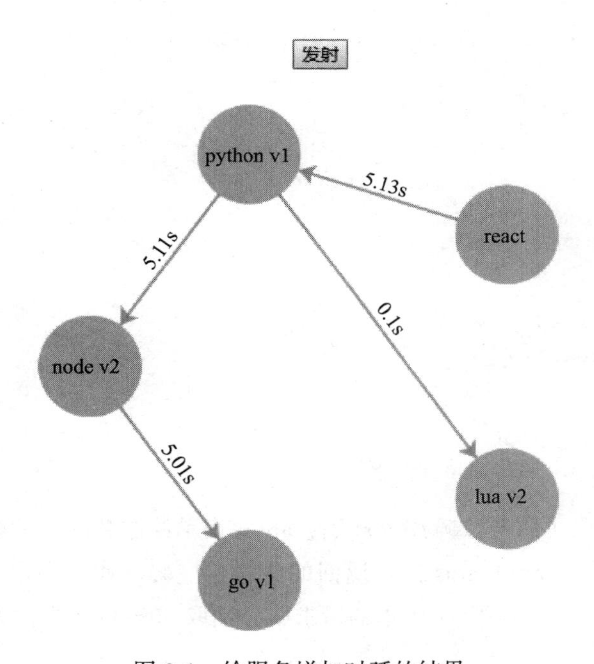

图 9-1　给服务增加时延的结果

4）清理：

```
$ kubectl delete -f istio/route/gateway-js-v1.yaml
$ kubectl delete -f istio/fault/virtual-service-go-delay.yaml
```

9.3　给服务注入错误

给服务注入错误，模拟后端服务出现故障，服务调用失败。

使用示例如下：

```
 1 apiVersion: networking.istio.io/v1alpha3
 2 kind: DestinationRule
 3 metadata:
 4   name: service-go
 5 spec:
 6   host: service-go
 7   subsets:
 8   - name: v1
 9     labels:
10       version: v1
11   - name: v2
12     labels:
13       version: v2
14 ---
15 apiVersion: networking.istio.io/v1alpha3
16 kind: VirtualService
17 metadata:
18   name: service-go
19 spec:
20   hosts:
21   - service-go
22   http:
23   - route:
24     - destination:
25         host: service-go
26         subset: v1
27     fault:
28       abort:
29         percent: 50
30         httpStatus: 500
```

第 27 ～ 30 行定义了服务故障注入规则，abort 表示给服务注入故障，percent 表示出错的百分比，默认值为 0，httpStatus 指定返回的错误响应码。代码中定义的故障注入规则表示，在所有调用 service-go 服务 v1 版本的请求中，抽取 50% 的请求注入出错响应码为 500。

【实验】

1）创建基础路由规则，创建 Gateway 应用访问入口，service-js 服务路由规则：

```
$ kubectl apply -f istio/route/gateway-js-v1.yaml
```

2）创建 service-go 服务的错误故障注入规则：

```
$ kubectl apply -f istio/fault/virtual-service-go-abort.yaml
```

3）浏览器访问。

浏览器访问地址为 http：//11.11.11.111：31380/，多次点击“发射”按钮，查看服务调用情况，可以看到，有部分请求出现调用 service-go 服务无法成功的现象，如图 9-2 所示。

Welcome to Istio

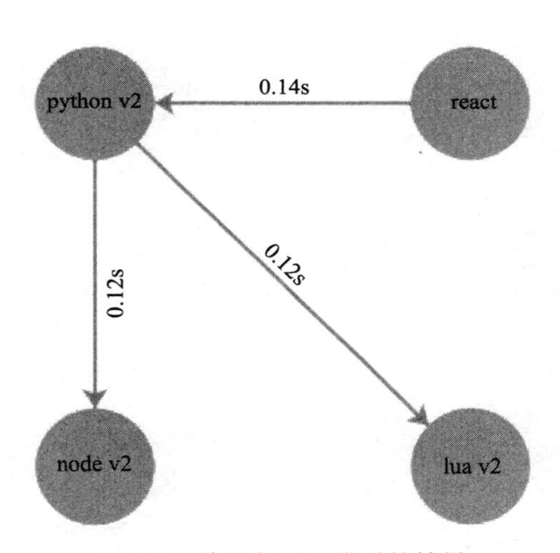

图 9-2　给服务注入错误的结果

4）清理：

```
$ kubectl delete -f istio/route/gateway-js-v1.yaml
$ kubectl delete -f istio/fault/virtual-service-go-abort.yaml
```

9.4　时延与错误配合使用

有时可能需要模拟服务时延和服务出错故障同时出现的场景，比如当服务出现过载情况时，可能会出现所有请求响应变慢，大部分请求失败，只有部分请求成功的情况。Istio 的时延和错误注入功能是相互独立工作的，可以同时设置，分别起作用，互不干扰，能够模拟服务过载出现的异常情况。

时延与错误配合使用示例如下：

```
 1 apiVersion: networking.istio.io/v1alpha3
 2 kind: DestinationRule
 3 metadata:
 4   name: service-go
 5 spec:
 6   host: service-go
 7   subsets:
 8   - name: v1
 9     labels:
10       version: v1
11   - name: v2
12     labels:
13       version: v2
14 ---
15 apiVersion: networking.istio.io/v1alpha3
16 kind: VirtualService
17 metadata:
18   name: service-go
19 spec:
20   hosts:
21   - service-go
22   http:
23   - route:
24     - destination:
25         host: service-go
26         subset: v1
27     fault:
28       delay:
29         percent: 100
30         fixedDelay: 5s
31       abort:
32         percent: 50
33         httpStatus: 500
```

第 27 ～ 33 行定义了服务故障注入规则，同时设置了时延和错误注入规则。代码中定义的规则表明要给调用 service-go 服务 v1 版本的所有请求增加 5 秒的时延，并且抽取 50% 的请求得到错误响应码为 500。

【实验】

1）创建基础路由规则，创建 Gateway 应用访问入口，service-js 服务路由规则：

```
$ kubectl apply -f istio/route/gateway-js-v1.yaml
```

2）创建 service-go 服务的时延和错误故障注入规则：

```
$ kubectl apply -f istio/fault/virtual-service-go-delay-abort.yaml
```

3）浏览器访问。

浏览器访问地址为 http://11.11.11.111:31380/，多次点击“发射”按钮，查看服务调用情况，可以看到，有部分请求出现 5 秒左右的调用时延但调用成功，如图 9-3 所示。

有部分请求出现 5 秒左右的调用时延，且调用失败，如图 9-4 所示。

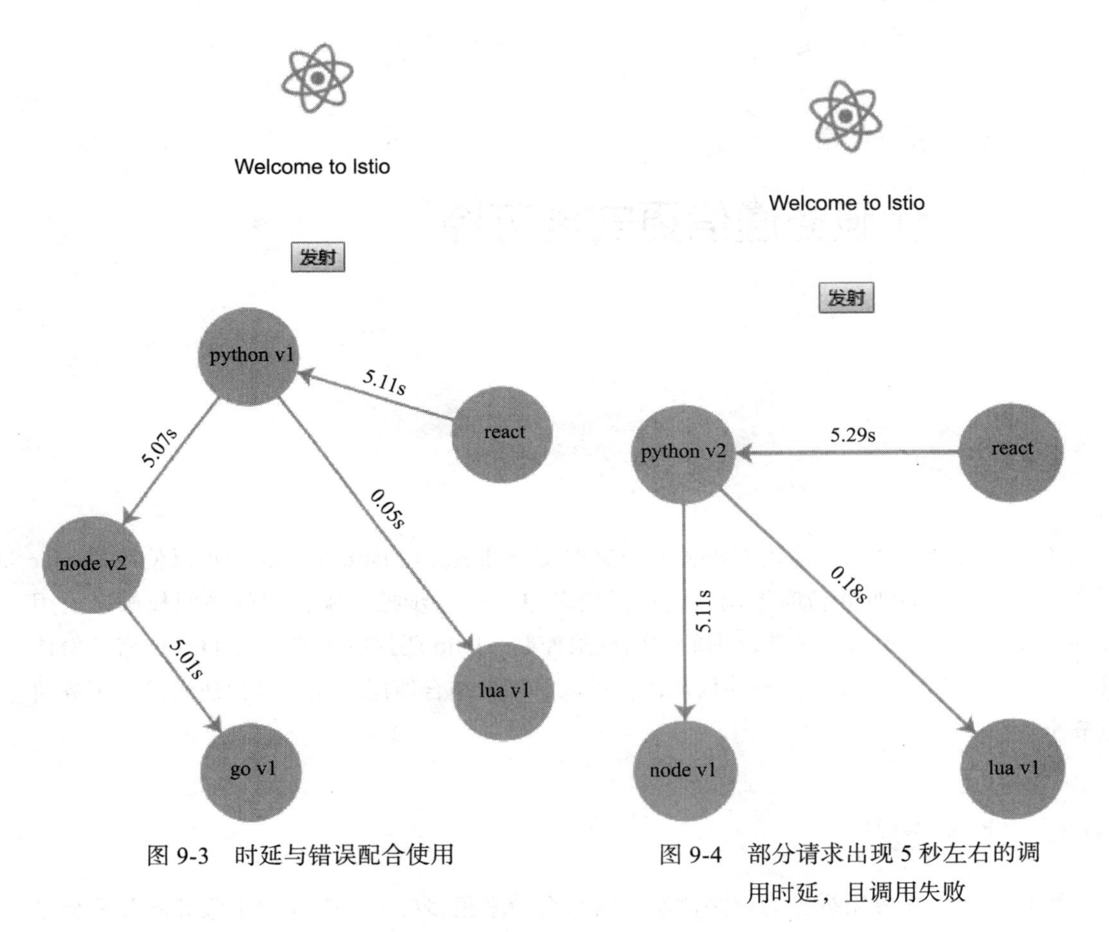

图 9-3　时延与错误配合使用

图 9-4　部分请求出现 5 秒左右的调用时延，且调用失败

4）清理：

```
$ kubectl delete -f istio/route/gateway-js-v1.yaml
$ kubectl delete -f istio/fault/virtual-service-go-delay-abort.yaml
```

9.5　本章小结

Istio 提供了开箱即用的故障注入功能，使我们可以方便地给服务注入故障，来测试服务的稳定性。这能帮助我们更早地处理线上可能会出现的各种故障问题，提前解决这些问题，提高服务的可用性与稳定性。

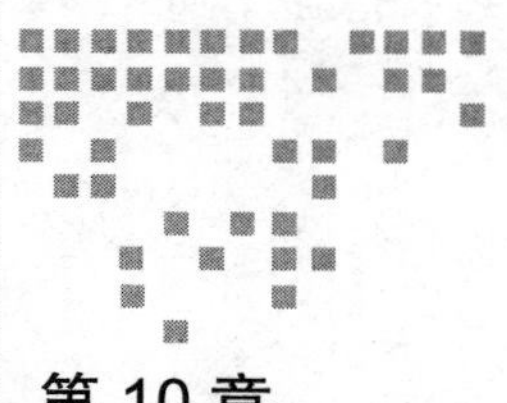

Chapter 10 第 10 章

让服务通信更安全可控

本章主要介绍如何让应用的服务通信更加安全可控。在 Istio 中，我们可以使用 Denier 适配器实现简单的服务访问控制，使用黑白名单更进一步控制服务间的访问权限，使用 RBAC 实现复杂的接口、函数级别的访问权限控制。Istio 还提供了双向的 TLS 加密身份认证，以及基于 JWT 的终端用户的认证，也可以将两者结合使用。Istio 极大地增强了服务间的安全性。

10.1 整体介绍

当把一个单体应用拆分为许多微服务时，会带来很多好处，比如单个服务的开发会更简单，服务更容易扩缩容，能更好地实现服务重用。但这也会带来很多安全问题，需要加强安全方面的管理，例如：

- 为了抵御中间人攻击，需要服务间的通信加密。
- 为了使服务间的通信控制更灵活，需要双向 TLS 加密以及更细粒度的访问控制。
- 需要审核谁在什么时候做了什么，需要审计工具。

Istio 尝试提供一个通用的安全解决方案来解决上述问题。本章主要介绍如何使服务间的通信更安全，以及如何控制服务间的访问权限。

Istio 安全功能架构如图 10-1 所示。

其中：

- Citadel 提供密钥和证书管理。

- Envoy 代理实现客户端和服务端之间的安全通信。
- Pilot 负责将认证策略和安全命名信息分发给 Envoy 代理。
- Mixer 负责管理授权和审计。

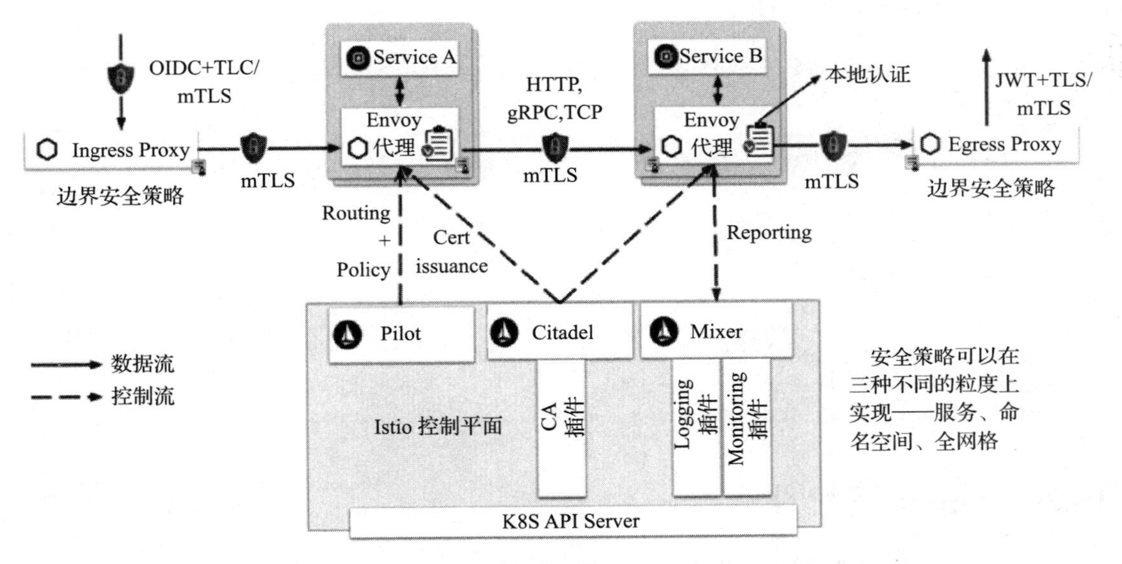

图 10-1　Istio 安全功能架构（图片来源：Istio 官方网站）

外部流量通过 Ingress 代理进入网格内，可以在 Ingress 代理上设置认证和加密策略。Ingress 代理与网格内部 Envoy 代理以及网格内部 Enovy 代理之间，通过 mTLS 认证和加密通信。需要访问外部服务的网格内的服务使用 Envoy 代理与 Egress 代理通过 mTLS 认证加密通信，Egress 代理代替服务访问外部服务。Egress 代理是可选的配置项，也可以不使用 Egress 代理，直接使用 Envoy 代理访问外部服务。上面这些认证加密规则是用户通过 Istio 的控制平面提供的 API 接口进行配置的。

【实验前的准备】

进行本章实验前，需要先执行如下的前置步骤。

1）下载实验时使用的源码仓库：

```
$ sudo yum install -y git

$ git clone https://github.com/mgxian/istio-lab
Cloning into 'istio-lab'...
remote: Enumerating objects: 247, done.
remote: Counting objects: 100% (247/247), done.
remote: Compressing objects: 100% (173/173), done.
remote: Total 774 (delta 153), reused 164 (delta 73), pack-reused 527
```

```
Receiving objects: 100% (774/774), 283.00 KiB | 229.00 KiB/s, done.
Resolving deltas: 100% (447/447), done.

$ cd istio-lab
```

2）开启 default 命名空间的自动注入功能：

```
$ kubectl label namespace default istio-injection=enabled
namespace/default labeled
```

3）部署用于测试的服务：

```
$ kubectl apply -f service/go/service-go.yaml
$ kubectl apply -f service/node/service-node.yaml
$ kubectl get pod
NAME                                READY     STATUS     RESTARTS     AGE
service-go-v1-7cc5c6f574-pd87x      2/2       Running    0            29s
service-go-v2-7656dcc478-5xdwc      2/2       Running    0            29s
service-node-v1-d44b9bf7b-vtzhq     2/2       Running    0            25s
service-node-v2-86545d9796-pxtsx    2/2       Running    0            25s
```

10.2 Denier 适配器

通过 Denier 适配器可以实现简单的服务访问控制，通过设置条件，可以拒绝部分服务间的访问。使用示例如下：

```
 1 apiVersion: config.istio.io/v1alpha2
 2 kind: denier
 3 metadata:
 4   name: deny-handler
 5 spec:
 6   status:
 7     code: 7
 8     message: Not allowed
 9 ---
10 apiVersion: config.istio.io/v1alpha2
11 kind: checknothing
12 metadata:
13   name: deny-request
14 spec:
15 ---
16 apiVersion: config.istio.io/v1alpha2
17 kind: rule
18 metadata:
19   name: deny-service-node-v2
20 spec:
21   match: destination.labels["app"] == "service-go" && source.labels["app"]
         == "service-node" && source.labels["version"] == "v2"
```

```
22    actions:
23    - handler: deny-handler.denier
24      instances:
25      - deny-request.checknothing
```

第 1 ～ 8 行定义了名为 deny-handler 的 denier 适配器，code 定义了请求被拒绝后返回的 gRPC 的响应码，code 为 7 表示没有访问权限，与 HTTP 中的 403 响应码意义相同，message 定义了请求被拒绝时返回的错误信息（面向开发者）。

第 10 ～ 14 行定义了名为 deny-request 的 checknothing 实例，直接返回空数据，一般用于测试。

第 16 ～ 26 行定义了名为 deny-service-node-v2 的 rule，该规则表明，当访问 service-node 服务的 v2 版本时，无法得到 service-go 服务的正常响应。

【实验】

1）创建测试 Pod：

```
$ kubectl apply -f kubernetes/dns-test.yaml
```

2）创建 Denier 规则：

```
$ kubectl apply -f istio/security/policy-rule-deny-simple.yaml
```

3）访问 service-node 服务：

```
$ kubectl exec dns-test -c dns-test -- curl -s http://service-node/env
{"message":"node v1","upstream":[{"message":"go v1","response_time":"0.50"}]}

$ kubectl exec dns-test -c dns-test -- curl -s http://service-node/env
{"message":"node v2","upstream":[]}
```

当访问 service-node 服务的 v1 版本时，可以得到 service-go 服务的正常响应；当访问 service-node 服务的 v2 版本时，无法得到 service-go 服务的正常响应。这正好符合我们设置的访问权限策略，说明我们设置的 Denier 适配器规则已经生效。

4）清理：

```
$ kubectl delete -f kubernetes/dns-test.yaml
$ kubectl delete -f istio/security/policy-rule-deny-simple.yaml
```

10.3　黑白名单

在 Istio 中可以通过黑白名单功能来实现服务访问控制，能实现比 Denier 适配器更强的服务访问控制策略。

1. 白名单

白名单规则是指，在默认情况下拒绝服务的所有请求，只允许符合条件的服务请求通过。使用示例如下：

```
 1 apiVersion: config.istio.io/v1alpha2
 2 kind: listchecker
 3 metadata:
 4   name: whitelist
 5 spec:
 6   overrides:
 7   - "v1"
 8   entryType: STRINGS
 9   blacklist: false
10 ---
11 apiVersion: config.istio.io/v1alpha2
12 kind: listentry
13 metadata:
14   name: app-version
15 spec:
16   value: source.labels["version"]
17 ---
18 apiVersion: config.istio.io/v1alpha2
19 kind: rule
20 metadata:
21   name: check-version
22 spec:
23   match: destination.labels["app"] == "service-go"
24   actions:
25   - handler: whitelist.listchecker
26     instances:
27     - app-version.listentry
```

第 1 ～ 9 行定义了名为 whitelist 的 listchecker 适配器。overrides 列出符合条件的字符串数组。blacklist 表示是否为黑名单，设置为 false 表示这是一个白名单。entryType 表示 overrides 列出的值的匹配类型，有 4 种类型可选：STRINGS 表示普通的字符串，CASE_INSENSITIVE_STRINGS 表示大小写敏感的字符串，IP_ADDRESSES 表示值为 IP 地址或者 IP 地址段，REGEX 表示使用正则匹配。

第 11 ～ 16 行定义了名为 app-version 的 listentry 实例，表示收集来源请求的 version 标签值。

第 18 ～ 27 行定义了名为 check-version 的 rule，只有请求目标服务为 service-go 时才应用此规则。该规则表明，在请求服务 service-go 的请求中，只有来源版本为 v1 的请求才允许通过。

2. 黑名单

黑名单规则是指，在默认情况下允许服务的所有请求，只拒绝符合条件的服务请求通过。使用示例如下：

```
 1 apiVersion: config.istio.io/v1alpha2
 2 kind: listchecker
 3 metadata:
 4   name: whitelist
 5 spec:
 6   overrides:
 7   - "v2"
 8   entryType: STRINGS
 9   blacklist: true
10 ---
11 apiVersion: config.istio.io/v1alpha2
12 kind: listentry
13 metadata:
14   name: app-version
15 spec:
16   value: source.labels["version"]
17 ---
18 apiVersion: config.istio.io/v1alpha2
19 kind: rule
20 metadata:
21   name: check-version
22 spec:
23   match: destination.labels["app"] == "service-go"
24   actions:
25   - handler: whitelist.listchecker
26     instances:
27     - app-version.listentry
```

黑名单的定义与白名单几乎一致，只需要把 blacklist 设置为 true 即可。

第 18 ～ 27 行定义的规则表明，在请求服务 service-go 的请求中，来源版本为 v2 的请求会被拒绝。

【实验】

1）创建测试 Pod：

```
$ kubectl apply -f kubernetes/dns-test.yaml
```

2）创建白名单规则：

```
$ kubectl apply -f istio/security/policy-rule-whitelist.yaml
```

3）访问 service-node 服务：

```
$ kubectl exec dns-test -c dns-test -- curl -s http://service-node/env
```

```
{"message":"node v1","upstream":[{"message":"go v2","response_time":"0.06"}]}

$ kubectl exec dns-test -c dns-test -- curl -s http://service-node/env
{"message":"node v2","upstream":[]}
```

当访问服务 service-node 的 v1 版本时，可以得到服务 service-go 的正常响应；当访问服务 service-node 的 v2 版本时，无法得到服务 service-go 的正常响应。在这里，我们通过设置白名单达到了和 Denier 适配器规则同样的效果。

4）删除白名单规则：

```
$ kubectl delete -f istio/security/policy-rule-whitelist.yaml
```

5）创建黑名单规则：

```
$ kubectl apply -f istio/security/policy-rule-blacklist.yaml
```

6）访问 service-node 服务：

```
$ kubectl exec dns-test -c dns-test -- curl -s http://service-node/env
{"message":"node v1","upstream":[{"message":"go v2","response_time":"0.06"}]}

$ kubectl exec dns-test -c dns-test -- curl -s http://service-node/env
{"message":"node v2","upstream":[]}
```

当访问服务 service-node 的 v1 版本时，可以得到服务 service-go 的正常响应；当访问服务 service-node 的 v2 版本时，无法得到服务 service-go 的正常响应。在这里，我们通过设置黑名单达到了和 Denier 适配器以及白名单规则同样的效果。

7）清理：

```
$ kubectl delete -f kubernetes/dns-test.yaml
$ kubectl delete -f istio/security/policy-rule-blacklist.yaml
```

10.4 服务与身份认证

Istio 提供了如下两种类型的身份认证：

- **传输认证**，也称为服务间的身份认证：验证直连的客户端，Istio 提供双向 TLS 作为传输认证的全栈解决方案。我们可以在不修改应用代码的情况下轻松使用双向 TLS，它有如下特点：
  - 为每一个服务提供强大的身份，表示其角色，以实现跨集群和云的互通性。
  - 为服务间和终端用户到服务的通信加密。
  - 提供密钥管理系统来自动化密钥和证书的生成、分发以及轮换。
- **来源认证**，也称为终端用户身份认证：验证请求来源客户端，把它们作为终端用户或设备。Istio 在请求级别使用 JSON Web Token（JWT）来验证并简化开发者对 Auth0、Firebase、Google Auth 以及自定义认证的对接工作。

Istio 认证架构如图 10-2 所示。

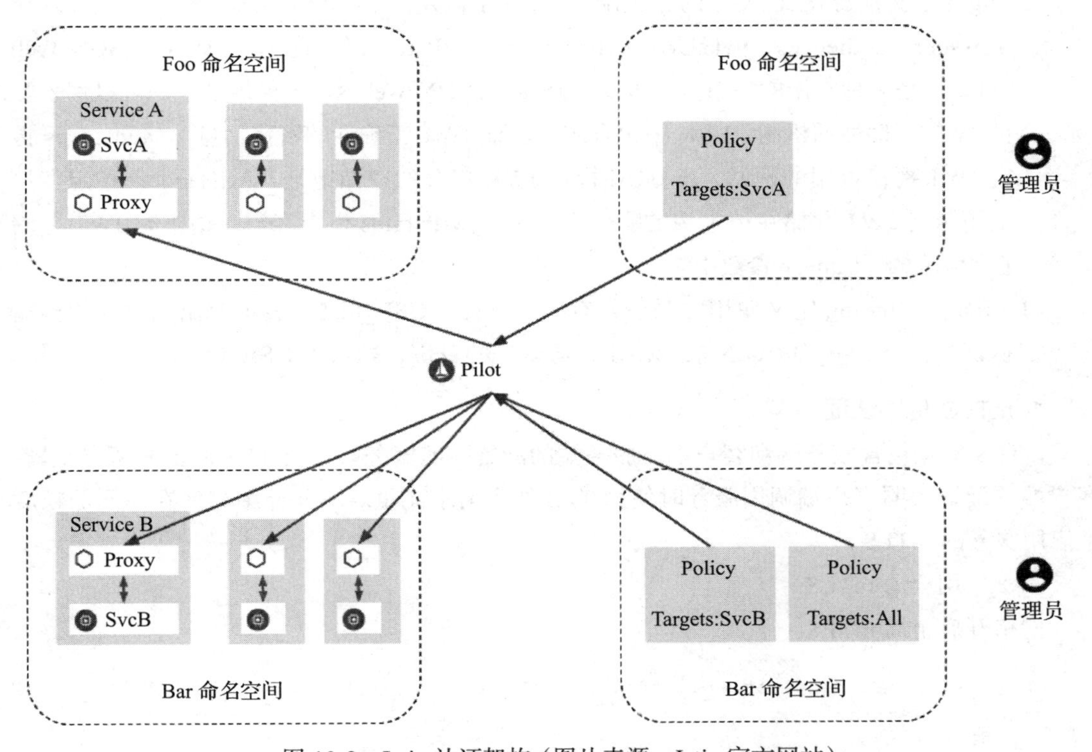

图 10-2　Istio 认证架构（图片来源：Istio 官方网站）

管理员可以通过 Pilot 提供的 API 为每个命名空间中的服务设置访问策略。

Istio 将传输认证和来源认证两种类型的身份验证以及凭证中的其他声明传输到下一层供授权使用。

通过创建 Policy 对象来控制 mTLS 和 JWT 认证策略，Policy 主要包含 target 列表、peer 列表、origin 列表、principalBinding 参数，含义如下：

- target 定义服务或目标地址。name 字段必须要有，且只能写服务的短域名。如：必须写 service-go 而不能写全域名 "service-go.default.svc.cluster.local"。ports 字段可以指定只对服务特定端口请求应用 Policy 认证策略，是一个 PortSelector 对象的列表，PortSelector 包含 name 和 number 字段，number 字段指定端口号，name 字段指定端口名，number 和 name 只能选择其中一个。
- peer 定义 mTLS 相关的配置，mtls 字段定义是否使用 mTLS。mtls 包含一个可以指定 TLS 模式的 mode 字段，当指定为 STRICT 时表示客户端必须使用 TLS 连接，当指定为 PERMISSIVE 时表示客户端既可以使用 TLS 连接，也可以使用普通连接。默认为 STRICT。

- origin 定义来源认证 JWT 的相关配置，jwt 字段定义 JWT 相关的参数配置，可以指定 issuer、audiences、jwksUri、jwtHeaders、jwtParams 的 配 置。JWT（JSON Web Token）是一种常用的简单认证方式，JWK（JSON Web Key）提供了一种可以用来验证 JWT 正确性的机制，JWK 中包含用来验证 JWT 正确性的密钥信息[1]。audiences 指定 JWT 授权可用的网站，jwksUri 指定 JWK 集合的存储地址，jwtHeaders 指定客户端请求时 JWT 存储在请求头的哪个字段中，jwtParams 指定客户端请求时 JWT 存储在请求的哪个 query 参数中。
- principalBinding 定义使用的身份，有两种选择：USE_PEER 表示使用 mTLS 的传输认证身份，USE_ORIGIN 表示使用来源认证的身份，默认为 USE_PEER。[2]

1. mTLS 传输认证

mTLS 配置包含服务端和客户端，服务端的配置表明服务端接收 TLS 加密的请求流量，客户端的配置表明客户端调用服务时使用 TLS 加密请求流量。下面介绍如何在不同的粒度上启用服务的 mTLS。

（1）启用全局 mTLS

网格开启全局 mTLS：

```
apiVersion: authentication.istio.io/v1alpha1
kind: MeshPolicy
metadata:
  name: default
spec:
  peers:
  - mtls: {}
```

上述策略表明网格中只接受使用 mTLS 加密的请求，MeshPolicy 类型的资源在整个网格内只能创建一个，且命名必须为 default。

客户端开启全局 mTLS：

```
apiVersion: networking.istio.io/v1alpha3
kind: DestinationRule
metadata:
  name: default
  namespace: default
spec:
```

---

[1] JWT 与 JWK 的区别和联系可以参考文章 https://redthunder.blog/2017/06/08/jwts-jwks-kids-x5ts-oh-my/。

[2] 详细说明可以参考官方文档 https://istio.io/docs/reference/config/istio.authentication.v1alpha1/。

```
  host: "*.local"
  trafficPolicy:
    tls:
      mode: ISTIO_MUTUAL
```

上述配置中 "*.local" 泛域名代表集群内的所有服务，此规则表明调用集群内的服务的请求都使用 mTLS 进行加密。

（2）启用命名空间级别 mTLS

```
 1 apiVersion: authentication.istio.io/v1alpha1
 2 kind: Policy
 3 metadata:
 4   name: default
 5   namespace: default
 6 spec:
 7   peers:
 8   - mtls: {}
 9 ---
10 apiVersion: networking.istio.io/v1alpha3
11 kind: DestinationRule
12 metadata:
13   name: default
14   namespace: default
15 spec:
16   host: "*.default.svc.cluster.local"
17   trafficPolicy:
18     tls:
19       mode: ISTIO_MUTUAL
```

第 1 ～ 8 行定义了名为 default 的 Policy 对象，表示对于 default 命名空间的服务默认情况下只允许 mTLS 加密的请求。

第 10 ～ 19 行定义了名为 default 的 DestinationRule 对象，表示对于所有 "*.default.svc.cluster.local" 服务的请求都使用 mTLS 进行加密。

（3）启用服务级别 mTLS

```
 1 apiVersion: authentication.istio.io/v1alpha1
 2 kind: Policy
 3 metadata:
 4   name: httpbin
 5   namespace: bar
 6 spec:
 7   targets:
 8   - name: httpbin
 9   peers:
10   - mtls: {}
11 ---
12 apiVersion: networking.istio.io/v1alpha3
```

```
13 kind: DestinationRule
14 metadata:
15   name: httpbin
16   namespace: bar
17 spec:
18   host: "httpbin.bar.svc.cluster.local"
19   trafficPolicy:
20     tls:
21       mode: ISTIO_MUTUAL
```

第1～10行定义了名为httpbin的Policy对象，表示对于bar命名空间的httpbin服务只允许mTLS加密的请求。

第12～21行定义了名为httpbin的DestinationRule对象，表示对于"httpbin.bar.svc.cluster.local"的服务请求都使用mTLS进行加密。

（4）启用端口级别mTLS

```
 1 apiVersion: authentication.istio.io/v1alpha1
 2 kind: Policy
 3 metadata:
 4   name: httpbin
 5   namespace: bar
 6 spec:
 7   targets:
 8   - name: httpbin
 9     ports:
10     - number: 1234
11   peers:
12   - mtls: {}
13 ---
14 apiVersion: networking.istio.io/v1alpha3
15 kind: DestinationRule
16 metadata:
17   name: httpbin
18   namespace: bar
19 spec:
20   host: httpbin.bar.svc.cluster.local
21   trafficPolicy:
22     tls:
23       mode: DISABLE
24     portLevelSettings:
25     - port:
26         number: 1234
27       tls:
28         mode: ISTIO_MUTUAL
```

第1～12行定义了名为httpbin的Policy对象，表示对于bar命名空间的httpbin服务的1234端口只允许mTLS加密的请求。

第 14 ～ 28 行定义了名为 httpbin 的 DestinationRule 对象，表示对于 "httpbin.bar.svc.cluster.local" 的服务，默认情况下不使用 mTLS 加密请求，但对于 1234 端口的请求使用 mTLS 进行加密。

优先级说明：服务级别和端口级别的 mTLS 策略比命名空间级别的 mTLS 策略的优先级更高，命名空间级别的 mTLS 策略比网格全局的 mTLS 策略的优先级更高。

2. 终端用户身份认证

使用 JWT 进行终端用户身份认证。

基本使用示例：

```
apiVersion: authentication.istio.io/v1alpha1
kind: Policy
metadata:
  name: jwt-example
  namespace: foo
spec:
  targets:
  - name: httpbin
  origins:
  - jwt:
    issuer: "testing@secure.istio.io"
    jwksUri: "https://raw.githubusercontent.com/istio/istio/release-1.0/security/
             tools/jwt/samples/jwks.json"
  principalBinding: USE_ORIGIN
```

自定义 JWT 头示例：

```
 1 apiVersion: authentication.istio.io/v1alpha1
 2 kind: Policy
 3 metadata:
 4   name: jwt-example
 5   namespace: foo
 6 spec:
 7   targets:
 8   - name: httpbin
 9   origins:
10   - jwt:
11     issuer: "testing@secure.istio.io"
12     jwksUri: "https://raw.githubusercontent.com/istio/istio/release-1.0/
              security/tools/jwt/samples/jwks.json"
13     jwtHeaders:
14     - X-Auth
15   principalBinding: USE_ORIGIN
```

第 13 ～ 14 行定义把 JWT 放在请求的 X-Auth 中。

自定义 query 参数示例：

```
 1 apiVersion: authentication.istio.io/v1alpha1
 2 kind: Policy
 3 metadata:
 4   name: jwt-example
 5   namespace: foo
 6 spec:
 7   targets:
 8   - name: httpbin
 9   origins:
10   - jwt:
11     issuer: "testing@secure.istio.io"
12     jwksUri: "https://raw.githubusercontent.com/istio/istio/release-1.0/
               security/tools/jwt/samples/jwks.json"
13     jwtParams:
14     - access_token
15   principalBinding: USE_ORIGIN
```

第13～14行定义把JWT放在请求query参数的access_token中。

自定义JWT头与query参数结合使用示例：

```
 1 apiVersion: authentication.istio.io/v1alpha1
 2 kind: Policy
 3 metadata:
 4   name: jwt-example
 5   namespace: foo
 6 spec:
 7   targets:
 8   - name: httpbin
 9   origins:
10   - jwt:
11     issuer: "testing@secure.istio.io"
12     jwksUri: "https://raw.githubusercontent.com/istio/istio/release-1.0/
               ecurity/tools/jwt/samples/jwks.json"
13     jwtHeaders:
14     - X-Auth
15     jwtParams:
16     - access_token
17   principalBinding: USE_ORIGIN
```

第13～16行定义可以把JWT放在请求的X-Auth中，也可以把JWT放在请求query参数的access_token中。

【实验一】 mTLS传输认证

（1）部署用于测试的服务

1）删除微服务的部署。

如果不暂时删除之前部署用于测试的微服务，可能会由于虚拟机集群资源和性能问题导致实验失败：

```
$ kubectl delete -f service/go/service-go.yaml
$ kubectl delete -f service/node/service-node.yaml
```

2）部署服务。

foo 和 bar 命名空间中部署的服务流量都由 Istio 管理，属于 Istio 集群中的服务。legacy 命名空间中部署的服务流量没有使用 Istio 管理，不属于 Istio 集群中的服务。使用如下命令部署服务：

```
$ kubectl create ns foo
$ kubectl apply -f <(istioctl kube-inject -f kubernetes/httpbin.yaml) -n foo
$ kubectl apply -f <(istioctl kube-inject -f kubernetes/sleep.yaml) -n foo

$ kubectl create ns bar
$ kubectl apply -f <(istioctl kube-inject -f kubernetes/httpbin.yaml) -n bar
$ kubectl apply -f <(istioctl kube-inject -f kubernetes/sleep.yaml) -n bar

$ kubectl create ns legacy
$ kubectl apply -f  kubernetes/httpbin.yaml -n legacy
$ kubectl apply -f  kubernetes/sleep.yaml -n legacy
```

3）查看部署服务的状态。

等待 Pod 全部为 Running 状态时再进行后面的实验步骤。使用如下命令查看服务状态：

```
$ kubectl get pod -n legacy
NAME                          READY     STATUS      RESTARTS     AGE
httpbin-b67975b8f-2z8v5       1/1       Running     0            67s
sleep-5875d97cd4-t4mdd        1/1       Running     0            66s

$ kubectl get pod -n foo
NAME                          READY     STATUS      RESTARTS     AGE
httpbin-6f9dfcf5d8-f8zwc      2/2       Running     0            85s
sleep-7584fcd546-sj2mn        2/2       Running     0            85s

$ kubectl get pod -n bar
NAME                          READY     STATUS      RESTARTS     AGE
httpbin-6f9dfcf5d8-tkp6n      2/2       Running     0            86s
sleep-7584fcd546-6rcv2        2/2       Running     0            85s
```

4）访问测试部署的服务是否正常。

有时可能由于虚拟机集群性能问题导致路由信息生效过慢，需要等一段时间再进行如下的测试。使用如下命令测试服务：

```
    $ for from in "foo" "bar" "legacy"; do for to in "foo" "bar" "legacy"; do
kubectl exec $(kubectl get pod -l app=sleep -n ${from} -o jsonpath={.items..
metadata.name}) -c sleep -n ${from} -- curl http://httpbin.${to}:8000/ip -s -o /
dev/null -w "sleep.${from} to httpbin.${to}: %{http_code}\n"; done; done
    sleep.foo to httpbin.foo: 200
    sleep.foo to httpbin.bar: 200
```

```
sleep.foo to httpbin.legacy: 200
sleep.bar to httpbin.foo: 200
sleep.bar to httpbin.bar: 200
sleep.bar to httpbin.legacy: 200
sleep.legacy to httpbin.foo: 200
sleep.legacy to httpbin.bar: 200
sleep.legacy to httpbin.legacy: 200
```

（2）测试全局 mTLS

1）服务端启用全局 mTLS：

```
$ kubectl apply -f istio/security/mtls-config-mesh-on.yaml
```

2）服务访问测试：

```
$ for from in "foo" "bar"; do for to in "foo" "bar"; do kubectl exec $(kubectl
get pod -l app=sleep -n ${from} -o jsonpath={.items..metadata.name}) -c sleep -n
${from} -- curl http://httpbin.${to}:8000/ip -s -o /dev/null -w "sleep.${from} to
httpbin.${to}: %{http_code}\n"; done; done
    sleep.foo to httpbin.foo: 503
    sleep.foo to httpbin.bar: 503
    sleep.bar to httpbin.foo: 503
    sleep.bar to httpbin.bar: 503
```

之前步骤的配置只是在服务端启用了全局 mTLS，此时服务端只接受 TLS 加密的请求，客户端由于没有配置 TLS 加密，仍然使用没有经过 TLS 加密的请求，所以 Istio 网格中的所有请求均会失败，得到 503 响应码。

3）客户端启用 mTLS：

```
$ kubectl apply -f istio/security/mtls-config-default-on.yaml
```

4）服务访问测试。

Istio 集群中的服务相互访问：

```
$ for from in "foo" "bar"; do for to in "foo" "bar"; do kubectl exec $(kubectl
get pod -l app=sleep -n ${from} -o jsonpath={.items..metadata.name}) -c sleep -n
${from} -- curl http://httpbin.${to}:8000/ip -s -o /dev/null -w "sleep.${from} to
httpbin.${to}: %{http_code}\n"; done; done
    sleep.foo to httpbin.foo: 200
    sleep.foo to httpbin.bar: 200
    sleep.bar to httpbin.foo: 200
    sleep.bar to httpbin.bar: 200
```

从非 Istio 网格中的服务访问 Istio 网格中的服务：

```
$ for from in "legacy"; do for to in "foo" "bar"; do kubectl exec $(kubectl
get pod -l app=sleep -n ${from} -o jsonpath={.items..metadata.name}) -c sleep -n
${from} -- curl http://httpbin.${to}:8000/ip -s -o /dev/null -w "sleep.${from} to
```

```
httpbin.${to}: %{http_code}\n"; done; done
    sleep.legacy to httpbin.foo: 000
    command terminated with exit code 56
    sleep.legacy to httpbin.bar: 000
    command terminated with exit code 56
```

从 Istio 网格中的服务访问非 Istio 网格中的服务：

```
    $ for from in "foo" "bar"; do for to in "legacy"; do kubectl exec $(kubectl
get pod -l app=sleep -n ${from} -o jsonpath={.items..metadata.name}) -c sleep -n
${from} -- curl http://httpbin.${to}:8000/ip -s -o /dev/null -w "sleep.${from} to
httpbin.${to}: %{http_code}\n"; done; done
    sleep.foo to httpbin.legacy: 503
    sleep.bar to httpbin.legacy: 503
```

上面步骤配置的规则表明，在客户端请求时也使用 TLS 进行加密，此时服务端与客户端均开启了 TLS 加密，所以 Istio 网格中的服务相互访问都会成功。默认情况下，非 Istio 网格中的服务访问 Istio 网格中的服务会失败，Istio 网格中的服务访问非 Istio 网格中的服务也会失败。

5）关闭访问非 Istio 网格中的服务的 TLS 功能：

```
    $ kubectl apply -f istio/security/mtls-httpbin-legacy-off.yaml
```

6）再次测试从 Istio 网格中的服务访问非 Istio 网格中的服务：

```
    $ for from in "foo" "bar"; do for to in "legacy"; do kubectl exec $(kubectl
get pod -l app=sleep -n ${from} -o jsonpath={.items..metadata.name}) -c sleep -n
${from} -- curl http://httpbin.${to}:8000/ip -s -o /dev/null -w "sleep.${from} to
httpbin.${to}: %{http_code}\n"; done; done
    sleep.foo to httpbin.legacy: 200
    sleep.bar to httpbin.legacy: 200
```

7）访问 Kubernetes API Server：

```
    $ TOKEN=$(kubectl describe secret $(kubectl get secrets | grep default | cut
-f1 -d ' ') | grep -E '^token' | cut -f2 -d':' | tr -d '\t')
    $ kubectl exec $(kubectl get pod -l app=sleep -n foo -o jsonpath={.items..
metadata.name}) -c sleep -n foo -- curl https://kubernetes.default/api --header
"Authorization: Bearer $TOKEN" --insecure -s -o /dev/null -w "%{http_code}\n"
    000
    command terminated with exit code 35
```

8）关闭访问 Kubernetes API Server 的 mTLS 功能：

```
    $ kubectl apply -f istio/security/mtls-apiserver-off.yaml
```

9）再次访问 Kubernetes API Server ：

```
    $ TOKEN=$(kubectl describe secret $(kubectl get secrets | grep default | cut
```

```
-f1 -d ' ') | grep -E '^token' | cut -f2 -d':' | tr -d '\t')
    $ kubectl exec $(kubectl get pod -l app=sleep -n foo -o jsonpath={.items..
metadata.name}) -c sleep -n foo -- curl https://kubernetes.default/api --header
"Authorization: Bearer $TOKEN" --insecure -s -o /dev/null -w "%{http_code}\n"
    200
```

10）清理相关 mTLS 规则：

```
$ kubectl delete -f istio/security/mtls-config-mesh-on.yaml
$ kubectl delete -f istio/security/mtls-config-default-on.yaml
$ kubectl delete -f istio/security/mtls-httpbin-legacy-off.yaml
$ kubectl delete -f istio/security/mtls-apiserver-off.yaml
```

（3）测试命名空间级别 mTLS

1）启用 foo 命名空间的 mTLS 功能：

```
$ kubectl apply -f istio/security/mtls-namespace-foo-on.yaml
```

2）访问服务：

```
    $ for from in "foo" "bar" "legacy"; do for to in "foo" "bar" "legacy"; do
kubectl exec $(kubectl get pod -l app=sleep -n ${from} -o jsonpath={.items..
metadata.name}) -c sleep -n ${from} -- curl http://httpbin.${to}:8000/ip -s -o /
dev/null -w "sleep.${from} to httpbin.${to}: %{http_code}\n"; done; done
    sleep.foo to httpbin.foo: 200
    sleep.foo to httpbin.bar: 200
    sleep.foo to httpbin.legacy: 200
    sleep.bar to httpbin.foo: 200
    sleep.bar to httpbin.bar: 200
    sleep.bar to httpbin.legacy: 200
    sleep.legacy to httpbin.foo: 000
    command terminated with exit code 56
    sleep.legacy to httpbin.bar: 200
    sleep.legacy to httpbin.legacy: 200
```

foo 命名空间开启了 mTLS 功能，并设置了去往 foo 命名空间的流量默认使用 mTLS 加密，当 legacy 命名空间请求 foo 命名空间时，由于 legacy 命名空间的请求没有使用 mTLS 加密，导致请求失败。

（4）测试服务级别 mTLS

1）启用 bar 命名空间的 httpbin 服务的 mTLS 功能：

```
$ kubectl apply -f istio/security/mtls-service-httpbin-on.yaml
```

2）访问服务：

```
    $ for from in "foo" "bar" "legacy"; do for to in "foo" "bar" "legacy"; do
kubectl exec $(kubectl get pod -l app=sleep -n ${from} -o jsonpath={.items..
```

```
metadata.name}) -c sleep -n ${from} -- curl http://httpbin.${to}:8000/ip -s -o /
dev/null -w "sleep.${from} to httpbin.${to}: %{http_code}\n"; done; done
    sleep.foo to httpbin.foo: 200
    sleep.foo to httpbin.bar: 200
    sleep.foo to httpbin.legacy: 200
    sleep.bar to httpbin.foo: 200
    sleep.bar to httpbin.bar: 200
    sleep.bar to httpbin.legacy: 200
    sleep.legacy to httpbin.foo: 000
    command terminated with exit code 56
    sleep.legacy to httpbin.bar: 000
    command terminated with exit code 56
    sleep.legacy to httpbin.legacy: 200
```

由于 bar 命名空间的 httpbin 服务开启了 mTLS 功能，导致 legacy 命名空间请求 bar 命名空间的 httpbin 服务也出现了失败的情况。

3）清理测试：

```
$ kubectl delete -f istio/security/mtls-namespace-foo-on.yaml
$ kubectl delete -f istio/security/mtls-service-httpbin-on.yaml
```

（5）测试服务端口级别 mTLS

1）启用 bar 命名空间的 httpbin 服务的 1234 端口的 mTLS 功能：

```
$ kubectl apply -f istio/security/mtls-service-httpbin-port-on.yaml
```

2）访问服务：

```
    $ kubectl exec $(kubectl get pod -l app=sleep -n legacy -o jsonpath={.items..
metadata.name}) -c sleep -n legacy -- curl http://httpbin.bar:8000/ip -s -o /dev/
null -w "%{http_code}\n"
    200
```

bar 命名空间的 httpbin 服务只在端口 1234 开启了 mTLS 功能，当 legacy 命名空间请求 bar 命名空间的 httpbin 服务时，由于使用的是 8000 端口，服务可以不使用 TLS 加密，正常通过。

3）清理：

```
$ kubectl delete -f istio/security/mtls-service-httpbin-port-on.yaml
```

【实验二】 终端用户身份认证

本实验需要部署实验一部署的 3 个命名空间的 httpbin 服务。

1）创建路由规则。创建 httpbin 的路由规则和 Gateway：

```
$ kubectl apply -f istio/route/gateway-httpbin-http.yaml -n foo
```

2）访问 httpbin 服务：

```
$ curl http://11.11.11.111:31380/headers -s -o /dev/null -w "%{http_code}\n"
200
```

3）创建 JWT 认证策略：

```
$ kubectl apply -f istio/security/jwt-httpbin.yaml
```

4）访问测试。由于开启了 JWT 认证，直接访问就会返回没有认证的 401 响应码：

```
$ curl http://11.11.11.111:31380/headers -s -o /dev/null -w "%{http_code}\n"
401
```

5）使用 JWT 访问测试。此时使用 JWT 的 token 进行访问，会得到正常的响应码：

```
$ TOKEN=$(curl https://raw.githubusercontent.com/istio/istio/release-1.0/
security/tools/jwt/samples/demo.jwt -s)
$ curl --header "Authorization: Bearer $TOKEN" http://11.11.11.111:31380/
headers -s -o /dev/null -w "%{http_code}\n"
200
```

6）配置 Python 环境。配置 Python 环境的同时下载用于测试的源码文件：

```
$ sudo yum install -y python-pip wget
$ sudo pip install jwcrypto -i http://pypi.douban.com/simple/ --trusted-host
pypi.douban.com
$ mkdir jwt-test
$ wget -O jwt-test/key.pem https://raw.githubusercontent.com/istio/istio/
release-1.0/security/tools/jwt/samples/key.pem
$ wget -O jwt-test/gen-jwt.py https://raw.githubusercontent.com/istio/istio/
release-1.0/security/tools/jwt/samples/gen-jwt.py
```

7）生成 token 并访问服务。生成有效期很短的 token，测试 token 过期特性：

```
$ TOKEN=$(python jwt-test/gen-jwt.py jwt-test/key.pem --expire 5)
$ for i in `seq 1 10`; do curl --header "Authorization: Bearer $TOKEN"
http://11.11.11.111:31380/headers -s -o /dev/null -w "%{http_code}\n"; sleep 1;
done
200
200
200
200
401
401
401
401
401
401
```

从以上结果可以看出，由于设置的 token 有效期仅为 5 秒，只有前几次的请求成功，后面的请求都返回没有认证的 401 响应码，说明 token 已经过期。注意：由于生成 token 到复制命令执行请求可能存在短暂时间间隔，真正的有效期可能已经不足 5 秒，导致看到的 200 响应码不足 5 次。

【实验三】 传输认证与终端认证配合使用

本实验需要部署实验一部署的 3 个命名空间的 httpbin 服务。

1）创建同时使用传输认证 mTLS 与终端认证 JWT 策略：

```
$ kubectl apply -f istio/security/jwt-mtls-httpbin.yaml
```

2）访问服务：

```
$ TOKEN=$(curl https://raw.githubusercontent.com/istio/istio/release-1.0/
security/tools/jwt/samples/demo.jwt -s)

$ curl --header "Authorization: Bearer $TOKEN" http://11.11.11.111:31380/
headers -s -o /dev/null -w "%{http_code}\n"
200

$ kubectl exec $(kubectl get pod -l app=sleep -n foo -o jsonpath={.items..
metadata.name}) -c sleep -n foo -- curl http://httpbin.foo:8000/ip -s -o /dev/null
-w "%{http_code}\n"
401

$ kubectl exec $(kubectl get pod -l app=sleep -n foo -o jsonpath={.items..
metadata.name}) -c sleep -n foo -- curl http://httpbin.foo:8000/ip -s -o /dev/null
-w "%{http_code}\n" --header "Authorization: Bearer $TOKEN"
200

$ kubectl exec $(kubectl get pod -l app=sleep -n legacy -o jsonpath={.items..
metadata.name}) -c sleep -n legacy -- curl http://httpbin.foo:8000/ip -s -o /dev/
null -w "%{http_code}\n" --header "Authorization: Bearer $TOKEN"
000
command terminated with exit code 56
```

对比测试结果可知，当没有 TLS 加密请求时，请求直接失败；当以 TLS 加密请求时，如果没有使用携带 token 会返回没有认证的 401 响应码；只有当以 TLS 加密请求并携带 token 时，请求才会成功，说明 mTLS 与 JWT 配合使用已经生效。

3）清理认证策略：

```
$ kubectl delete -f istio/security/jwt-mtls-httpbin.yaml
```

4）创建自定义认证头规则：

```
$ kubectl apply -f istio/security/jwt-httpbin-custom-header.yaml
```

5）访问服务：

```
$ curl --header "X-Auth: $TOKEN" http://11.11.11.111:31380/headers -s -o /dev/
null -w "%{http_code}\n"
200
```

6）创建自定义认证query参数规则：

```
$ kubectl apply -f istio/security/jwt-httpbin-custom-query.yaml
```

7）访问服务：

```
$ curl http://11.11.11.111:31380/headers?access_token=$TOKEN -s -o /dev/null
-w "%{http_code}\n"
200
```

8）创建自定义认证头结合自定义认证query参数规则：

```
$ kubectl apply -f istio/security/jwt-httpbin-custom-header-query.yaml
```

9）访问服务：

```
$ curl --header "X-Auth: $TOKEN" http://11.11.11.111:31380/headers -s -o /dev/
null -w "%{http_code}\n"
200

$ curl http://11.11.11.111:31380/headers?access_token=$TOKEN -s -o /dev/null
-w "%{http_code}\n"
200
```

10）清理：

```
$ kubectl delete ns foo bar legacy
```

10.5 RBAC访问控制

Istio的授权功能使用的是基于角色的访问权限控制，在服务网格中提供了命名空间级别、服务级别以及方法级别的访问控制功能，具有如下特点：

- 基于角色的语法语义，简单易用。
- 支持服务到服务和终端用户到服务的授权功能。
- 可以自定义属性，这样更灵活，例如在角色和角色绑定上使用条件。
- 高性能，Istio授权是在Envoy代理本地实施的。

授权架构如图10-3所示。

管理员通过Istio的配置创建服务访问策略，Pilot接收策略规则后，生成Envoy代理需要的配置格式并分发给对应的Envoy代理，Envoy代理根据访问策略在本地执行服务访问权限的检查，判断服务的请求能否被通过。

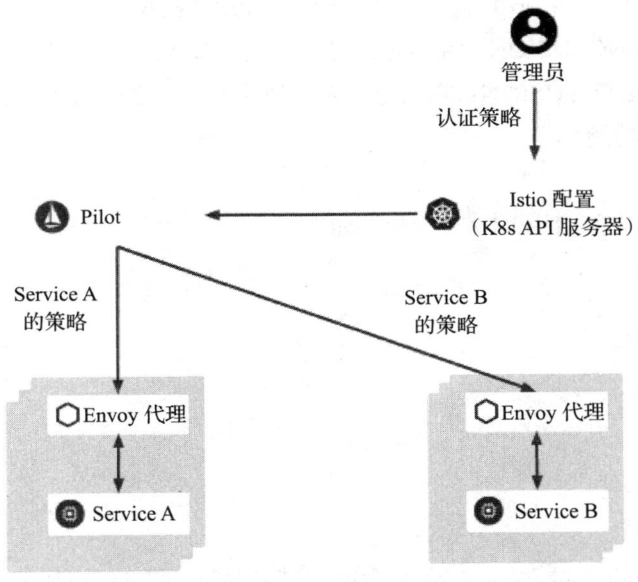

图 10-3　授权架构（图片来源：Istio 官方网站）

管理人员通过编写 yaml 文件来配置管理授权规则，然后提交配置存储到 IstioConfigStore 中。Pilot 实时监控授权策略，当有授权策略发生改变时，Pilot 会立即更新授权策略，然后把授权策略分发给服务实例相关的 Envoy 代理。每一个 Envoy 代理会在运行时启动授权引擎，以便在请求到来时进行授权验证，当一个请求到达 Envoy 代理时，授权引擎以当前的授权策略评估请求，返回通过或者拒绝的授权验证结果。

在 Istio 中，RBAC 访问控制使用 Kubernetes 中的 serviceaccount 作为服务的身份标识，用于分配 servicerole 对象，因此在 Kubernetes 中部署服务时需要创建对应的 serviceaccount，并给部署的 deployment 指定创建的 serviceaccount。

在使用 RBAC 访问控制，并且使用某些特殊的属性时，比如 source.namespace，由于会用到 source.principal 属性来实现匹配 source.namespace 属性的功能，因此分配角色的主体对象有时必须是认证的用户或者服务身份标识。当服务没有开启 mTLS 时，没有经过身份认证，就不能正常应用 RBAC 访问控制策略。如果使用其他属性，如 source.ip、request.headers 等，由于不会使用到 source.principal 属性来实现，就不需要给服务启用 mTLS 功能。

1. 启用授权

通过创建 RbacConfig 对象，可以启用 RBAC 授权功能。RbacConfig 对象是整个网络中唯一的，且必须命名为 default。在 RbacConfig 对象中可以通过指定 mode 参数来指定 RBAC 访问控制启用的范围，mode 可取值如下：

- OFF：完全关闭 RBAC 授权功能。
- ON：为所有的服务和命名空间启用 RBAC 授权功能。
- ON_WITH_INCLUSION：在指定命名空间或者服务上启用 RBAC 授权功能。
- ON_WITH_EXCLUSION：在指定命名空间或者服务之外的命名空间或者服务中启用 RBAC 授权功能。

使用示例如下。

在 default 命名空间启用 RBAC 授权功能：

```
apiVersion: rbac.istio.io/v1alpha1
kind: RbacConfig
metadata:
  name: default
spec:
  mode: 'ON_WITH_INCLUSION'
  inclusion:
    namespaces:
    - default
```

在 istio-system 和 kube-system 之外的命名空间中启用 RBAC 授权功能：

```
apiVersion: rbac.istio.io/v1alpha1
kind: RbacConfig
metadata:
  name: default
spec:
  mode: 'ON_WITH_EXCLUSION'
  exclusion:
    namespaces:
    - istio-system
    - kube-system
```

2. 授权策略

RBAC 授权策略由 ServiceRole 和 ServiceRoleBinding 组成，ServiceRole 中包含了服务访问权限的集合，ServiceRoleBinding 把 ServiceRole 分配给指定的对象，如用户和服务。

ServiceRole 对象中包含了字段为 rules 的权限列表，每一个 rule 有如下字段：

- services：服务列表，可以设置为 * 表示指定命名空间的所有服务。
- methods：HTTP 的方法列表，在 gRPC 请求中该字段会被忽略，因为 gRPC 中只有 POST 方法，可以设置为 * 表示不限定方法。
- paths：HTTP 的请求 URI 列表或者 gRPC 请求中的方法列表，gRPC 方法必须是 /packageName.serviceName/methodName 这种格式，并且是大小写敏感的。URI 支持前缀、后缀和精确匹配，例如：/books/review 精确匹配 /books/review，前缀匹配 /

books/，后缀匹配 /review。如果没有指定此字段表示对任何 URI 或者 gRPC 请求中的方法都生效。

- constraints：表示特殊限制列表，比如限制服务版本、请求头等信息[1]。

rule 中所有的字段均支持前缀和后缀匹配。

定义命名空间级别的 ServiceRole 示例如下：

```
 1 apiVersion: rbac.istio.io/v1alpha1
 2 kind: ServiceRole
 3 metadata:
 4   name: service-viewer
 5   namespace: default
 6 spec:
 7   rules:
 8   - services: ["*"]
 9     methods: ["GET"]
10     constraints:
11     - key: "destination.labels[app]"
12       values:
13       - service-js
14       - service-python
15       - service-lua
16       - service-node
17       - service-go
```

如上的 ServiceRole 定义表示 default 命名空间的所有服务都有 GET 方法的权限，但是第 10 ～ 17 行又定义了特殊的限制，限制请求的目标地址只能为 values 中指定的服务，访问其他服务时没有权限。

使用前缀和后缀匹配示例如下：

```
apiVersion: rbac.istio.io/v1alpha1
kind: ServiceRole
metadata:
  name: tester
  namespace: default
spec:
  rules:
  - services: ["test-*"]
    methods: ["*"]
  - services: ["bookstore.default.svc.cluster.local"]
    paths: ["*/reviews"]
    methods: ["GET"]
```

ServiceRoleBinding 的定义包含如下两个字段：

[1] 可以使用的值参考官方文档 https://istio.io/docs/reference/config/authorization/constraints-and-properties/#constraints。

- roleRef 指定某个同命名空间的 ServiceRole 名字。
- subjects 指定角色分配的主体对象列表。主体可以是 user，也可以是 properties，还可以是两者的组合。user 的值一般是用户的标识或者服务的身份标识，也就是 Pod 使用的 serviceaccount 名称，使用如 "cluster.local/ns/default/sa/service-python" 这样的形式。properties 与 ServiceRole 中的 constraints 字段的使用方法类似[⊖]。

定义命名空间级别的 ServiceRoleBinding 示例如下：

```
apiVersion: rbac.istio.io/v1alpha1
kind: ServiceRoleBinding
metadata:
  name: bind-service-viewer
  namespace: default
spec:
  subjects:
  - properties:
      source.namespace: "istio-system"
  - properties:
      source.namespace: "default"
  roleRef:
    kind: ServiceRole
    name: "service-viewer"
```

如上代码中的 ServiceRoleBinding 定义，将名为 service-viewer 的 ServiceRole 分配给请求的来源命名空间为 istio-system 和 default 的请求主体。

分配角色给所有用户示例：

```
apiVersion: rbac.istio.io/v1alpha1
kind: ServiceRoleBinding
metadata:
  name: bind-service-js-service-python-viewer
  namespace: default
spec:
  subjects:
  - user: "*"
  roleRef:
    kind: ServiceRole
    name: "service-js-service-python-viewer"
```

分配角色给所有认证的用户示例：

```
apiVersion: rbac.istio.io/v1alpha1
kind: ServiceRoleBinding
metadata:
  name: bind-service-js-service-python-viewer
```

⊖ 具体可以使用的值参考官方文档 https://istio.io/docs/reference/config/authorization/constraints-and-properties/#properties。

```
  namespace: default
spec:
  subjects:
  - properties:
      source.principal: "*"
  roleRef:
    kind: ServiceRole
    name: "service-js-service-python-viewer"
```

基于服务级别的访问控制策略使用示例：

```
 1 apiVersion: rbac.istio.io/v1alpha1
 2 kind: ServiceRole
 3 metadata:
 4   name: service-js-service-python-viewer
 5   namespace: default
 6 spec:
 7   rules:
 8   - services:
 9     - "service-js.default.svc.cluster.local"
10     - "service-python.default.svc.cluster.local"
11     methods: ["GET"]
12 ---
13 apiVersion: rbac.istio.io/v1alpha1
14 kind: ServiceRoleBinding
15 metadata:
16   name: bind-service-js-service-python-viewer
17   namespace: default
18 spec:
19   subjects:
20   - user: "*"
21   roleRef:
22     kind: ServiceRole
23     name: "service-js-service-python-viewer"
24 ---
25 apiVersion: rbac.istio.io/v1alpha1
26 kind: ServiceRole
27 metadata:
28   name: service-lua-service-node-viewer
29   namespace: default
30 spec:
31   rules:
32   - services:
33     - "service-lua.default.svc.cluster.local"
34     - "service-node.default.svc.cluster.local"
35     methods: ["GET"]
36 ---
37 apiVersion: rbac.istio.io/v1alpha1
38 kind: ServiceRoleBinding
39 metadata:
```

```
40   name: bind-service-lua-service-node-viewer
41   namespace: default
42 spec:
43   subjects:
44   - user: "cluster.local/ns/default/sa/service-python"
45   roleRef:
46     kind: ServiceRole
47     name: "service-lua-service-node-viewer"
48 ---
49 apiVersion: rbac.istio.io/v1alpha1
50 kind: ServiceRole
51 metadata:
52   name: service-go-viewer
53   namespace: default
54 spec:
55   rules:
56   - services: ["service-go.default.svc.cluster.local"]
57     methods: ["GET"]
58 ---
59 apiVersion: rbac.istio.io/v1alpha1
60 kind: ServiceRoleBinding
61 metadata:
62   name: bind-service-go-viewer
63   namespace: default
64 spec:
65   subjects:
66   - user: "cluster.local/ns/default/sa/service-node"
67   roleRef:
68     kind: ServiceRole
69     name: "service-go-viewer"
```

第1～23行定义的访问控制策略表明，服务service-js和服务service-python可以被任何用户以GET方法访问。

第25～47行定义的访问控制策略表明，服务service-node和服务service-lua只可以被default命名空间中使用了名为service-python的serviceaccount的服务实例以GET方法访问。

第49～69行定义的访问控制策略表明，服务service-go只可以被default命名空间中使用了名为service-node的serviceaccount的服务实例以GET方法访问。

【实验】

1）部署服务：

```
$ kubectl apply -f service/js/service-go.yaml
$ kubectl apply -f service/js/service-node.yaml
$ kubectl apply -f service/lua/service-lua.yaml
$ kubectl apply -f service/python/service-python.yaml
$ kubectl apply -f service/js/service-js.yaml
```

2）启用 default 命名空间的 mTLS 功能。

RBAC 的部分实验需要服务身份认证，需要开启 mTLS：

```
$ kubectl apply -f istio/security/mtls-namespace-default-on.yaml
```

3）创建 Gateway 和 service-js 路由规则。

创建 Gateway 和服务 service-js 的 mTLS 路由规则，并路由到 v1 版本：

```
$ kubectl apply -f istio/route/gateway-js-v1-mtls.yaml
```

4）替换 service-node 服务和 service-python 服务的部署为带有 serviceaccount 的 Deployment：

```
$ kubectl apply -f service/node/service-node-with-serviceaccount.yaml
$ kubectl apply -f service/python/service-python-with-serviceaccount.yaml
```

5）创建测试 Pod：

```
$ kubectl apply -f kubernetes/dns-test.yaml
```

6）访问服务：

```
$ kubectl exec dns-test -c dns-test -- curl -s http://service-go/env
{"message":"go v2"}

$ kubectl exec dns-test -c dns-test -- curl -s http://service-node/env
{"message":"node v1","upstream":[{"message":"go v2","response_time":"0.40"}]}

$ kubectl exec dns-test -c dns-test -- curl -s http://service-lua/env
{"message":"lua v2"}

$ kubectl exec dns-test -c dns-test -- curl -s http://service-python/env
{"message":"python v1","upstream":[{"message":"lua v1","response_
time":0.22},{"message":"node v1","response_time":0.47,"upstream":[{"message":"go
v1","response_time":"0.24"}]}]}
```

此时服务开启了 mTLS，所有服务访问正常。

7）浏览器访问。

浏览器访问地址为 http：//11.11.11.112：31380/，服务正常如图 10-4 所示。

8）启动 default 命名空间的 RBAC 访问控制：

```
$ kubectl apply -f istio/security/rbac-config-on.yaml
```

9）访问服务：

```
$ kubectl exec dns-test -c dns-test -- curl -s http://service-go/env
RBAC: access denied

$ kubectl exec dns-test -c dns-test -- curl -s http://service-node/env
```

```
RBAC: access denied

$ kubectl exec dns-test -c dns-test -- curl -s http://service-lua/env
RBAC: access denied

$ kubectl exec dns-test -c dns-test -- curl -s http://service-python/env
RBAC: access denied

$ kubectl exec dns-test -c dns-test -- curl -s http://service-js/env
RBAC: access denied
```

Welcome to Istio

发射

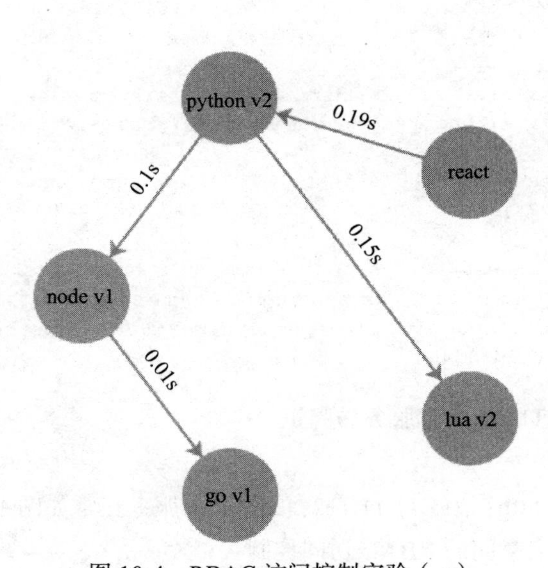

图 10-4　RBAC 访问控制实验（一）

此时由于开启了 RBAC 访问控制，而没有创建授权策略，所有服务访问都会被拒绝。

10）创建 RBAC 命名空间级别规则：

```
$ kubectl apply -f istio/security/rbac-namespace-policy.yaml
```

11）访问服务：

```
$ kubectl exec dns-test -c dns-test -- curl -s http://service-go/env
{"message":"go v1"}

$ kubectl exec dns-test -c dns-test -- curl -s http://service-node/env
{"message":"node v2","upstream":[{"message":"go v1","response_time":"0.02"}]}

$ kubectl exec dns-test -c dns-test -- curl -s http://service-lua/env
{"message":"lua v1"}

$ kubectl exec dns-test -c dns-test -- curl -s http://service-python/env
{"message":"python v1","upstream":[{"message":"lua v2","response_
time":0.06},{"message":"node v2","response_time":0.13,"upstream":[{"message":"go
v1","response_time":"0.02"}]}]}
```

此时已经创建了授权策略，所有服务访问正常。

12）浏览器访问。

浏览器访问地址为 http：//11.11.11.112：31380/，服务正常如图 10-5 所示。

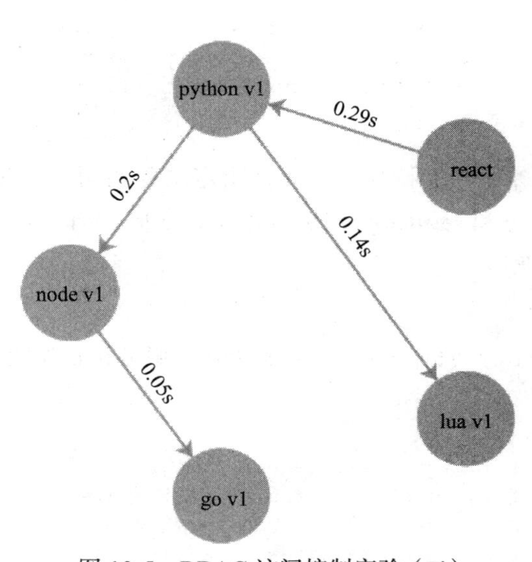

图 10-5　RBAC 访问控制实验（二）

13）删除命名空间级别的授权：

```
$ kubectl delete -f istio/security/rbac-namespace-policy.yaml
```

14）创建服务级别的授权：

```
$ kubectl apply -f istio/security/rbac-service-policy.yaml
```

15）访问服务：

```
$ kubectl exec dns-test -c dns-test -- curl -s http://service-go/env
RBAC: access denied

$ kubectl exec dns-test -c dns-test -- curl -s http://service-node/env
RBAC: access denied

$ kubectl exec dns-test -c dns-test -- curl -s http://service-lua/env
RBAC: access denied

$ kubectl exec dns-test -c dns-test -- curl -s http://service-python/env
{"message":"python v2","upstream":[{"message":"lua v1","response_
time":0.3},{"message":"node v2","upstream":[{"message":"go v2","response_
time":"0.02"}],"response_time":0.25}]}

$ kubectl exec dns-test -c dns-test -- curl -s http://service-js/
<!doctype html><html lang="en"><head><meta charset="utf-8"><meta
name="viewport" content="width=device-width,initial-scale=1,shrink-to-
fit=no"><meta name="theme-color" content="#000000"><link rel="manifest" href="/
manifest.json"><link rel="shortcut icon" href="/favicon.ico"><title>React
App</title><link href="/static/css/main.c17080f1.css" rel="stylesheet"></
head><body><noscript>You need to enable JavaScript to run this app.</noscript><div
id="root"></div><script type="text/javascript" src="/static/js/main.bc334d90.
js"></script></body></html>
```

以上步骤创建了服务级别的授权，只允许指定服务调用，因此我们使用 dns-test 访问时只有 service-python 服务和 service-js 服务可以正常访问，因为服务 service-python 与服务 service-js 允许任何用户调用。

16）浏览器访问。

浏览器访问地址为 http：//11.11.11.112：31380/，服务正常如图 10-6 所示。

17）清理：

```
$ kubectl delete -f kubernetes/dns-test.yaml
$ kubectl delete -f istio/security/rbac-config-on.yaml
$ kubectl delete -f istio/security/rbac-service-policy.yaml
$ kubectl delete -f istio/route/gateway-js-v1-mtls.yaml
$ kubectl delete -f istio/security/mtls-namespace-default-on.yaml
```

Welcome to Istio

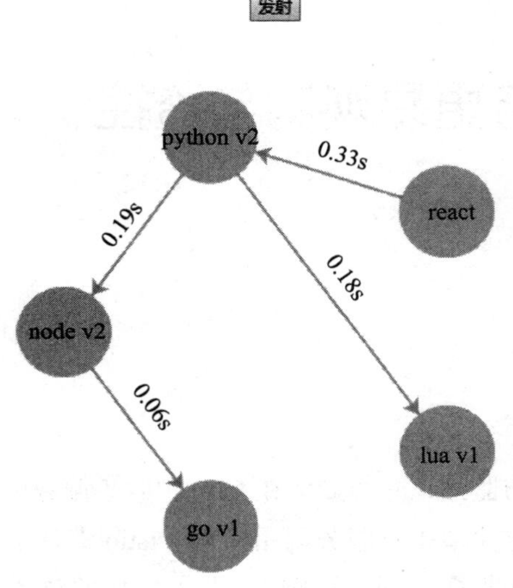

图 10-6　RBAC 访问控制实验（三）

10.6　本章小结

借助 Istio 提供的服务访问安全功能，我们可以轻松地实现服务间的通信加密，以及服务访问权限控制，这会大大地增强服务的安全性。而使用这些增强服务安全性相关的功能特性，却几乎不需要修改已有服务的代码，这也给服务迁移到服务网格中带来极大的便利。

Chapter 11 第 11 章

让服务更易观测与监控

本章介绍服务的观测与监控，介绍如何在 Istio 中收集服务的指标数据、日志数据，如何借助调用链追踪功能，更方便地排查服务调用问题。Istio 默认情况下已经收集了许多服务指标数据，并使用 Grafana 创建了许多图表来展示服务网格的整体运行状态，利用收集到的日志数据和服务指标数据，Istio 也生成了服务的调用依赖树，让我们对服务依赖关系一目了然。

11.1 整体介绍

为了获取服务运行时的行为，可观测性很有必要。为了让服务更易观测，我们需要收集服务的日志（logging)，统计服务的运行指标数据（metrics)，记录服务的调用链追踪(tracing)。这三者都有互有重叠的功能部分，关系如图 11-1 所示。

图 11-1 服务的日志、指标数据、调用链追踪之间的关系（图片来源：Peter Bourgon 的文章“Metrics，tracing，and logging”）

日志是事件驱动的并根据时间变化不断存储的不可变数据。日志一般有如下几种格式：

- 纯文本：这是最常用的日志格式。
- 结构化数据：一般使用 JSON 格式存储。
- 二进制：一般是为节省空间或有特殊用途的格式，比如：MySQL 的二进制日志等。

一般会将日志收集到一个集中的日志中心，用于存储、搜索和分析。通常情况下使用开源的日志收集处理技术栈是个不错的选择。比如开源的 ELK/EFK 日志系统。

指标数据是指一段时间内对系统的测量数值，是一段时间内数据的聚合。通过指标数据通常可以计算出一段时间内的平均值、最大值和最小值，可以了解数据在某段时间内的变化趋势，指标数据占用的存储空间一般也会比日志系统小得多，大多数的监控系统都是基于指标数据做出来的，我们可以聚合多个指标数据组成一个监控面板，观测过去某段时间内数据的变化。监控面板一般有如下几种类型：

- 计数器（Counter）：只可以增加不可以减少，一个简单的计数功能。例如：统计请求次数。
- 仪表盘（Gauge）：可以增加也可以减少，仪表盘只代表当前的数据状态。例如：统计 CPU 使用率。
- 柱状图（Histogram）：用于统计一些数据的分布情况，计算在一定范围内的分布情况。例如：统计一段时间内响应时间大于某个数值的请求数。
- 概要（Summary）：与柱状图类似，计算一段时间内指标对象总数以及指标的总和，提供百分位的功能。例如：请求响应时间的 99 百分位。

通过指标数据，我们可以了解一段时间内整体数据的变化情况。收集指标数据的工具现在比较多，例如 Prometheus、Influxdb、Grafana，可用于指标数据的可视化。

调用链追踪是请求级别的数据记录，包含系统中一个请求在生命周期内所涉及的所有数据的集合。比如：调用远程服务的耗时，查询数据库时所用的 SQL 语句等。一般情况下，调用链涉及从请求进入服务到服务响应数据的所有调用过程。调用链系统会在请求进入服务时生成唯一的请求 ID，这个 ID 会一直传递给将要调用的其他服务，直到本次调用成功或者失败，调用链系统会记录每一次调用的耗时等信息。方便查看一个请求的完整调用过程。调用链有如下几个重要概念：

- Span：是服务调用信息的存储单位，每一次服务调用就会生成一个 Span 用于存储本次服务调用的信息。
- Trace：请求的整个服务调用链，包含多个 Span。

Trace 与 Span 的关系如图 11-2 所示。

通过调用链追踪，我们能清楚地知道一个用户请求在整个系统的服务调用情况，知道每一个请求的每一个调用的耗时，看到系统中真实的服务调用情况，在 OpenTracing 组织的努力下，服务调用链追踪有了统一的标准。调用链追踪可以使用 Jaeger、Zipkin 等来收集、存储、查看和分析。

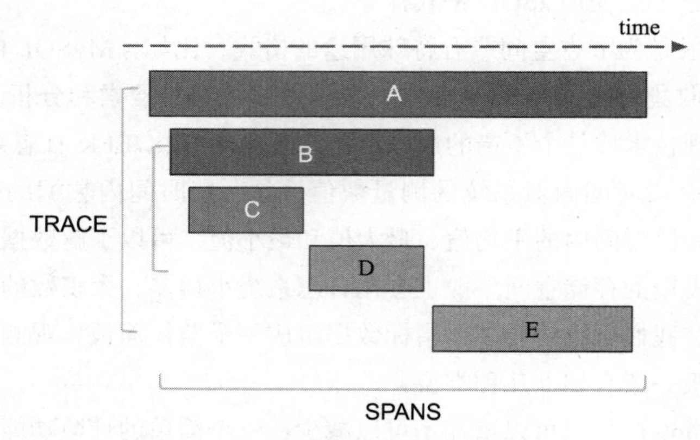

图 11-2 Trace 与 Span 的关系

日志、指标数据与调用链并不是孤立存在的，可以配合使用来更好地排查服务的问题。当我们观测到服务指标数据出现问题时，可以通过调用链来查看服务的调用情况，找到调用链上有问题的服务，然后再配合日志找到服务的具体问题。例如：当我们收到监控系统的报警提示服务 A 调用耗时过长，我们可以通过调用链追踪系统，找到服务调用链上响应时间过长的是服务 B，然后通过日志系统，搜索服务 B 的相关日志，进而找到问题的原因，修复服务 B 的问题，最终解决服务指标异常的问题。

【实验前的准备】

进行本章实验前，需要先执行如下的前置步骤。

1）部署 Istio 的官方示例完整功能版本：

```
# 配置
$ cp /usr/local/istio/install/kubernetes/istio-demo.yaml /usr/local/istio/install/kubernetes/istio-demo.yaml.ori
$ sed -i 's@2048Mi@500Mi@g' /usr/local/istio/install/kubernetes/istio-demo.yaml
$ sed -i 's@quay.io/coreos/hyperkube:v1.7.6_coreos.0@registry.cn-shanghai.aliyuncs.com/gcr-k8s/hyperkube:v1.7.6_coreos.0@g' /usr/local/istio/install/kubernetes/istio-demo.yaml

# 部署
$ kubectl apply -f /usr/local/istio/install/kubernetes/helm/istio/templates/crds.yaml
$ kubectl apply -f /usr/local/istio/install/kubernetes/istio-demo.yaml

# 查看状态
$ kubectl get svc -n istio-system
$ kubectl get pod -n istio-system
```

等待所有 pod 启动完成，再进行后续的实验步骤。

2）下载实验时用到的源码仓库：

```
$ sudo yum install -y git

$ git clone https://github.com/mgxian/istio-lab
Cloning into 'istio-lab'...
remote: Enumerating objects: 247, done.
remote: Counting objects: 100% (247/247), done.
remote: Compressing objects: 100% (173/173), done.
remote: Total 774 (delta 153), reused 164 (delta 73), pack-reused 527
Receiving objects: 100% (774/774), 283.00 KiB | 229.00 KiB/s, done.
Resolving deltas: 100% (447/447), done.

$ cd istio-lab
```

3）开启 default 命名空间的自动注入功能：

```
$ kubectl label namespace default istio-injection=enabled
namespace/default labeled
```

4）部署用于测试的服务：

```
$ kubectl apply -f service/go/service-go.yaml
$ kubectl apply -f service/node/service-node.yaml
$ kubectl apply -f service/lua/service-lua.yaml
$ kubectl apply -f service/python/service-python.yaml
$ kubectl apply -f service/js/service-js.yaml

$ kubectl get pod
NAME                                   READY     STATUS     RESTARTS    AGE
service-go-v1-7cc5c6f574-chmc4         2/2       Running    0           3m43s
service-go-v2-7656dcc478-kqwg5         2/2       Running    0           3m43s
service-js-v1-55756d577-qvkkn          2/2       Running    0           3m36s
service-js-v2-86bdfc86d9-hhhs8         2/2       Running    0           3m36s
service-lua-v1-5c9bcb7778-n515t        2/2       Running    0           3m41s
service-lua-v2-75cb5cdf8-bv7kp         2/2       Running    0           3m40s
service-node-v1-d44b9bf7b-5qsnq        2/2       Running    0           3m42s
service-node-v2-86545d9796-tg5vx       2/2       Running    0           3m42s
service-python-v1-79fc5849fd-kt7ls     2/2       Running    0           3m39s
service-python-v2-7b6864b96b-4hln6     2/2       Running    0           3m39s
```

11.2　指标收集

Istio 提供统一的“服务指标”收集功能，可以通过简单的配置就能收集服务指标，比如，收集服务的响应时间、请求次数、响应流量等。Istio 使用 Prometheus 获取指标数据。

统计服务的请求次数示例：

```
 1 apiVersion: config.istio.io/v1alpha2
 2 kind: metric
 3 metadata:
 4   name: myrequestcount
 5   namespace: istio-system
 6 spec:
 7   value: "1"
 8   dimensions:
 9     destination_service: destination.service.name | "unknown"
10     destination_namespace: destination.service.namespace | "unknown"
11     destination_version: destination.labels["version"] | "unknown"
12     response_code: response.code | 200
13   monitored_resource_type: '"UNSPECIFIED"'
14 ---
15 apiVersion: config.istio.io/v1alpha2
16 kind: prometheus
17 metadata:
18   name: myrequestcounthandler
19   namespace: istio-system
20 spec:
21   metrics:
22   - name: request_count
23     instance_name: myrequestcount.metric.istio-system
24     kind: COUNTER
25     label_names:
26     - destination_service
27     - destination_namespace
28     - destination_version
29     - response_code
30 ---
31 apiVersion: config.istio.io/v1alpha2
32 kind: rule
33 metadata:
34   name: request-count
35   namespace: istio-system
36 spec:
37   actions:
38   - handler: myrequestcounthandler.prometheus
39     instances:
40     - myrequestcount.metric
```

第1～13行定义了名为myrequestcount的metric实例。value定义表示每次请求收集的指标数据计数值为1，dimensions定义了被收集指标的身份标识，用于区分不同的指标数据。monitored_resource_type定义了被监控资源的类型为"UNSPECIFIED"，仅对支持该字段的后端有用，此处可以忽略。

第15～29行定义了名为myrequestcounthandler的prometheus适配器。metrics定义要存储在Prometheus中的指标数据，name定义了指标名称，默认情况下在Prometheus中以istio后接name的形式存储，如"istio_request_count"，也可以通过指定namespace字段改变指标名称前缀。instance_name 定义了使用 metric 的名称，必须使用全名称，kind 表示存储在 Prometheus 时的指标类型，label_names 定义了 metric 存储到 Prometheus 时要存储的 label 名称，必须与 metric 中的定义相对应。

第 31 ～ 40 行定义了名为 request-count 的 rule 规则。表明把 myrequestcount 实例收集到的指标数据发送给 myrequestcounthandler 适配器处理。

统计 TCP 服务数据传输大小示例：

```
 1 apiVersion: config.istio.io/v1alpha2
 2 kind: metric
 3 metadata:
 4   name: mytcpsentbytes
 5   namespace: default
 6 spec:
 7   value: connection.sent.bytes | 0
 8   dimensions:
 9     source_service: source.labels["app"] | source.workload.name | "unknown"
10     source_version: source.labels["version"] | "unknown"
11     destination_service: destination.service.name | "unknown"
12     destination_version: destination.labels["version"] | "unknown"
13   monitoredResourceType: '"UNSPECIFIED"'
14 ---
15 apiVersion: config.istio.io/v1alpha2
16 kind: metric
17 metadata:
18   name: mytcpreceivedbytes
19   namespace: default
20 spec:
21   value: connection.received.bytes | 0
22   dimensions:
23     source_service: source.labels["app"] | source.workload.name | "unknown"
24     source_version: source.labels["version"] | "unknown"
25     destination_service: destination.service.name | "unknown"
26     destination_version: destination.labels["version"] | "unknown"
27   monitoredResourceType: '"UNSPECIFIED"'
28 ---
29 apiVersion: config.istio.io/v1alpha2
30 kind: prometheus
31 metadata:
32   name: tcphandler
33   namespace: default
34 spec:
35   metrics:
36   - name: tcp_sent_bytes
37     instance_name: mytcpsentbytes.metric.default
38     kind: COUNTER
39     label_names:
40     - source_service
41     - source_version
```

```
42      - destination_service
43      - destination_version
44    - name: tcp_received_bytes
45      instance_name: mytcpreceivedbytes.metric.default
46      kind: COUNTER
47      label_names:
48      - source_service
49      - source_version
50      - destination_service
51      - destination_version
52 ---
53 apiVersion: config.istio.io/v1alpha2
54 kind: rule
55 metadata:
56   name: tcp-sent-received-bytes
57   namespace: default
58 spec:
59   match: context.protocol == "tcp"
60          && destination.service.namespace == "default"
61   actions:
62   - handler: tcphandler.prometheus
63     instances:
64     - mytcpreceivedbytes.metric
65     - mytcpsentbytes.metric
```

第1～13行定义了名为mytcpsentbytes的metric实例，value的定义表示使用TCP发送数据的大小作为收集的指标数据，dimensions定义了被收集指标的身份标识。monitored_resource_type定义了被监控资源的类型为"UNSPECIFIED"。

第15～27行定义了名为mytcpreceivedbytes的metric实例，value的定义表示使用TCP接收数据的大小作为收集的指标数据，其他定义均与之前统计服务请求次数的定义一样，在此不再赘述。

第29～51行定义了名为tcphandler的prometheus适配器。其他定义均与之前统计服务请求次数的定义一样，在此不再赘述。

第53～65行定义了名为tcp-sent-received-bytes的rule规则。表明当请求的协议为TCP且目标服务的命名空间为default时，把mytcpsentbytes实例和mytcpreceivedbytes实例收集到的指标数据发送给tcphandler适配器处理。

【实验】

1）查看是否部署了Prometheus：

```
$ kubectl get deploy prometheus -n istio-system
NAME         DESIRED   CURRENT   UP-TO-DATE   AVAILABLE   AGE
prometheus   1         1         1            1           21m

$ kubectl get svc prometheus -n istio-system
```

```
NAME         TYPE        CLUSTER-IP      EXTERNAL-IP   PORT(S)    AGE
prometheus   ClusterIP   10.99.177.169   <none>        9090/TCP   21m
```

上面命令的结果表示 Prometheus 已经部署。

2）创建用于请求的 Pod：

```
$ kubectl apply -f kubernetes/fortio.yaml
```

3）创建服务 HTTP 指标收集规则：

```
$ kubectl apply -f istio/telemetry/metric-http-request-count.yaml
```

4）暴露 Prometheus 查询 Web 服务：

```
$ kubectl apply -f kubernetes/istio-prometheus-service.yaml
```

5）并发请求服务：

```
$ kubectl exec fortio -c fortio /usr/local/bin/fortio -- load -curl http://
service-python/env
HTTP/1.1 200 OK
content-type: application/json
content-length: 177
server: envoy
date: Fri, 18 Jan 2019 10:56:41 GMT
x-envoy-upstream-service-time: 1014

{"message":"python v2","upstream":[{"message":"lua v2","response_
time":0.13},{"message":"node v2","response_time":1.0,"upstream":[{"message":"go
v1","response_time":"0.32"}]}]}

$ kubectl exec fortio -c fortio /usr/local/bin/fortio -- load -qps 30 -n 300
-loglevel Error http://service-python/env
10:59:53 I logger.go:97> Log level is now 4 Error (was 2 Info)
Fortio 1.0.1 running at 30 queries per second, 2->2 procs, for 300 calls:
http://service-python/env
Aggregated Sleep Time : count 296 avg -50.177311 +/- 36.53 min -130.220001356
max -0.261570142 sum -14852.4839
# range, mid point, percentile, count
>= -130.22 <= -0.26157 , -65.2408 , 100.00, 296
# target 50% -65.4611
WARNING 100.00% of sleep were falling behind
Aggregated Function Time : count 300 avg 1.8540739 +/- 1.99 min 0.075693251
max 7.331838352 sum 556.222183
# target 50% 0.45
# target 75% 3.13333
# target 90% 5.07522
# target 99% 7.10618
# target 99.9% 7.30927
Sockets used: 6 (for perfect keepalive, would be 4)
```

```
Code 200 : 298 (99.3 %)
Code 503 : 2 (0.7 %)
All done 300 calls (plus 0 warmup) 1854.074 ms avg, 2.1 qps
```

6）在 Prometheus UI 上查看指标数据。

访问地址 http://11.11.11.111:32141/，使用 'istio_request_count' 条件查询创建的指标数据，可以在 Console 中看到如下的指标数据：

```
    istio_request_count{destination_namespace="default",destination_service="service-go",destination_version="v1",instance="10.244.2.6:42422",job="istio-mesh",response_code="200"}           272
    istio_request_count{destination_namespace="default",destination_service="service-go",destination_version="v2",instance="10.244.2.6:42422",job="istio-mesh",response_code="200"}           282
    ...
    istio_request_count{destination_namespace="default",destination_service="service-python",destination_version="v2",instance="10.244.2.6:42422",job="istio-mesh",response_code="200"}       256
    istio_request_count{destination_namespace="istio-system",destination_service="istio-policy",destination_version="unknown",instance="10.244.2.6:42422",job="istio-mesh",response_code="200"}      38
    istio_request_count{destination_namespace="istio-system",destination_service="istio-telemetry",destination_version="unknown",instance="10.244.2.6:42422",job="istio-mesh",response_code="200"}  862
```

查看图表数据，如图 11-3 所示。

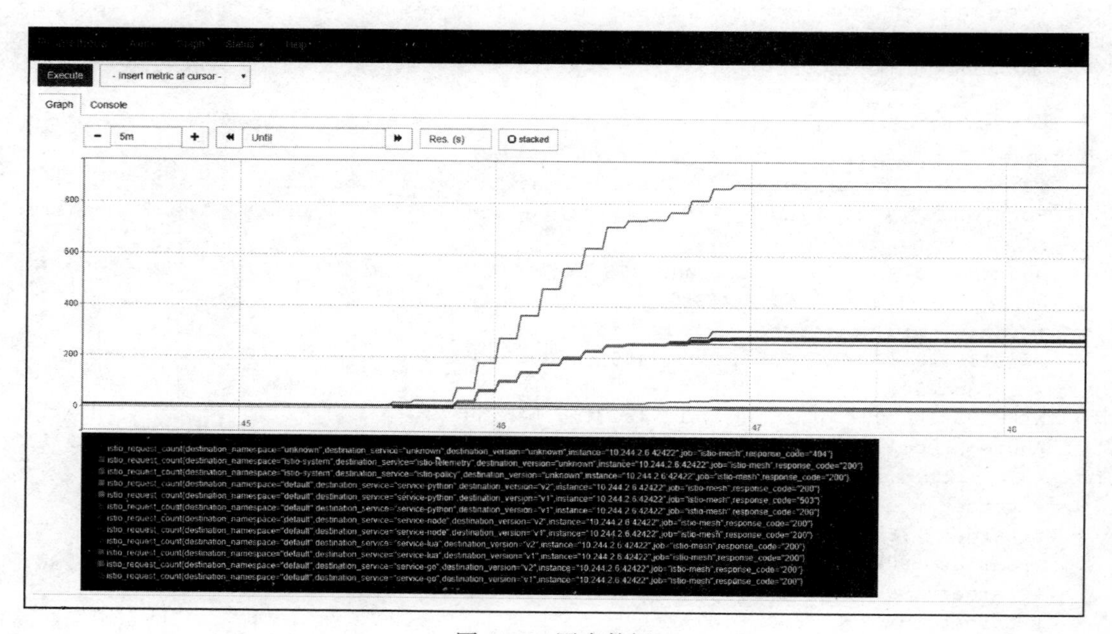

图 11-3　图表数据

使用如下的条件查询 service-go 服务的请求数据：

```
istio_request_count{destination_service="service-python", destination_version="v1",response_code="200"}
```

可以在 Console 看到如下的数据：

```
istio_request_count{destination_namespace="default",destination_service="service-python",destination_version="v1",instance="10.244.2.6:42422",job="istio-mesh",response_code="200"}  307
```

7）清理：

```
$ kubectl delete -f istio/telemetry/metric-http-request-count.yaml
```

8）创建服务 TCP 指标收集规则：

```
$ kubectl apply -f istio/telemetry/metric-tcp-data-size.yaml
```

9）部署 Redis 服务：

```
$ kubectl apply -f kubernetes/redis-server.yaml
```

10）部署 service-redis 服务：

```
$ kubectl apply -f service/redis/service-redis.yaml
```

11）并发请求服务：

```
$ kubectl exec fortio -c fortio /usr/local/bin/fortio -- load -curl http://service-redis/env
HTTP/1.1 200 OK
content-type: text/plain; charset=utf-8
date: Fri, 18 Jan 2019 13:47:17 GMT
x-envoy-upstream-service-time: 661
server: envoy
transfer-encoding: chunked

800
# Server
redis_version:5.0.1
redis_git_sha1:00000000
redis_git_dirty:0
redis_build_id:8a9d320088384235
redis_mode:standalone
os:Linux 3.10.0-693.5.2.el7.x86_64 x86_64
arch_bits:64
...
# Clients
connected_clients:1
client_recent_max_input_buffer:0
client_recent_max_output_buffer:0
```

```
blocked_clients:0

# Memory
used_memory:853752
...

$ kubectl exec fortio -c fortio /usr/local/bin/fortio -- load -qps 10 -n 100 -loglevel Error http://service-redis/env
13:47:28 I logger.go:97> Log level is now 4 Error (was 2 Info)
Fortio 1.0.1 running at 10 queries per second, 2->2 procs, for 100 calls: http://service-redis/env
Aggregated Function Time : count 100 avg 0.010159774 +/- 0.007213 min 0.004142955 max 0.043500525 sum 1.01597743
# target 50% 0.00816667
# target 75% 0.0105
# target 90% 0.0148
# target 99% 0.0426254
# target 99.9% 0.043413
Sockets used: 4 (for perfect keepalive, would be 4)
Code 200 : 100 (100.0 %)
All done 100 calls (plus 0 warmup) 10.160 ms avg, 10.0 qps
```

12）在 Prometheus UI 上查看指标数据。

访问地址 http://11.11.11.111:32141/，使用 'istio_tcp_sent_bytes' 条件查询创建的指标数据，可以在 Console 中看到如下的指标数据：

```
istio_tcp_sent_bytes{destination_service="redis",destination_version="v1",instance="10.244.2.6:42422",job="istio-mesh",source_service="service-redis",source_version="v1"}  326355
```

使用 'istio_tcp_received_bytes' 条件查询创建的指标数据，可以在 Console 中看到如下的指标数据：

```
istio_tcp_received_bytes{destination_service="redis",destination_version="v1",instance="10.244.2.6:42422",job="istio-mesh",source_service="service-redis",source_version="v1"}  1400
```

13）清理：

```
$ kubectl delete -f service/redis/service-redis.yaml
$ kubectl delete -f kubernetes/redis-server.yaml
$ kubectl delete -f istio/telemetry/metric-tcp-data-size.yaml
$ kubectl delete -f kubernetes/fortio.yaml
$ kubectl delete -f kubernetes/istio-prometheus-service.yaml
```

11.3 日志收集

Istio 提供了开箱即用的日志收集组件，可以把日志输出到 Mixer 组件上的标准输出或者存储到 Mixer 组件上的文件里。也可以把日志发送到 Fluentd 组件，然后收集存储到日志中心，用于检索分析日志，生产环境推荐使用这种方式来收集日志。

收集 HTTP 协议服务的请求日志输出到标准输出示例：

```
 1 apiVersion: config.istio.io/v1alpha2
 2 kind: logentry
 3 metadata:
 4   name: myaccesslog
 5   namespace: istio-system
 6 spec:
 7   severity: '"Default"'
 8   timestamp: request.time
 9   variables:
10     source_ip: source.ip | ip("0.0.0.0")
11     destination_ip: destination.ip | ip("0.0.0.0")
12     source_user: source.principal | ""
13     method: request.method | ""
14     url: request.path | ""
15     protocol: request.scheme | "http"
16     response_code: response.code | 0
17     response_size: response.size | 0
18     request_size: request.size | 0
19     latency: response.duration | "0ms"
20   monitored_resource_type: '"UNSPECIFIED"'
21 ---
22 apiVersion: config.istio.io/v1alpha2
23 kind: stdio
24 metadata:
25   name: myaccessloghandler
26   namespace: istio-system
27 spec:
28  severity_levels:
29    info: 1
30  outputAsJson: true
31 ---
32 apiVersion: config.istio.io/v1alpha2
33 kind: rule
34 metadata:
35   name: myaccesslog-logstdio
36   namespace: istio-system
37 spec:
38   match: "true"
39   actions:
40    - handler: myaccessloghandler.stdio
41      instances:
42    - myaccesslog.logentry
```

第 1 ～ 20 行定义了名为了 myaccesslog 的 logentry 实例。severity 表示收集的日志级别为 Default，此字段供支持日志级别的适配器使用。timestamp 表示使用请求的时间作为日志的时间。variables 定义了日志收集的数据。monitored_resource_type 定义了被监控资源的类型为 "UNSPECIFIED"。

第 22 ～ 30 行定义了名为 myaccessloghandler 的 stdio 适配器。stdio 适配器会把收集到的数据输出到 Mixer 组件的标准输出，severity_levels 定义了日志的级别为 info，outputAsJson 设置为 true 表示以 JSON 的格式输出日志，可以通过设置 outputLevel 字段指定输出的级别。可以通过更多的配置选项输出日志到文件中并自动轮转日志，此次实验并没有使用这种方法，有兴趣的读者可以查看官方文档[⊖]。

第 32 ～ 42 行定义了名为 myaccesslog-logstdio 的规则，表明把 myaccesslog 实例收集到的日志发送给 myaccessloghandler 适配器处理。

收集服务请求日志发送到 Fluentd 组件示例：

```
 1 apiVersion: config.istio.io/v1alpha2
 2 kind: logentry
 3 metadata:
 4   name: newlog
 5   namespace: istio-system
 6 spec:
 7   severity: '"info"'
 8   timestamp: request.time
 9   variables:
10     source: source.labels["app"] | source.workload.name | "unknown"
11     user: source.principal | "unknown"
12     destination:destination.labels["app"] |destination.service.name | "unknown"
13     response_code: response.code | 0
14     response_size: response.size | 0
15     latency: response.duration | "0ms"
16   monitored_resource_type: '"UNSPECIFIED"'
17 ---
18 apiVersion: config.istio.io/v1alpha2
19 kind: fluentd
20 metadata:
21   name: fluentdhandler
22   namespace: istio-system
23 spec:
24   address: "fluentd-es.logging:24224"
25 ---
26 apiVersion: config.istio.io/v1alpha2
27 kind: rule
28 metadata:
29   name: newlogtofluentd
30   namespace: istio-system
31 spec:
32   match: "true"
```

⊖ 官方文档地址为 https://istio.io/docs/reference/config/policy-and-telemetry/adapters/stdio/。

```
33  actions:
34   - handler: fluentdhandler.fluentd
35     instances:
36     - newlog.logentry
```

第 1 ～ 16 行定义了名为 newlog 的 logentry 实例。与收集 HTTP 协议服务的请求日志定义类似，在此不再赘述。

第 18 ～ 24 行定义了名为 fluentdhandler 的 fluentd 适配器。address 定义了 fluentd 服务地址。

第 26 ～ 36 行定义了名为 newlogtofluentd 的规则，表明把 newlog 实例收集到的日志数据发送到 fluentdhandler 适配器。

【实验一】 测试日志输出到 Mixer 标准输出

1）创建用于请求的 Pod：

```
$ kubectl apply -f kubernetes/fortio.yaml
```

2）创建服务日志收集到 Mixer 标准输出的规则：

```
$ kubectl apply -f istio/telemetry/log-http-access-log.yaml
```

3）打开新终端查看收集到的服务日志：

```
$ kubectl -n istio-system logs -f $(kubectl -n istio-system get pods -l istio-mixer-type=telemetry -o jsonpath='{.items[0].metadata.name}') -c mixer | grep \"instance\":\"myaccesslog.logentry.istio-system\"
```

4）另外打开一个终端并发请求服务：

```
$ kubectl exec fortio -c fortio /usr/local/bin/fortio -- load -curl http://service-python/env
HTTP/1.1 200 OK
content-type: application/json
content-length: 176
server: envoy
date: Fri, 18 Jan 2019 13:53:26 GMT
x-envoy-upstream-service-time: 595

{"message":"python v2","upstream":[{"message":"lua v2","response_time":0.1},{"message":"node v1","upstream":[{"message":"go v2","response_time":"0.30"}],"response_time":0.76}]}

$ kubectl exec fortio -c fortio /usr/local/bin/fortio -- load -qps 10 -n 100 -loglevel Error http://service-python/env
14:01:11 I logger.go:97> Log level is now 4 Error (was 2 Info)
Fortio 1.0.1 running at 10 queries per second, 2->2 procs, for 100 calls: http://service-python/env
Aggregated Sleep Time : count 96 avg -3.8814565 +/- 2.688 min -9.370066353 max
```

```
0.046896141 sum -372.619823
    # range, mid point, percentile, count
    >= -9.37007 <= -0.001 , -4.68553 , 98.96, 95
    > 0.044 <= 0.0468961 , 0.0454481 , 100.00, 1
    # target 50% -4.68553
    WARNING 98.96% of sleep were falling behind
    Aggregated Function Time : count 100 avg 0.73897862 +/- 0.4143 min 0.097562604
max 1.980383317 sum 73.8978625
    # target 50%  0.669231
    # target 75%  1
    # target 90%  1.58823
    # target 99%  1.94117
    # target 99.9% 1.97646
    Sockets used: 5 (for perfect keepalive, would be 4)
    Code 200 : 99 (99.0 %)
    Code 503 : 1 (1.0 %)
    All done 100 calls (plus 0 warmup) 738.979 ms avg, 5.1 qps
```

5）清理：

```
$ kubectl delete -f istio/telemetry/log-http-access-log.yaml
```

【实验二】 测试日志输出到 Fluentd 标准输出

由于机器性能问题，本次实验环境只使用Fluentd收集日志，并输出到Fluentd的标准输出上，在生产环境中，把日志通过Fluentd收集，然后保存到ElasticSearch集群中，通过Kibana在Web中搜索查看分析日志。在机器性能足够的情况下，可以使用loging-stack.yaml部署EFK日志收集平台来模拟生产环境中的日志收集场景。

1）部署Fluentd：

```
$ kubectl apply -f kubernetes/loging-fluentd-stdout.yaml

$ kubectl get pod -n logging
NAME                          READY   STATUS    RESTARTS   AGE
fluentd-es-6cd547b4bc-sjqmk   1/1     Running   0          3m13s
```

2）创建服务日志发送到Fluentd的日志收集规则：

```
$ kubectl apply -f istio/telemetry/log-fluentd.yaml
```

3）打开终端查看服务日志：

```
$ kubectl -n logging logs -f $(kubectl -n logging get pods -l app=fluentd-es
-o jsonpath='{.items[0].metadata.name}') | grep newlog.logentry.istio-system
```

4）另外打开一个终端并发请求服务：

```
$ kubectl exec fortio -c fortio /usr/local/bin/fortio -- load -curl http://
service-python/env
```

```
HTTP/1.1 200 OK
content-type: application/json
content-length: 178
server: envoy
date: Fri, 18 Jan 2019 14:08:10 GMT
x-envoy-upstream-service-time: 667

{"message":"python v1","upstream":[{"message":"lua v1","response_time":1.03},{"message":"node v2","response_time":1.95,"upstream":[{"message":"go v2","response_time":"1.04"}]}]}

$ kubectl exec fortio -c fortio /usr/local/bin/fortio -- load -qps 10 -n 100 -loglevel Error http://service-python/env
14:08:51 I logger.go:97> Log level is now 4 Error (was 2 Info)
Fortio 1.0.1 running at 10 queries per second, 2->2 procs, for 100 calls: http://service-python/env
Aggregated Sleep Time : count 96 avg -3.8814565 +/- 2.688 min -9.370066353 max 0.046896141 sum -372.619823
# range, mid point, percentile, count
>= -9.37007 <= -0.001 , -4.68553 , 98.96, 95
> 0.044 <= 0.0468961 , 0.0454481 , 100.00, 1
# target 50% -4.68553
WARNING 98.96% of sleep were falling behind
Aggregated Function Time : count 100 avg 0.73897862 +/- 0.4143 min 0.097562604 max 1.980383317 sum 73.8978625
# target 50% 0.669231
# target 75% 1
# target 90% 1.58823
# target 99% 1.94117
# target 99.9% 1.97646
Sockets used: 5 (for perfect keepalive, would be 4)
Code 200 : 99 (99.0 %)
Code 503 : 1 (1.0 %)
All done 100 calls (plus 0 warmup) 738.979 ms avg, 5.1 qps
```

5）清理：

```
$ kubectl delete -f kubernetes/loging-fluentd-stdout.yaml
$ kubectl delete -f istio/telemetry/log-fluentd.yaml
```

【实验三】 测试日志输出到 EFK

进行此实验步骤时，每台虚拟机分配了 4 核 CPU 和 4G 内存，资源不够的读者可以跳过此步骤实验。

1）部署 EFK 日志收集平台：

```
$ kubectl apply -f kubernetes/loging-stack.yaml
```

2）创建服务日志发送到 Fluentd 日志收集规则：

```
$ kubectl apply -f istio/telemetry/log-fluentd.yaml
```

3）并发请求服务：

```
$ kubectl exec fortio -c fortio /usr/local/bin/fortio -- load -curl http://
service-python/env
HTTP/1.1 200 OK
content-type: application/json
content-length: 178
server: envoy
date: Fri, 18 Jan 2019 14:10:25 GMT
x-envoy-upstream-service-time: 179

{"message":"python v1","upstream":[{"message":"lua v1","response_
time":0.15},{"message":"node v2","response_time":0.51,"upstream":[{"message":"go
v1","response_time":"0.40"}]}]}

$ kubectl exec fortio -c fortio /usr/local/bin/fortio -- load -qps 10 -n 300
-loglevel Error http://service-python/env
14:11:50 I logger.go:97> Log level is now 4 Error (was 2 Info)
Fortio 1.0.1 running at 10 queries per second, 2->2 procs, for 300 calls:
http://service-python/env
Aggregated Sleep Time : count 296 avg -33.639106 +/- 23.5 min -84.517839391
max -0.225980787 sum -9957.17528
# range, mid point, percentile, count
>= -84.5178 <= -0.225981 , -42.3719 , 100.00, 296
# target 50% -42.5148
WARNING 100.00% of sleep were falling behind
Aggregated Function Time : count 300 avg 1.4654142 +/- 1.06 min 0.009331473
max 7.796983286 sum 439.624248
# target 50%   1.26667
# target 75%   1.98095
# target 90%   2.81132
# target 99%   6.5
# target 99.9% 7.70789
Sockets used: 10 (for perfect keepalive, would be 4)
Code 200 : 293 (97.7 %)
Code 503 : 7 (2.3 %)
All done 300 calls (plus 0 warmup) 1465.414 ms avg, 2.6 qps
```

4）在Kibana上查看日志数据。

访问地址http://11.11.11.111:32142/，设置并查看日志。选择Management导航栏，点击Index Patterns创建Index Pattern，如图11-4所示。

填入logstash-*匹配ElasticSearch中存储的日志索引，如图11-5所示。

选择@timestamp作为时间字段，如图11-6所示。

在Discover导航栏查看日志，如图11-7所示。

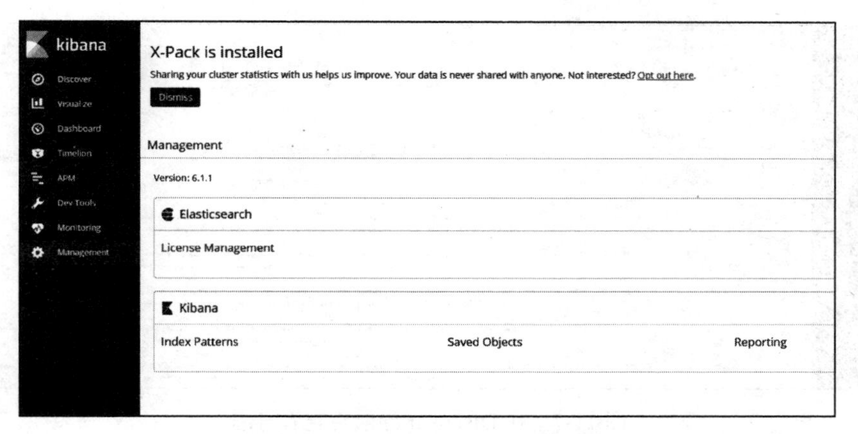

图 11-4　在 Kibana 上查看日志数据

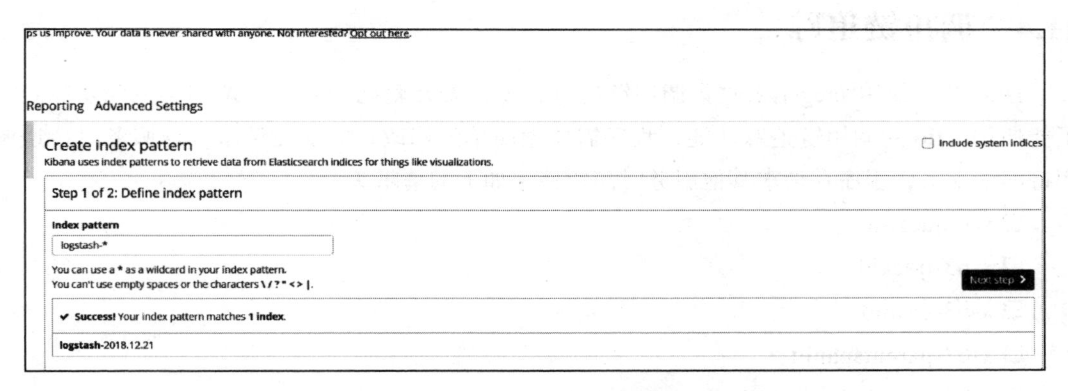

图 11-5　匹配 ElasticSearch 中存储的日志索引

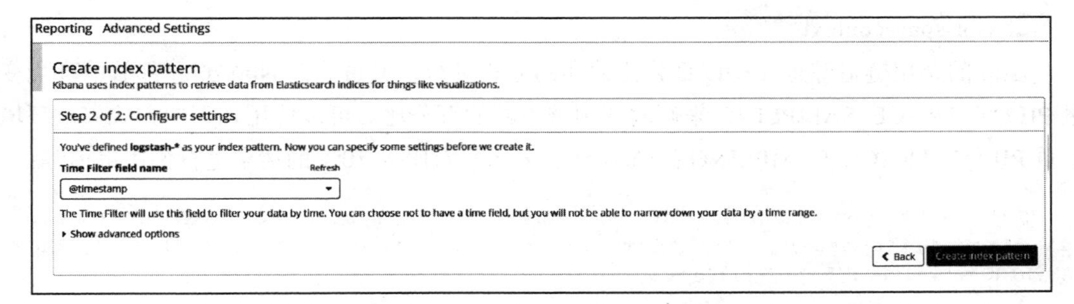

图 11-6　选择时间字段

5）清理：

```
$ kubectl delete -f istio/telemetry/log-fluentd.yaml
$ kubectl delete -f kubernetes/loging-stack.yaml
$ kubectl delete -f kubernetes/fortio.yaml
```

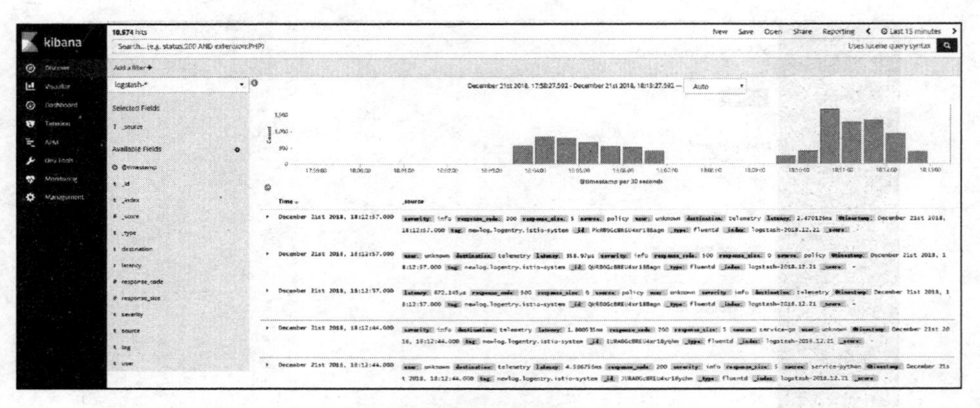

图 11-7　在 Discover 导航栏查看日志

11.4　调用链追踪

Istio 默认使用 Jaeger 来收集调用链信息，默认安装就已经启用了调用链追踪功能。为了能使用 Istio 的调用链追踪功能，程序需要做简单的修改，需要在请求其他服务时传递调用链的请求头，程序在请求其他服务时需要传递如下的请求头：

- x-request-id
- x-b3-traceid
- x-b3-spanid
- x-b3-parentspanid
- x-b3-sampled
- x-b3-flags
- x-ot-span-context

Istio 的调用链追踪也支持以百分比抽样的方式进行，通过修改 istio-pilot 部署的环境变量 PILOT_TRACE_SAMPLING 来修改调用链追踪的百分比。可以使用如下的方式修改环境变量 PILOT_TRACE_SAMPLING 的值为 50，表示只对服务 50% 的请求进行调用链追踪：

```
$ kubectl patch deploy istio-pilot -n istio-system -p '{"spec":{"template":{"spec":{"containers":[{"name":"discovery","env":[{"name":"PILOT_TRACE_SAMPLING","value":"50"}]}]}}}}'
```

Istio 的调用链追踪功能支持使用 HTTP/1.1、HTTP/2.0 和 gRPC 协议的服务，这已经能够覆盖大部分服务使用的协议，如果使用的是其他的协议，Istio 将无法实现自动的调用链追踪功能。

虽然 Istio 能自动实现调用链追踪功能，但是你无法在调用链中的 span 中添加自定义的数据，比如用户相关的信息。如果你想在 span 中添加自定义的数据，或者在程序中添加对其他协议的支持，可以参考如下两篇文章：

❑ https://aspenmesh.io/2018/07/distributed-tracing-istio-and-your-applications/

❑ https://aspenmesh.io/2018/04/tracing-grpc-with-istio/

【实验】

1）查看是否已经开启调用链追踪功能：

```
$ kubectl get deploy istio-tracing -n istio-system
NAME             DESIRED      CURRENT        UP-TO-DATE     AVAILABLE     AGE
istio-tracing    1            1              1              1             21d

$ kubectl get svc tracing -n istio-system
NAME             TYPE         CLUSTER-IP     EXTERNAL-IP    PORT(S)       AGE
tracing          ClusterIP    10.103.169.41  <none>         80/TCP        21d

$ kubectl get svc zipkin -n istio-system
NAME             TYPE         CLUSTER-IP     EXTERNAL-IP    PORT(S)       AGE
zipkin           ClusterIP    10.109.10.212  <none>         9411/TCP      21d
```

上面的命令结果表示已经开启了调用链追踪功能。

2）创建用于请求的 Pod：

```
$ kubectl apply -f kubernetes/fortio.yaml
```

3）暴露 Jaeger 查询的 Web 服务：

```
$ kubectl apply -f kubernetes/istio-tracing-service.yaml
```

4）并发请求服务：

```
    $ kubectl exec fortio -c fortio /usr/local/bin/fortio -- load -curl http://
service-python/env
    HTTP/1.1 200 OK
    content-type: application/json
    content-length: 177
    server: envoy
    date: Fri, 18 Jan 2019 14:17:32 GMT
    x-envoy-upstream-service-time: 804

    {"message":"python v2","upstream":[{"message":"lua v2","response_
time":0.11},{"message":"node v1","upstream":[{"message":"go v1","response_
time":"0.01"}],"response_time":0.11}]}

    $ kubectl exec fortio -c fortio /usr/local/bin/fortio -- load -qps 10 -n 100
-loglevel Error http://service-python/env
    14:18:01 I logger.go:97> Log level is now 4 Error (was 2 Info)
    Fortio 1.0.1 running at 10 queries per second, 2->2 procs, for 100 calls:
http://service-python/env
    Aggregated Sleep Time : count 96 avg -29.406556 +/- 17.84 min -71.182683879
max -0.663793146 sum -2823.02941
    # range, mid point, percentile, count
    >= -71.1827 <= -0.663793 , -35.9232 , 100.00, 96
```

```
# target 50% -36.2944
WARNING 100.00% of sleep were falling behind
Aggregated Function Time : count 100 avg 2.9197529 +/- 1.781 min 0.081594508
max 6.892183204 sum 291.975288
# target 50%   2.5625
# target 75%   4.38462
# target 90%   5.77913
# target 99%   6.78088
# target 99.9% 6.88105
Sockets used: 8 (for perfect keepalive, would be 4)
Code 200 : 96 (96.0 %)
Code 503 : 4 (4.0 %)
All done 100 calls (plus 0 warmup) 2919.753 ms avg, 1.2 qps
```

5）创建 Web 访问的路由规则：

```
$ kubectl apply -f istio/route/gateway-js-v1.yaml
```

6）浏览器访问。访问地址 http://11.11.11.112:31380/，多次刷新并点击“发射”按钮，如图 11-8 所示。

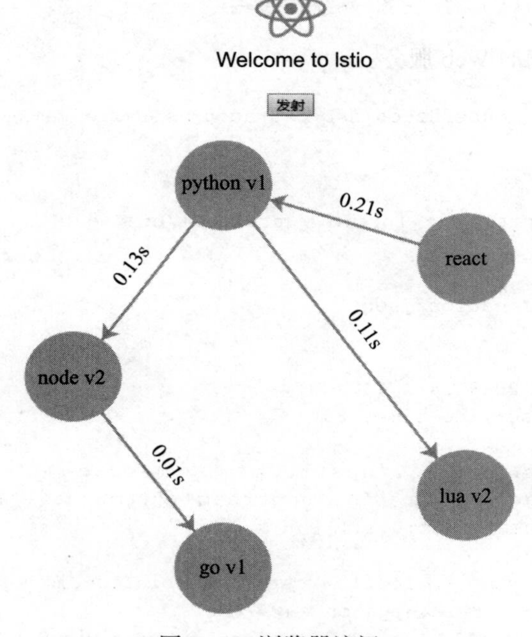

图 11-8 浏览器访问

7）在 Jaeger UI 上查看指标数据。访问地址 http://11.11.11.111:32144/，查看 service-python 服务调用链，如图 11-9 所示。

查询 service-python 服务响应时间大于 3s 的请求调用链，如图 11-10 所示。

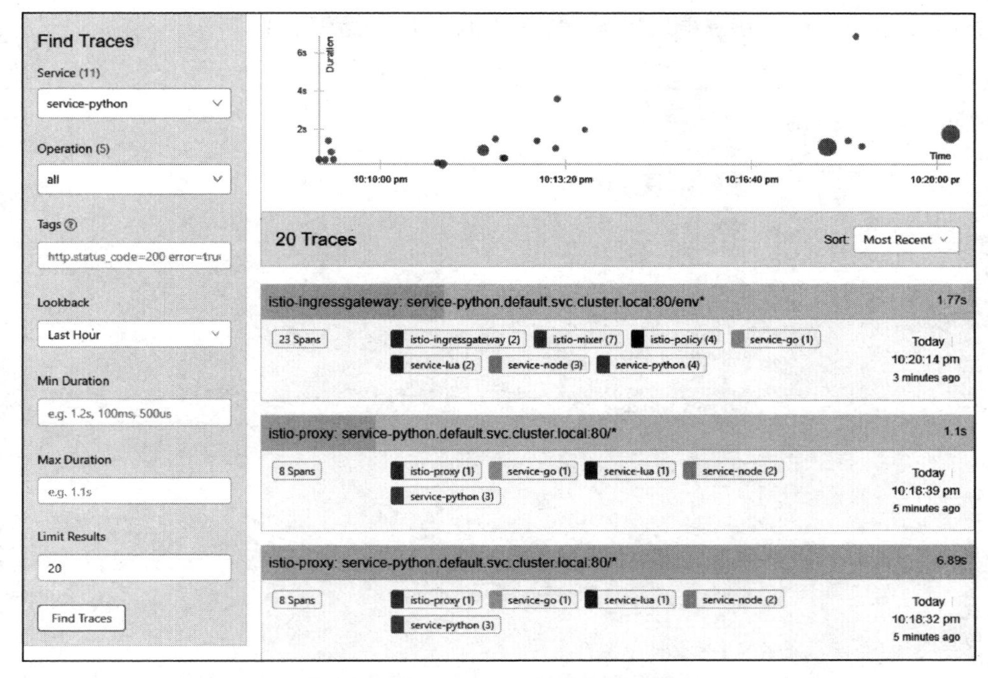

图 11-9　在 Jaeger UI 上查看指标数据

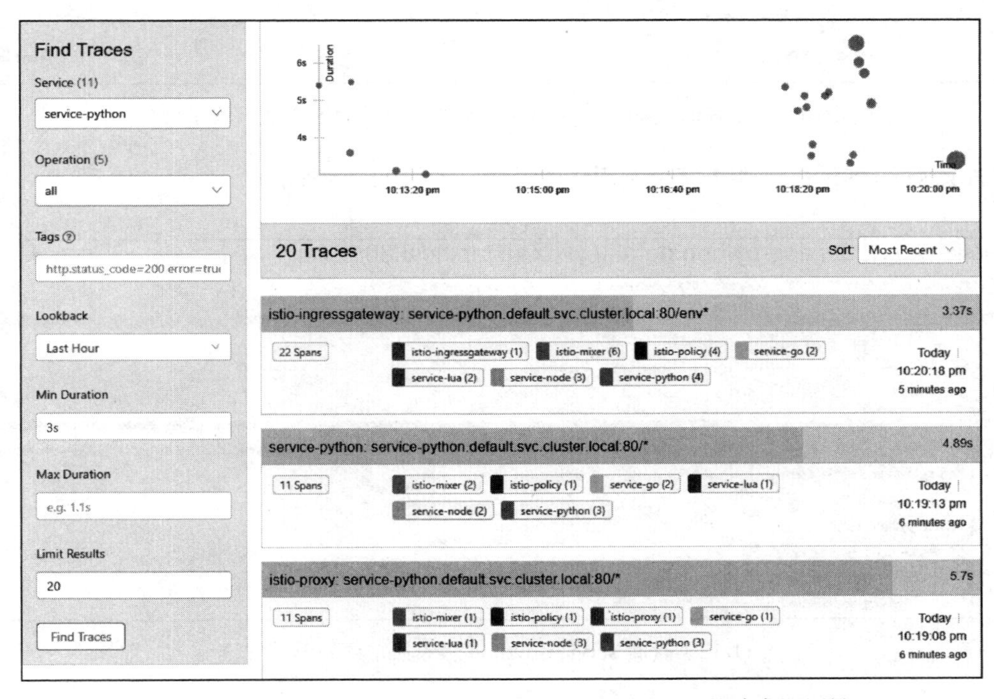

图 11-10　查询 service-python 服务响应时间大于 3s 的请求调用链

查询 service-python 服务响应码为 503 请求调用链，如图 11-11 所示。

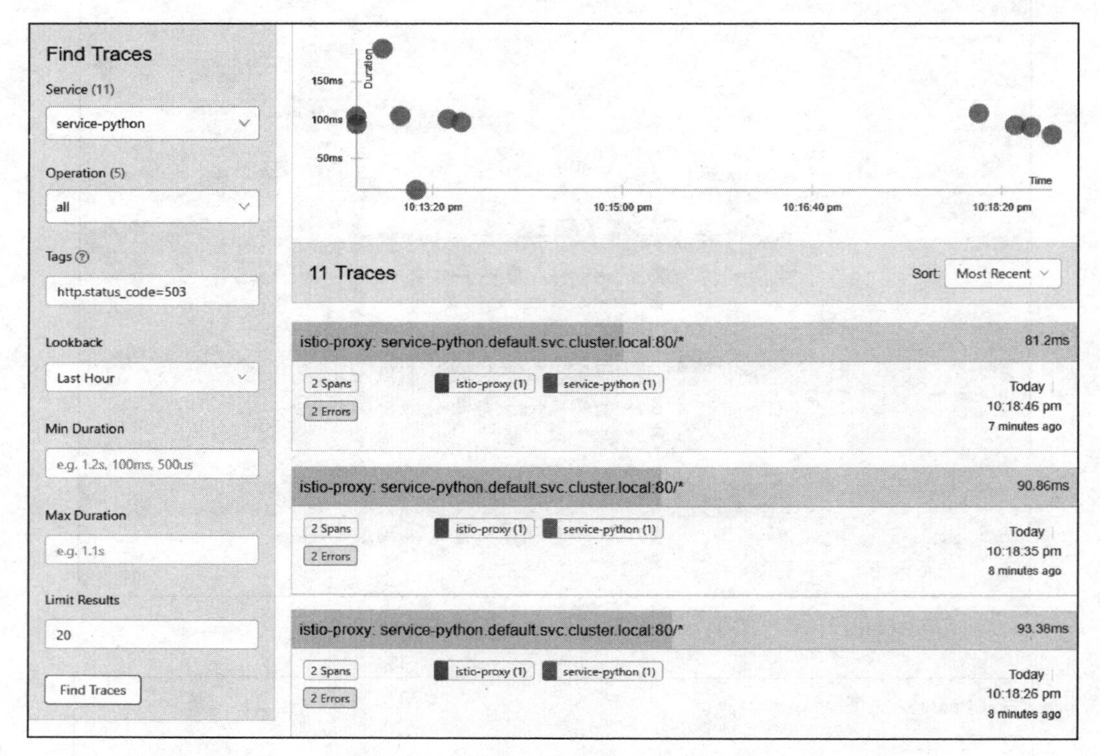

图 11-11 查询 service-python 服务响应码为 503 请求调用链

查看 service-python 服务某个调用链的详情，如图 11-12 所示。

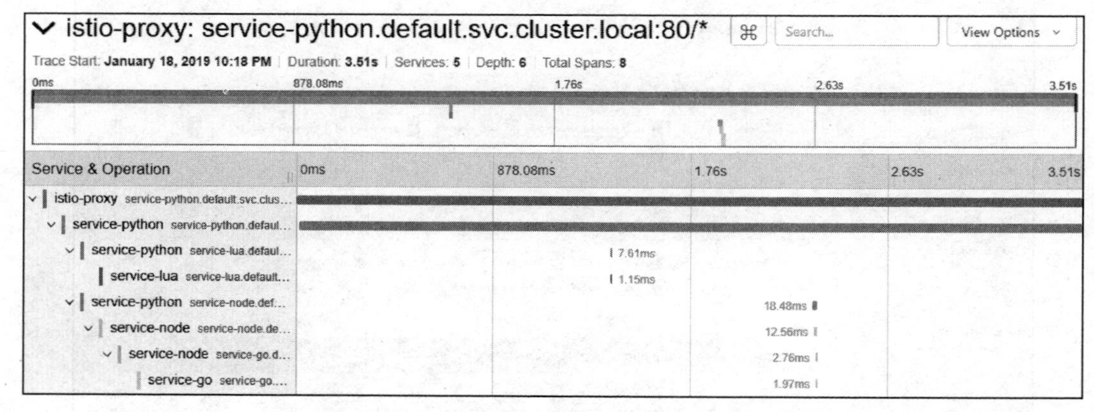

图 11-12 查看 service-python 服务某个调用链的详情

查看服务调用链中服务的请求详情，如图 11-13 所示。

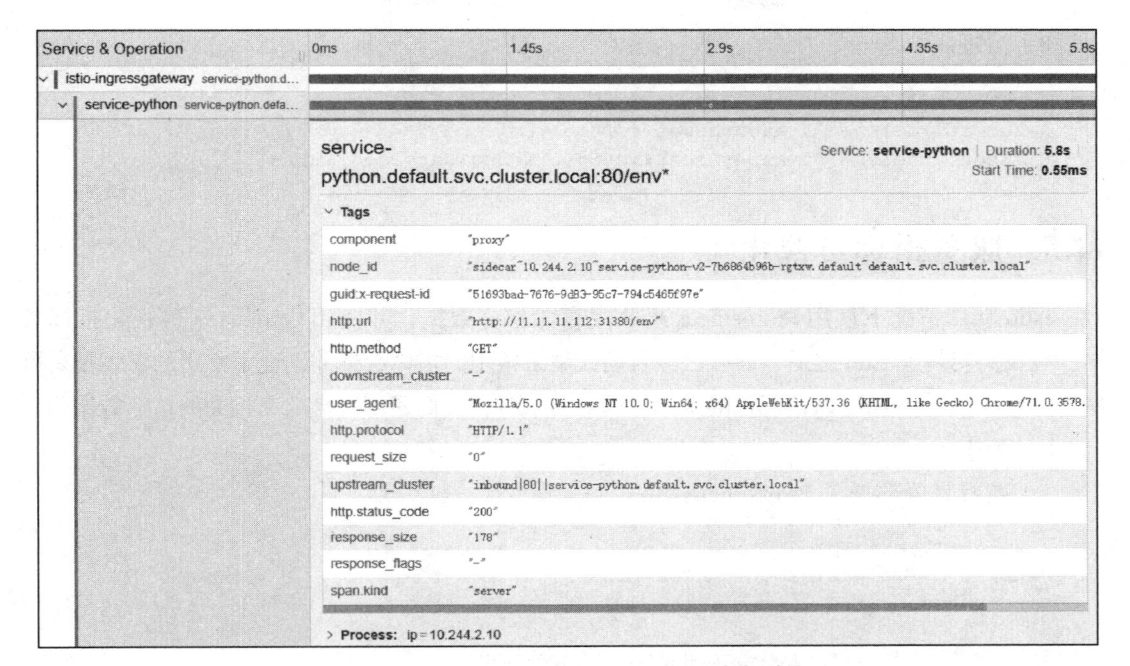

图 11-13　查看服务调用链中服务的请求详情

查看网格中的服务调用树，如图 11-14 所示。

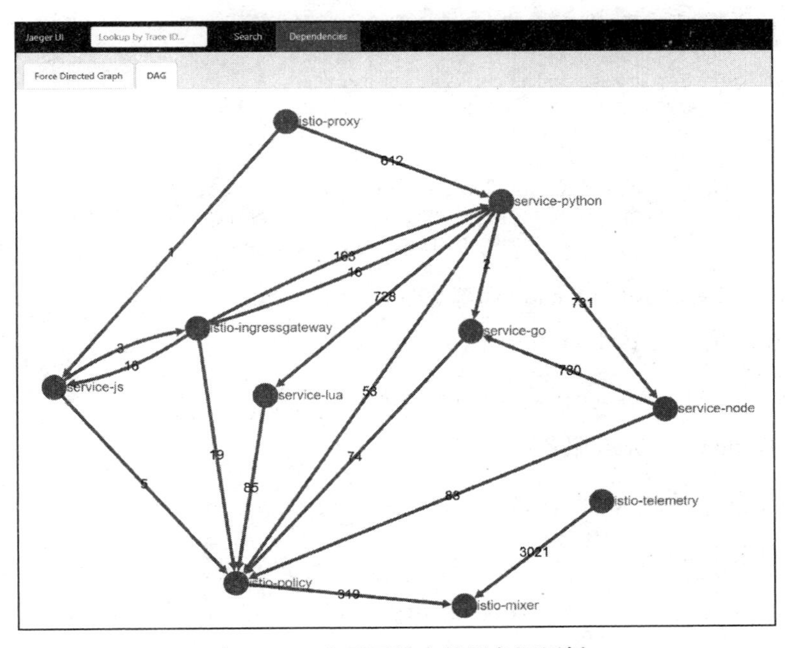

图 11-14　查看网格中的服务调用树

8）清理：

```
$ kubectl delete -f istio/route/gateway-js-v1.yaml
$ kubectl delete -f kubernetes/fortio.yaml
$ kubectl delete -f kubernetes/istio-tracing-service.yaml
```

11.5 服务指标可视化

Istio默认情况下使用Prometheus来收集存储指标数据，并使用Grafana来可视化指标数据。当我们安装好Istio之后，Istio已经自动收集了常用的指标数据并创建了常用的仪表板来可视化指标数据。当Istio默认配置的图无法满足需求时，我们还可以自定义相关的指标数据图。

【实验】

1）查看是否部署了Prometheus：

```
$ kubectl get deploy prometheus -n istio-system
NAME          DESIRED     CURRENT         UP-TO-DATE    AVAILABLE    AGE
prometheus    1           1               1             1            21d

$ kubectl get svc prometheus -n istio-system
NAME          TYPE        CLUSTER-IP      EXTERNAL-IP   PORT(S)      AGE
prometheus    ClusterIP   10.99.177.169   <none>        9090/TCP     21d
```

上面的命令结果表示Prometheus已经部署。

2）查看是否部署了Grafana：

```
$ kubectl get deploy grafana -n istio-system
NAME          DESIRED   CURRENT   UP-TO-DATE   AVAILABLE   AGE
grafana       1         1               1             1            21d

$ kubectl get svc grafana -n istio-system
NAME          TYPE        CLUSTER-IP      EXTERNAL-IP     PORT(S)     AGE
grafana       ClusterIP   10.99.125.166   <none>          3000/TCP    21d
```

如上的命令结果表示Grafana已经部署。

3）创建用于请求的Pod：

```
$ kubectl apply -f kubernetes/fortio.yaml
```

4）暴露Grafana的Web服务：

```
$ kubectl apply -f kubernetes/istio-grafana-service.yaml
```

5）并发请求服务：

```
$ kubectl exec fortio -c fortio /usr/local/bin/fortio -- load -curl http://
service-python/env
HTTP/1.1 200 OK
```

```
content-type: application/json
content-length: 177
server: envoy
date: Fri, 18 Jan 2019 14:36:22 GMT
x-envoy-upstream-service-time: 2317

{"message":"python v1","upstream":[{"message":"lua v2","response_time":0.03},{"message":"node v2","response_time":0.1,"upstream":[{"message":"go v1","response_time":"0.02"}]}]}

$ kubectl exec fortio -c fortio /usr/local/bin/fortio -- load -qps 10 -n 100 -loglevel Error http://service-python/env
14:36:54 I logger.go:97> Log level is now 4 Error (was 2 Info)
Fortio 1.0.1 running at 10 queries per second, 2->2 procs, for 100 calls: http://service-python/env
Aggregated Sleep Time : count 96 avg -57.921022 +/- 33.4 min -131.227050538 max -2.920655649 sum -5560.41808
# range, mid point, percentile, count
>= -131.227 <= -2.92066 , -67.0739 , 100.00, 96
# target 50% -67.7491
WARNING 100.00% of sleep were falling behind
Aggregated Function Time : count 100 avg 5.493663 +/- 2.894 min 0.01501075 max 13.482176793 sum 549.366298
# target 50%      4.78947
# target 75%      7.77778
# target 90%      9.86111
# target 99%     13.0953
# target 99.9%   13.4435
Sockets used: 6 (for perfect keepalive, would be 4)
Code 200 : 98 (98.0 %)
Code 503 : 2 (2.0 %)
All done 100 calls (plus 0 warmup) 5493.663 ms avg, 0.7 qps
```

6）在 Grafana UI 上查看指标数据。访问地址 http://11.11.11.111:32143/，首页如图 11-15 所示。

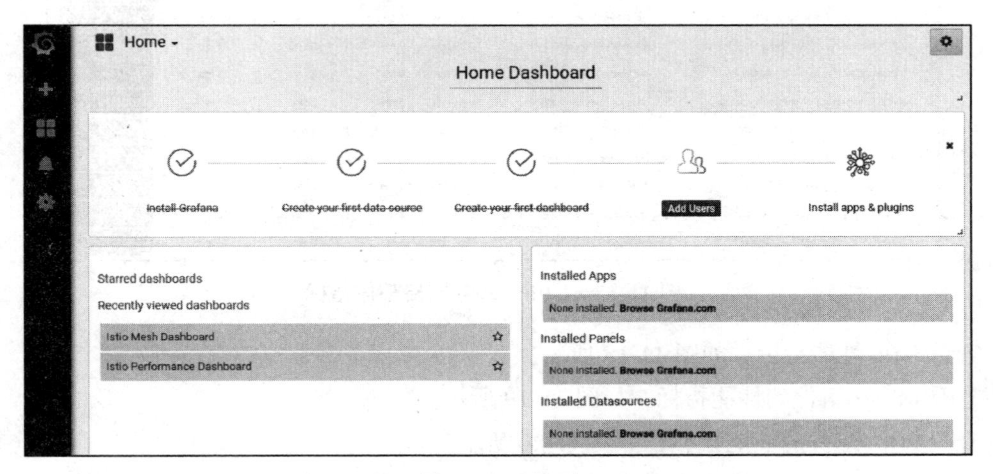

图 11-15　首页

查看 Istio 默认创建的仪表板，Home 下拉菜单的 istio 文件夹下 Dashboard 均为 Istio 默认创建，如图 11-16 所示。

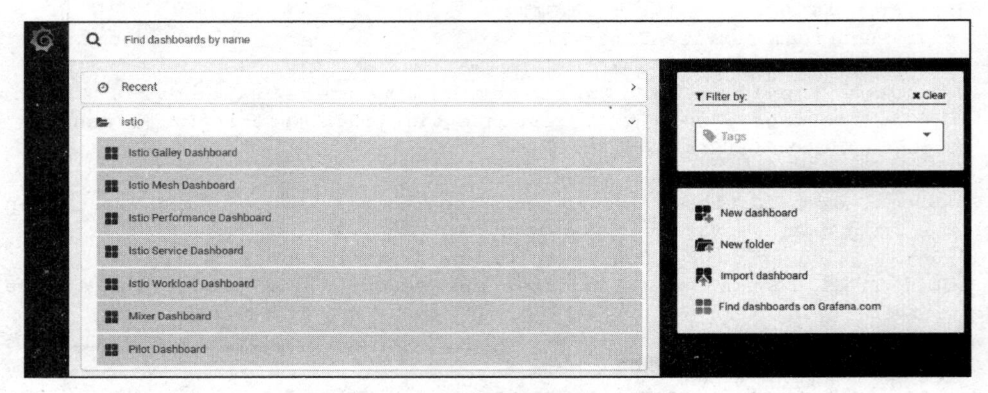

图 11-16 Istio 默认创建的仪表板

查看 Istio 服务网格整体指标，如图 11-17 所示。

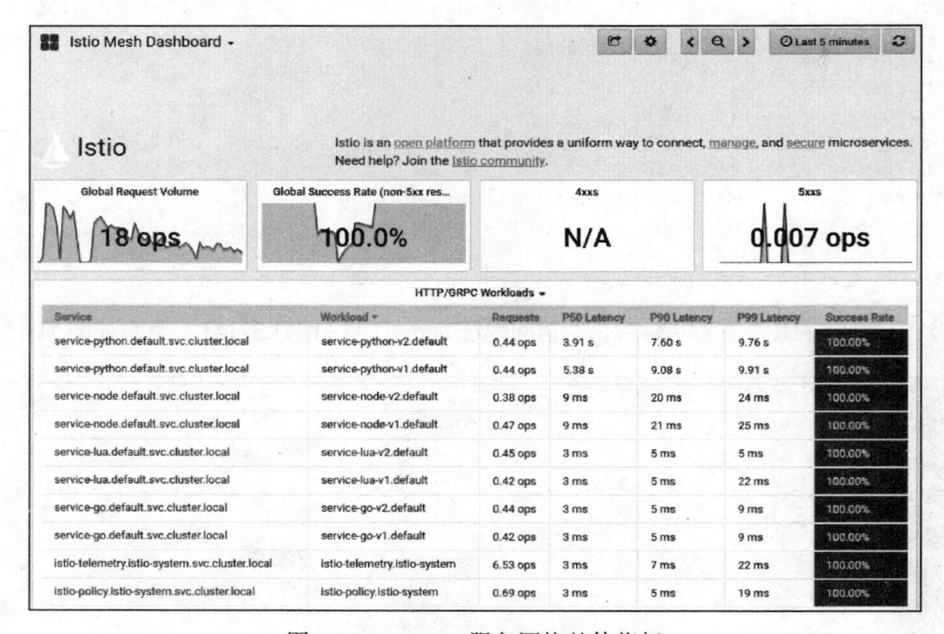

Service	Workload	Requests	P50 Latency	P90 Latency	P99 Latency	Success Rate
service-python.default.svc.cluster.local	service-python-v2.default	0.44 ops	3.91 s	7.60 s	9.76 s	100.00%
service-python.default.svc.cluster.local	service-python-v1.default	0.44 ops	5.38 s	9.08 s	9.91 s	100.00%
service-node.default.svc.cluster.local	service-node-v2.default	0.38 ops	9 ms	20 ms	24 ms	100.00%
service-node.default.svc.cluster.local	service-node-v1.default	0.47 ops	9 ms	21 ms	25 ms	100.00%
service-lua.default.svc.cluster.local	service-lua-v2.default	0.45 ops	3 ms	5 ms	5 ms	100.00%
service-lua.default.svc.cluster.local	service-lua-v1.default	0.42 ops	3 ms	5 ms	22 ms	100.00%
service-go.default.svc.cluster.local	service-go-v2.default	0.44 ops	3 ms	5 ms	9 ms	100.00%
service-go.default.svc.cluster.local	service-go-v1.default	0.42 ops	3 ms	5 ms	9 ms	100.00%
istio-telemetry.istio-system.svc.cluster.local	istio-telemetry.istio-system	6.53 ops	3 ms	7 ms	22 ms	100.00%
istio-policy.istio-system.svc.cluster.local	istio-policy.istio-system	0.69 ops	3 ms	5 ms	19 ms	100.00%

图 11-17 Istio 服务网格整体指标

查看 Istio 性能指标，如图 11-18 所示。

查看 Mixer 组件的性能指标，如图 11-19 所示。

查看 Pilot 组件的性能指标，如图 11-20 所示。

查看 service-python 服务的请求指标，如图 11-21 所示。

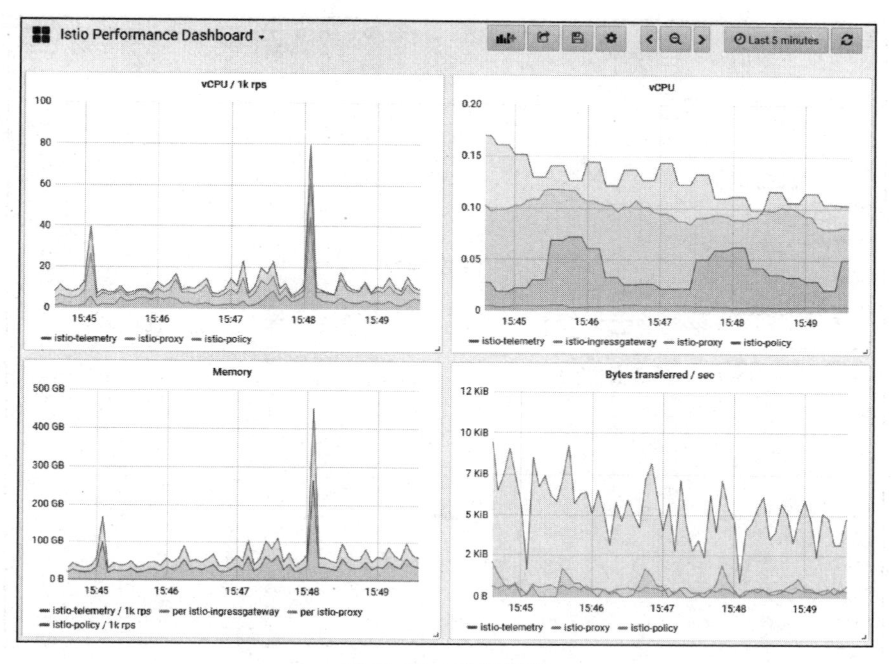

图 11-18　性能指标

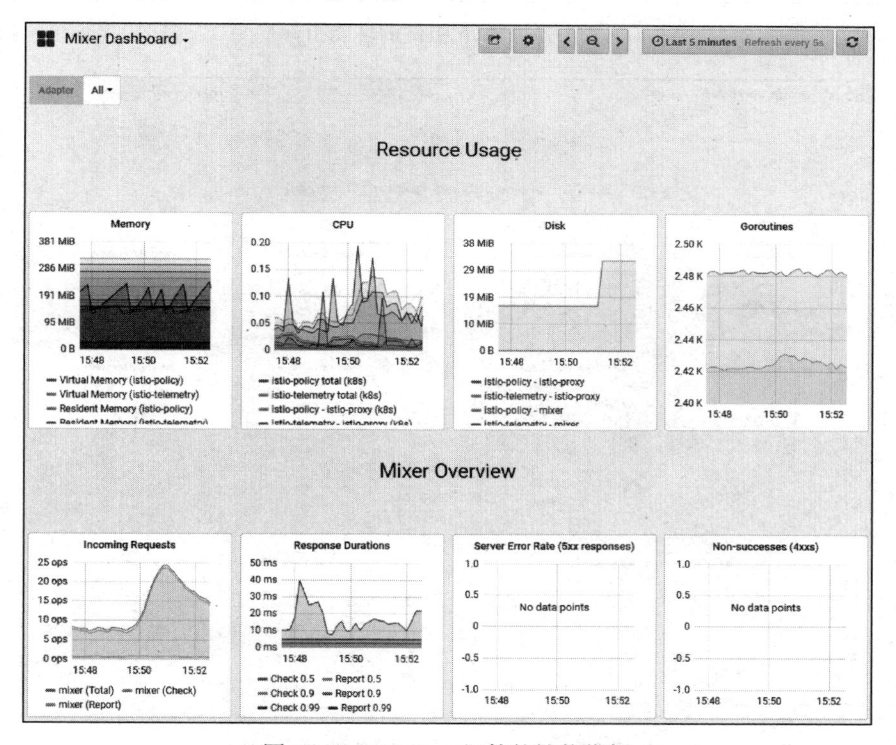

图 11-19　Mixer　组件的性能指标

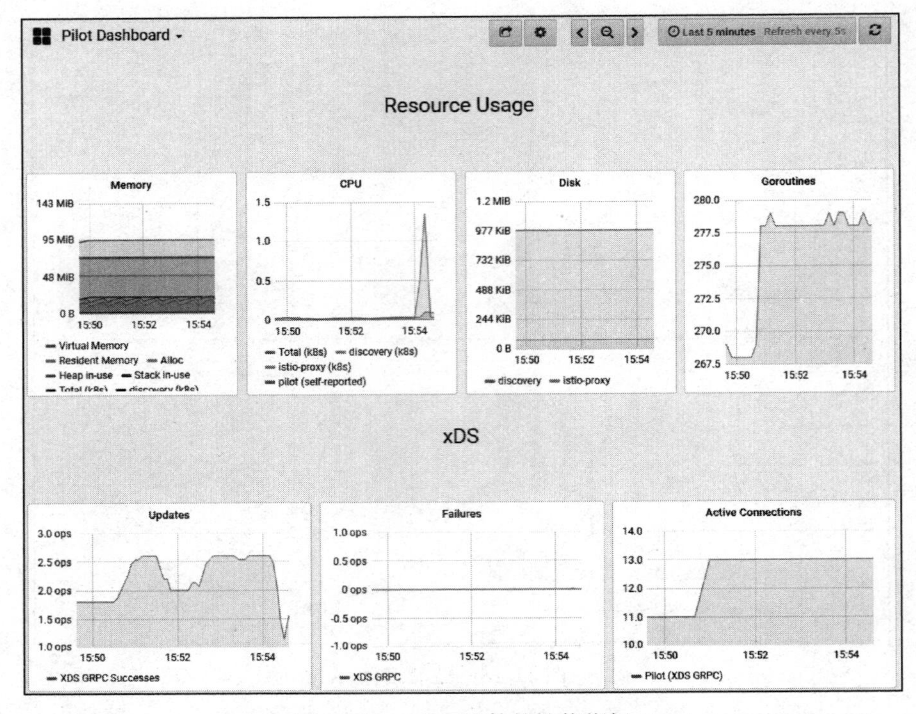

图 11-20 Pilot 组件的性能指标

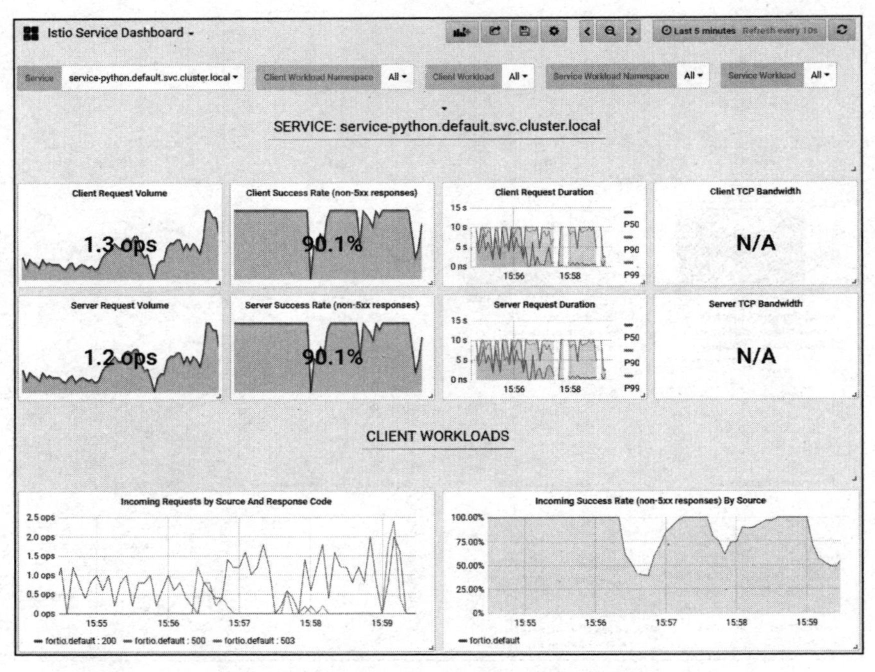

图 11-21 service-python 服务的请求指标

7）清理：

```
$ kubectl delete -f kubernetes/fortio.yaml
$ kubectl delete -f kubernetes/istio-grafana-service.yaml
```

11.6　服务调用树

服务调用树可以帮助我们更好地了解服务间的调用关系，并优化服务间的调用关系，减少服务间不需要的依赖关系，从而使应用更稳定。Istio 默认情况下使用 ServiceGraph 根据 Prometheus 中存储的指标数据来绘制网格内服务调用树。

【实验】

1）查看是否部署了 Prometheus：

```
$ kubectl get deploy prometheus -n istio-system
NAME          DESIRED       CURRENT           UP-TO-DATE      AVAILABLE     AGE
prometheus    1             1                 1               1             21d

$ kubectl get svc prometheus -n istio-system
NAME          TYPE          CLUSTER-IP        EXTERNAL-IP     PORT(S)       AGE
prometheus    ClusterIP     10.99.177.169     <none>          9090/TCP      21d
```

上面的命令结果表示 Prometheus 已经部署。

2）查看是否部署了 ServiceGraph：

```
$ kubectl get deploy servicegraph -n istio-system
NAME          DESIRED       CURRENT           UP-TO-DATE      AVAILABLE     AGE
servicegraph  1             1                 1               1             21d

$ kubectl get svc servicegraph -n istio-system
NAME          TYPE          CLUSTER-IP        EXTERNAL-IP     PORT(S)       AGE
servicegraph  ClusterIP     10.101.107.109    <none>          8088/TCP      21d
```

如上的命令结果表示 ServiceGraph 已经部署。

3）创建用于请求的 Pod：

```
$ kubectl apply -f kubernetes/fortio.yaml
```

4）暴露 ServiceGraph 的 Web 服务：

```
$ kubectl apply -f kubernetes/istio-servicegraph-service.yaml
```

5）并发请求服务：

```
$ kubectl exec fortio -c fortio /usr/local/bin/fortio -- load -curl http://
service-python/env
HTTP/1.1 200 OK
```

```
content-type: application/json
content-length: 178
server: envoy
date: Fri, 18 Jan 2019 14:41:18 GMT
x-envoy-upstream-service-time: 2621

{"message":"python v1","upstream":[{"message":"lua v2","response_time":0.03},{"message":"node v2","response_time":0.1,"upstream":[{"message":"go v1","response_time":"0.02"}]}]}

$ kubectl exec fortio -c fortio /usr/local/bin/fortio -- load -qps 10 -n 100 -loglevel Error http://service-python/env
14:46:20 I logger.go:97> Log level is now 4 Error (was 2 Info)
Fortio 1.0.1 running at 10 queries per second, 2->2 procs, for 100 calls: http://service-python/env
Aggregated Sleep Time : count 96 avg -14.174398 +/- 9.139 min -31.539875778 max -0.121480823 sum -1360.74219
# range, mid point, percentile, count
>= -31.5399 <= -0.121481 , -15.8307 , 100.00, 96
# target 50% -15.996
WARNING 100.00% of sleep were falling behind
Aggregated Function Time : count 100 avg 1.5912442 +/- 1.105 min 0.011235438 max 4.197832121 sum 159.124416
# target 50%   1.17647
# target 75%   2.52381
# target 90%   3.38462
# target 99%   4.09892
# target 99.9% 4.18794
Sockets used: 6 (for perfect keepalive, would be 4)
Code 200 : 98 (98.0 %)
Code 503 : 2 (2.0 %)
All done 100 calls (plus 0 warmup) 1591.244 ms avg, 2.4 qps
```

6）创建 Web 访问的路由规则：

```
$ kubectl apply -f istio/route/gateway-js-v1.yaml
```

7）浏览器访问。访问地址 http://11.11.11.112:31380/，多次刷新并点击“发射”按钮。

8）在 ServiceGraph UI 上查看服务调用树。访问地址 http://11.11.11.111:32145/force/forcegraph.html，查看服务调用树，如图 11-22 所示。

访问地址 http://11.11.11.111:32145/force/forcegraph.html?timehorizon=15m&filterempty=true，查看最近 15 分钟内的服务调用树，如图 11-23 所示。

9）清理：

```
$ kubectl delete -f kubernetes/fortio.yaml
$ kubectl delete -f istio/route/gateway-js-v1.yaml
$ kubectl delete -f kubernetes/istio-servicegraph-service.yaml
```

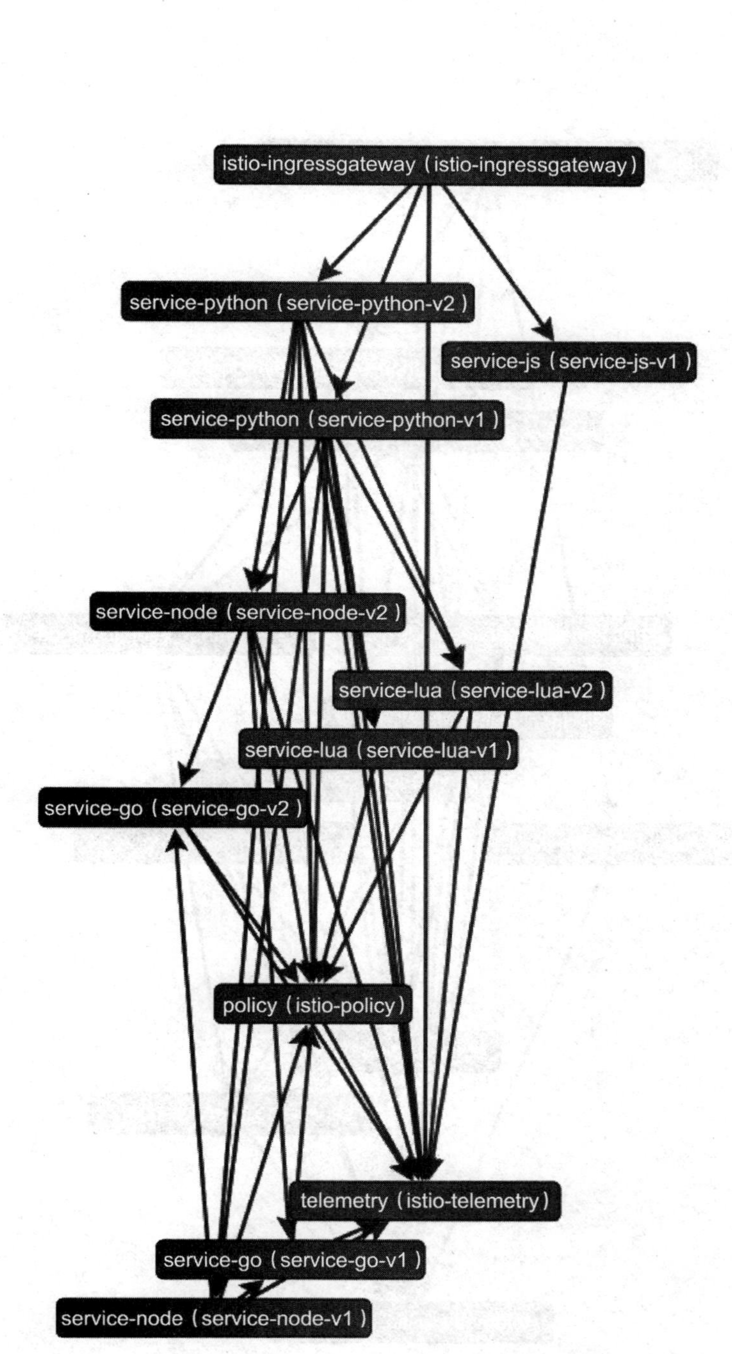

图 11-22　在 ServiceGraph UI 上查看服务调用树

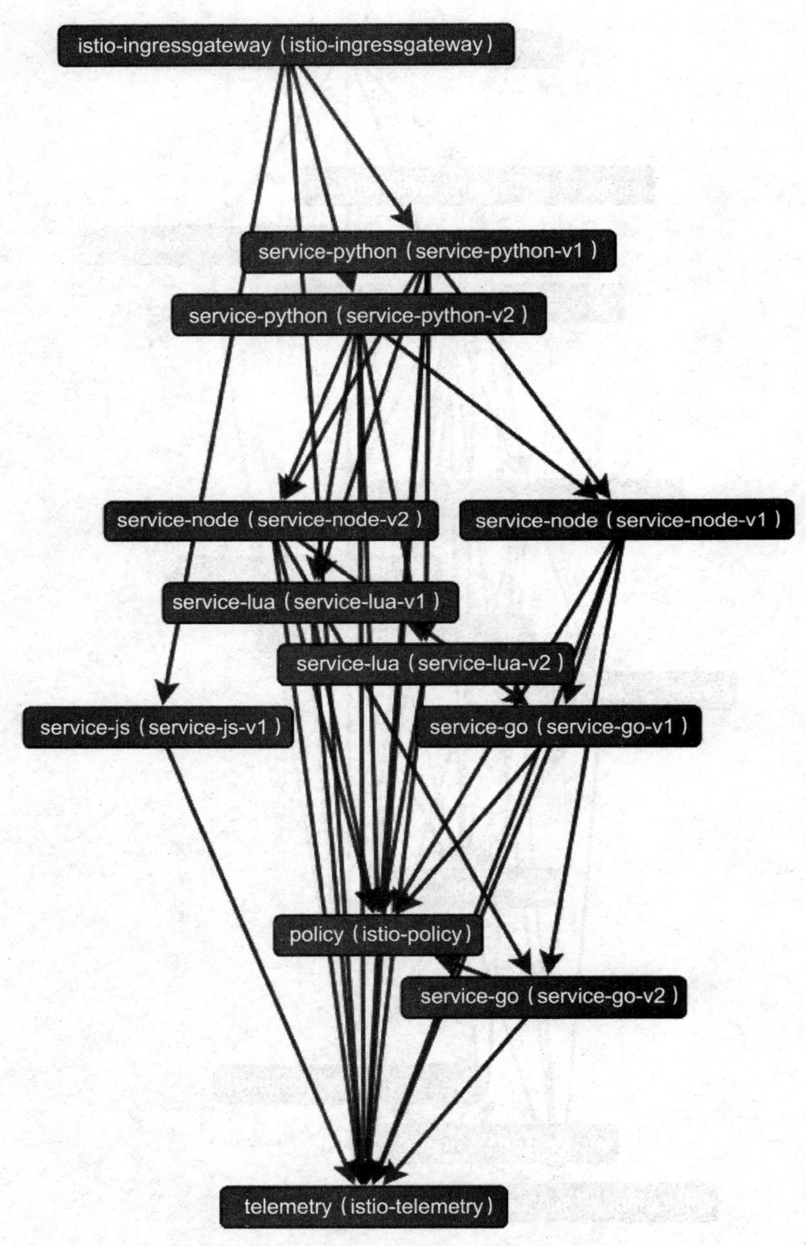

图 11-23　最近 15 分钟内的服务调用树

11.7　本章小结

Istio 在默认情况下已经为我们收集了许多指标数据和日志数据，当我们需要收集更细粒度的数据时，可以通过添加相应的收集规则来满足需求，非常方便。只需要传递简单的服务请求头信息就可以使用 Istio 提供的调用链追踪功能。通过网格内的服务调用树，可以实时地查看服务的调用情况，分析依赖关系，这可以在很大程度上帮助我们更好地理解服务架构，看清楚服务间的依赖关系，去除服务间不合理的依赖关系，增强整个应用的可用性和稳定性。

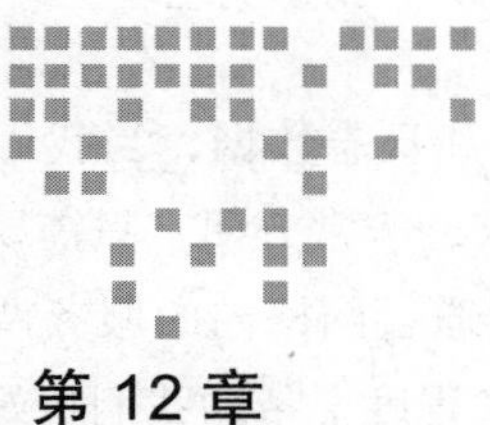

Chapter 12 第12章

Istio 维护

本章主要介绍 Istio 维护，包括如何可视化管理 Istio，如何在不中断业务的情况下升级 Istio，如何定制部署 Istio，当出现故障时如何排查问题。本章还将展示服务调用从请求发出到收到响应的整个流程，这对深入学习 Istio 是很有帮助的。

12.1 整体介绍

Istio 的维护管理内容包括：

- 可视化，通过 Istio 仪表板更好地观测 Istio 网络中的流量情况。
- 升级 Istio。
- 定制部署，使用 Helm 定制化地部署 Istio。
- 故障排除。

【实验前的准备】

进行本章实验前，需要先执行如下所示的前置步骤。

1）下载实验时用到的源码仓库：

```
$ sudo yum install -y git

$ git clone https://github.com/mgxian/istio-lab
Cloning into 'istio-lab'...
remote: Enumerating objects: 247, done.
remote: Counting objects: 100% (247/247), done.
remote: Compressing objects: 100% (173/173), done.
```

```
remote: Total 774 (delta 153), reused 164 (delta 73), pack-reused 527
Receiving objects: 100% (774/774), 283.00 KiB | 229.00 KiB/s, done.
Resolving deltas: 100% (447/447), done.

$ cd istio-lab
```

2）开启 default 命名空间的自动注入功能：

```
$ kubectl label namespace default istio-injection=enabled
namespace/default labeled
```

12.2 Istio 服务网格仪表板

Istio 服务网格仪表板（Dashboard）可以用来更好地观测 Istio 的运行状态以及服务网格内服务的流量情况，甚至可以直接在 Web 上修改 Istio 的路由等配置。在目前的 Istio 服务网格仪表板中，Kiali 功能最多，本节介绍如何安装 Kiali 并使用其功能。

注意 本节实验在以官方示例的方式部署 Istio 的集群中进行，并且部署所有用于实验的微服务。

（1）安装 Kiali

暴露 Jaeger 查询 Web 服务：

```
$ kubectl apply -f kubernetes/istio-tracing-service.yaml
```

部署 Kiali：

```
# 部署
$ export GRAFANA_URL='http://grafana.istio-system:3000'
$ export JAEGER_URL='http://11.11.11.111:32144'
$ export KIALI_USERNAME='admin'
$ export KIALI_PASSPHRASE='admin'
$ curl -sL http://git.io/getLatestKialiKubernetes | sed 's/create/apply/g' | bash

# 查看状态
$ kubectl get pod -n istio-system -l app=kiali
NAME                          READY       STATUS       RESTARTS       AGE
kiali-86bf676785-g9vht        1/1         Running      0              81s
```

（2）访问服务

创建用于请求的 Pod：

```
$ kubectl apply -f kubernetes/fortio.yaml
```

并发请求服务：

```
$ kubectl exec fortio -c fortio /usr/local/bin/fortio -- load -curl http://
service-python/env
HTTP/1.1 200 OK
content-type: application/json
content-length: 178
server: envoy
date: Sat, 19 Jan 2019 03:45:15 GMT
x-envoy-upstream-service-time: 1778

{"message":"python v1","upstream":[{"message":"lua v1","response_
time":0.38},{"message":"node v1","response_time":1.11,"upstream":[{"message":"go
v2","response_time":"0.42"}]}]}

$ kubectl exec fortio -c fortio /usr/local/bin/fortio -- load -qps 10 -n 100
-loglevel Error http://service-python/env
03:46:03 I logger.go:97> Log level is now 4 Error (was 2 Info)
Fortio 1.0.1 running at 10 queries per second, 2->2 procs, for 100 calls:
http://service-python/env
Aggregated Sleep Time : count 96 avg -3.857372 +/- 1.741 min -7.105141974 max
-0.501128479 sum -370.307712
# range, mid point, percentile, count
>= -7.10514 <= -0.501128 , -3.80314 , 100.00, 96
# target 50% -3.83789
WARNING 100.00% of sleep were falling behind
Aggregated Function Time : count 100 avg 0.68609566 +/- 0.4486 min 0.03083198
max 2.099653766 sum 68.6095656
# target 50% 0.633333
# target 75%   1.14815
# target 90%   1.7037
# target 99%   2.04983
# target 99.9% 2.09467
Sockets used: 6 (for perfect keepalive, would be 4)
Code 200 : 98 (98.0 %)
Code 503 : 2  (2.0 %)
All done 100 calls (plus 0 warmup) 686.096 ms avg, 5.8 qps
```

创建 Web 访问的路由规则：

```
$ kubectl apply -f istio/route/gateway-js-v1.yaml
```

浏览器访问。

访问地址 http://11.11.11.112:31380/，多次刷新并点击“发射”按钮。

（3）访问 Kiali

获取访问地址：

```
$ KIALI_NODEPORT=$(kubectl get svc kiali -n istio-system -o jsonpath='{.spec.
ports[0].nodePort}')
$ INGRESS_HOST=$(kubectl get node -o jsonpath='{.items[0].status.addresses[0].
address}')
```

```
$ echo "https://$INGRESS_HOST:$KIALI_NODEPORT/kiali/"
https://11.11.11.111:32072/kiali/
```

访问上一步骤中获取的 kialia 地址 https://11.11.11.111:32072/kiali/，用户名和密码均为 admin。查看服务调用树，如图 12-1 所示。

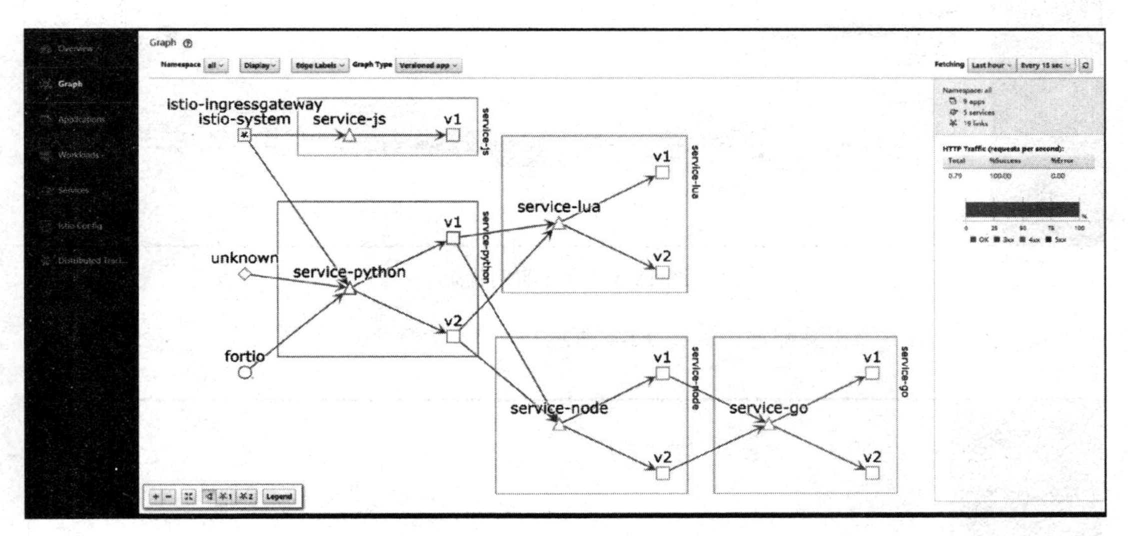

图 12-1　服务调用树

查看应用列表，如图 12-2 所示。

Applications				
service-go default	Health:	Error Rate: No requests	
service-js default	Health:	Error Rate: No requests	
service-lua default	Health:	Error Rate: No requests	
service-node default	Health:	Error Rate: No requests	
service-python default	Health:	Error Rate: No requests	
grafana istio-system	Health:	Error Rate: No requests	Missing Sidecar
istio-egressgateway istio-system	Health:	Error Rate: No requests	Missing Sidecar
istio-grafana istio-system	Health:	Error Rate: No requests	Missing Sidecar
istio-ingressgateway istio-system	Health:	Error Rate: No requests	Missing Sidecar

图 12-2　应用列表

查看服务实例状态，如图 12-3 所示。

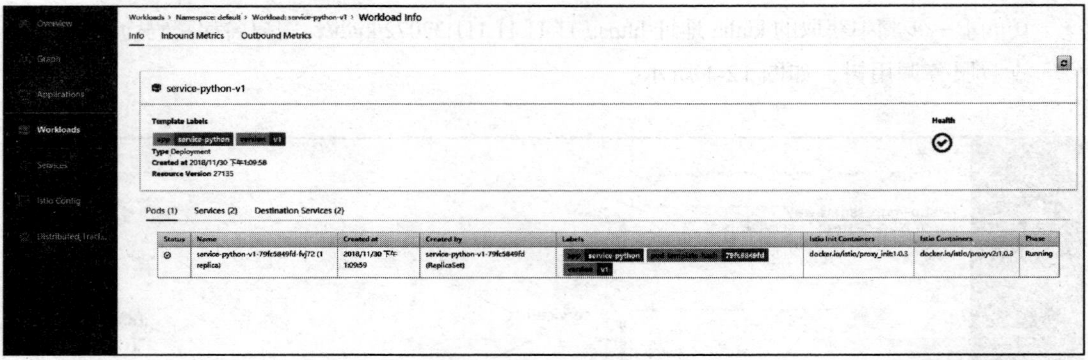

图 12-3　服务实例状态

查看 Istio 路由等相关配置，如图 12-4 所示。

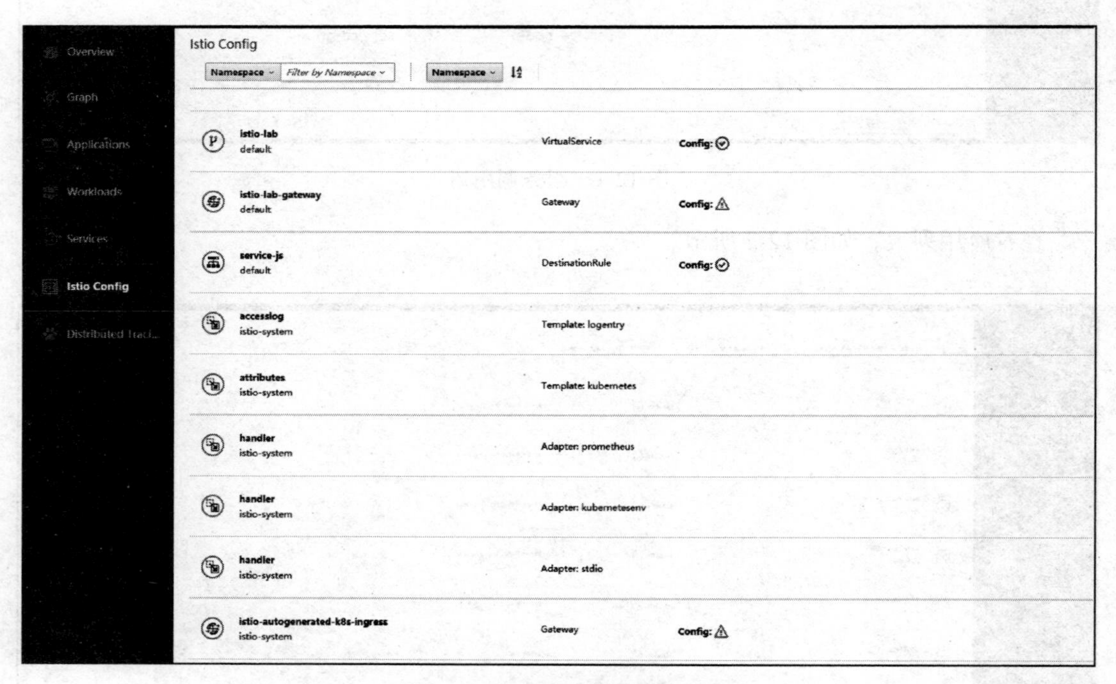

图 12-4　Istio 路由等相关配置

（4）清理

```
$ kubectl delete -f kubernetes/fortio.yaml
$ kubectl delete -f istio/route/gateway-js-v1.yaml
$ kubectl delete -f kubernetes/istio-tracing-service.yaml
```

```
$ kubectl delete all,secrets,sa,configmaps,deployments,ingresses,clusterrole
s,clusterrolebindings,virtualservices,destinationrules,customresourcedefinitions
--selector=app=kiali -n istio-system
```

12.3　升级 Istio

升级 Istio 的过程中可能会导致服务不可用，为了减少服务不可用时间，需要保证 Istio 控制平面组件和网格内的服务以多副本的方式部署。如果是大版本升级可能还会涉及改变新的配置，调整新的 API 等。本次实验仅以小版本升级作为演示，如下的实验步骤把 Istio 从 1.0.3 版本升级到 1.0.5 版本。

注意　本节实验在以官方示例的方式部署 Istio 的集群中进行。

（1）下载安装 Istio

下载 Istio 安装包：

```
$ sudo yum install -y wget
$ wget https://github.com/istio/istio/releases/download/1.0.5/istio-1.0.5-
linux.tar.gz
```

解压安装：

```
$ tar xf istio-1.0.5-linux.tar.gz
$ sudo mv istio-1.0.5 /usr/local
$ sudo rm -f /usr/local/istio
$ sudo ln -sv /usr/local/istio-1.0.5/ /usr/local/istio
```

添加到 PATH 路径中：

```
$ echo 'export PATH=/usr/local/istio/bin:$PATH' | sudo tee /etc/profile.d/
istio.sh
```

验证安装：

```
$ source /etc/profile.d/istio.sh
$ istioctl version
Version: 1.0.5
GitRevision: c1707e45e71c75d74bf3a5dec8c7086f32f32fad
User: root@6f6ea1061f2b
Hub: docker.io/istio
GolangVersion: go1.10.4
BuildStatus: Clean
```

（2）配置 Istio

复制源文件，便于出错时恢复：

```
$ cp /usr/local/istio/install/kubernetes/istio-demo.yaml /usr/local/istio/
install/kubernetes/istio-demo.yaml.ori
```

修改Istio部署配置。由于实验使用的虚拟机每台只有2G内存，默认情况下pilot的Deployment请求2G内存，为了使实验顺利进行。把13531行左右关于istio-pilot的内存配置修改成如下内容：

```
containers:
  - name: discovery
    image: "docker.io/istio/pilot:1.0.5"
    imagePullPolicy: IfNotPresent
    args:
    - "discovery"
    ports:
      ...
        initialDelaySeconds: 5
        periodSeconds: 30
        timeoutSeconds: 5
      ...
        resources:
          requests:
            cpu: 500m
            memory: 500Mi

        volumeMounts:
        - name: config-volume
          mountPath: /etc/istio/config
        - name: istio-certs
          mountPath: /etc/certs
          readOnly: true
```

配置istio-egressgateway启动多副本：

```
apiVersion: extensions/v1beta1
kind: Deployment
metadata:
  name: istio-egressgateway
  namespace: istio-system
...
spec:
  replicas: 2
  template:
    metadata:
      labels:
        app: istio-egressgateway
        istio: egressgateway
```

配置istio-ingressgateway启动多副本：

```
apiVersion: extensions/v1beta1
kind: Deployment
metadata:
  name: istio-ingressgateway
  namespace: istio-system
  labels:
...
spec:
  replicas: 2
  template:
    metadata:
      labels:
        app: istio-ingressgateway
        istio: ingressgateway
```

配置 istio-pilot 启动多副本：

```
apiVersion: extensions/v1beta1
kind: Deployment
metadata:
  name: istio-pilot
  namespace: istio-system
  # TODO: default template doesn't have this, which one is right ?
  labels:
...
spec:
  replicas: 2
  template:
    metadata:
      labels:
        istio: pilot
        app: pilot
```

配置 istio-policy 启动多副本：

```
apiVersion: extensions/v1beta1
kind: Deployment
metadata:
  name: istio-policy
  namespace: istio-system
...
spec:
  replicas: 2
  template:
    metadata:
      labels:
        app: policy
        istio: mixer
        istio-mixer-type: policy
```

配置 istio-telemetry 启动多副本：

```
apiVersion: extensions/v1beta1
kind: Deployment
metadata:
  name: istio-telemetry
  namespace: istio-system
...
spec:
  replicas: 2
  template:
    metadata:
      labels:
        app: telemetry
        istio: mixer
        istio-mixer-type: telemetry
```

使用命令行修改配置：

```
$ sed -i '13531 s/memory: 2048Mi/memory: 500Mi/g' /usr/local/istio/install/kubernetes/istio-demo.yaml
$ sed -i '12838 s/replicas: 1/replicas: 2/g' /usr/local/istio/install/kubernetes/istio-demo.yaml
$ sed -i '12977 s/replicas: 1/replicas: 2/g' /usr/local/istio/install/kubernetes/istio-demo.yaml
$ sed -i '13482 s/replicas: 1/replicas: 2/g' /usr/local/istio/install/kubernetes/istio-demo.yaml
$ sed -i '13228 s/replicas: 1/replicas: 2/g' /usr/local/istio/install/kubernetes/istio-demo.yaml
$ sed -i '13368 s/replicas: 1/replicas: 2/g' /usr/local/istio/install/kubernetes/istio-demo.yaml
```

查看配置：

```
$ grep -B5 'memory: 200Mi' /usr/local/istio/install/kubernetes/istio-demo.yaml
          - name: PILOT_TRACE_SAMPLING
            value: "100"
          resources:
            requests:
              cpu: 500m
              memory: 200Mi

$ grep -A5 'replicas: 2' /usr/local/istio/install/kubernetes/istio-demo.yaml
  replicas: 2
  template:
    metadata:
      labels:
        app: istio-egressgateway
        istio: egressgateway
--
  replicas: 2
  template:
    metadata:
```

```
        labels:
          app: istio-ingressgateway
          istio: ingressgateway
--
  replicas: 2
  template:
    metadata:
      labels:
        app: policy
        istio: mixer
--
  replicas: 2
  template:
    metadata:
      labels:
        app: telemetry
        istio: mixer
--
  replicas: 2
  template:
    metadata:
      labels:
        istio: pilot
        app: pilot
```

修改镜像使用国内镜像，加速部署，执行以下命令修改镜像：

```
$ sed -i 's@quay.io/coreos/hyperkube:v1.7.6_coreos.0@registry.cn-shanghai.
aliyuncs.com/gcr-k8s/hyperkube:v1.7.6_coreos.0@g' /usr/local/istio/install/
kubernetes/istio-demo.yaml
```

（3）配置 Istio 控制平面组件多副本

把 Istio 控制平面组件都设置为两副本，实现 Istio 控制平面高可用。Citadel 负责证书和密码的管理，Galley 负责 Istio 其他组件的配置验证，Sidecar-injector 负责 Pod 的 Envoy 代理自动注入功能。Istio 的更新一般都在业务的低流量时期进行，并且会禁止对网格内服务的相关操作，上述组件在更新时暂时不可用，并不会影响服务间的访问，因此本次实验不把上述组件进行高可用配置：

```
$ kubectl scale --replicas=2 -n istio-system deployment istio-egressgateway
$ kubectl scale --replicas=2 -n istio-system deployment istio-ingressgateway
$ kubectl scale --replicas=2 -n istio-system deployment istio-pilot
$ kubectl scale --replicas=2 -n istio-system deployment istio-policy
$ kubectl scale --replicas=2 -n istio-system deployment istio-telemetry

$ kubectl get pod -n istio-system
NAME                                READY   STATUS    RESTARTS   AGE
grafana-546d9997bb-4xs5s            1/1     Running   0          25d
istio-citadel-6955bc9cb7-dsl78      1/1     Running   0          25d
```

```
istio-cleanup-secrets-ntxn8              0/1     Completed   0          25d
istio-egressgateway-7dc5cbbc56-lpkqn     1/1     Running     0          24s
istio-egressgateway-7dc5cbbc56-rc5pl     1/1     Running     0          25d
istio-galley-545b6b8f5b-5pd4n            1/1     Running     0          25d
istio-grafana-post-install-97s5m         0/1     Completed   0          25d
istio-ingressgateway-7958d776b5-ccxxq    1/1     Running     0          24s
istio-ingressgateway-7958d776b5-qfpg7    1/1     Running     0          25d
istio-pilot-64958c46fc-5ct7q             2/2     Running     0          24s
istio-pilot-64958c46fc-cdk62             2/2     Running     0          25d
istio-policy-5c689f446f-l87kl            2/2     Running     0          25d
istio-policy-5c689f446f-zbl2r            2/2     Running     0          23s
istio-security-post-install-j8xr7        0/1     Completed   0          25d
istio-sidecar-injector-99b476b7b-58lk7   1/1     Running     0          25d
istio-telemetry-55d68b5dfb-4xjlp         2/2     Running     0          23s
istio-telemetry-55d68b5dfb-6v27b         2/2     Running     0          25d
istio-tracing-6445d6dbbf-p5nt5           1/1     Running     0          25d
prometheus-65d6f6b6c-2lf6n               1/1     Running     0          25d
servicegraph-57c8cbc56f-c2g2j            1/1     Running     3          25d
```

（4）升级期间的服务可用性测试

部署并扩容 service-go 服务：

```
$ kubectl apply -f service/go/service-go.yaml
$ kubectl scale --replicas=2 deployment service-go-v1
$ kubectl scale --replicas=2 deployment service-go-v2

$ kubectl get pod
NAME                            READY     STATUS      RESTARTS   AGE
service-go-v1-7cc5c6f574-4xjlp  2/2       Running     0          26s
service-go-v1-7cc5c6f574-sj2mn  2/2       Running     0          25d
service-go-v2-7656dcc478-f8zwc  2/2       Running     0          25d
service-go-v2-7656dcc478-zbl2r  2/2       Running     0          26s
```

创建 service-go 服务路由规则。在升级 Istio 时，可能会出现服务瞬时不可用的现象，配置服务 service-go 重试规则，可以减少服务瞬时不可用的现象，添加服务 service-go 的 RBAC 和 mTLS 配置，测试升级过程中 mTLS 和 RBAC 能否正常工作：

```
$ kubectl apply -f istio/upgrade/virtual-service-go-rbac-mtls-retry.yaml
```

创建测试的 Pod：

```
$ kubectl create ns testing
$ kubectl apply -f <(istioctl kube-inject -f kubernetes/sleep.yaml) -n testing
$ kubectl get pod -n testing
```

打开一个新终端，持续请求服务：

```
$ SLEEP_POD=$(kubectl get pod -l app=sleep -n testing -o jsonpath={.items..
metadata.name})
```

```
$ kubectl exec -n testing $SLEEP_POD -c sleep -- sh -c 'while true;do date
| tr -s "\n" " ";curl -s http://service-go.default/env -o /dev/null -w "%{http_
code}\n";sleep 1;done'
Sat Jan 19 04:16:33 UTC 2019 200
Sat Jan 19 04:16:34 UTC 2019 200
Sat Jan 19 04:16:36 UTC 2019 200
...
```

（5）升级 Istio

1）更新 Istio CRD：

```
$ kubectl apply -f /usr/local/istio/install/kubernetes/helm/istio/templates/
crds.yaml
```

2）查看 Istio CRD：

```
$ kubectl get crd
NAME                                        CREATED AT
adapters.config.istio.io                    2019-01-28T06:44:40Z
...
rules.config.istio.io                       2019-01-28T06:44:39Z
servicecontrolreports.config.istio.io       2019-01-28T06:44:40Z
...
tracespans.config.istio.io                  2019-01-28T06:44:40Z
virtualservices.networking.istio.io         2019-01-28T06:44:39Z
```

3）升级 Istio 相关组件：

```
$ kubectl apply -f /usr/local/istio/install/kubernetes/istio-demo.yaml
```

由于 Job 类型有部分字段不可修改，可能会出现 3 个关于 Job 的错误，由于不影响升级，可以直接忽略。

4）查看 Istio 控制平面组件状态：

```
$ kubectl get svc -n istio-system
NAME              TYPE      CLUSTER-IP       EXTERNAL-IP      PORT(S)            AGE
...
istio-egressgateway  ClusterIP    108.226.165      <none>    80/TCP,443/TCP      2m
istio-galley         ClusterIP    10.106.189.177  <none>    443/TCP,9093/TCP    72m
istio-ingressgateway      LoadBalancer    10.105.194.10     <pending>
80:31380/TCP,443:31390/TCP,31400:31400/TCP,15011:31606/TCP,8060:30116/
TCP,853:31117/TCP,15030:31051/TCP,15031:30994/TCP    72m
istio-pilot     ClusterIP     10.111.192.14     <none>     15010/TCP,15011/TCP,8080/
TCP,9093/TCP     72m
...

$ kubectl get pod -n istio-system
grafana-7ffdd5fb74-sjqgg                 1/1      Running      0             7m58s
istio-citadel-55cdfdd57c-qr8ts           1/1      Running      0             7m54s
```

```
istio-egressgateway-7798845f5d-ckqcf     1/1     Running     0     7m57s
istio-egressgateway-7798845f5d-xbrl5     1/1     Running     0     7m58s
istio-galley-76bbb946c8-n6fts            1/1     Running     0     7m59s
istio-grafana-post-install-d7qwp         0/1     Completed   0     62m
istio-ingressgateway-78c6d8b8d7-jkt9c    1/1     Running     0     7m58s
istio-ingressgateway-78c6d8b8d7-rj67m    1/1     Running     0     7m57s
istio-pilot-865fd9c96-2zfbh              2/2     Running     0     7m53s
istio-pilot-865fd9c96-qvtrw              2/2     Running     0     7m56s
istio-policy-7b6cc95d7b-9tw2h            2/2     Running     0     7m55s
istio-policy-7b6cc95d7b-ldmjc            2/2     Running     0     7m57s
istio-sidecar-injector-9c6698858-9b7j6   1/1     Running     0     7m52s
istio-telemetry-bfc9ff784-gs9gg          2/2     Running     0     7m56s
istio-telemetry-bfc9ff784-kh4cj          2/2     Running     0     7m55s
istio-tracing-6445d6dbbf-4cbbj           1/1     Running     0     62m
prometheus-65d6f6b6c-xgfkn               1/1     Running     0     62m
servicegraph-5c6f47859-vq7wx             1/1     Running     0     7m53s
```

5）观察整个控制平面组件升级过程中的服务访问日志：

```
...
Sat Jan 19 04:19:39 UTC 2019 200
Sat Jan 19 04:19:40 UTC 2019 200
...
Sat Jan 19 04:19:41 UTC 2019 200
Sat Jan 19 04:19:43 UTC 2019 200
...
Sat Jan 19 04:19:44 UTC 2019 503
Sat Jan 19 04:19:45 UTC 2019 200
Sat Jan 19 04:19:46 UTC 2019 200
...
```

从服务访问日志中可以看出，只有极少数异常服务响应码（503），这表明整个控制平面组件升级过程中，网格内的服务几乎不受影响，可以正常提供服务。

6）升级服务的 Envoy 代理：

```
$ TERMINATION_SECONDS=$(kubectl get deploy service-go-v1 -o jsonpath='{.spec.
template.spec.terminationGracePeriodSeconds}')
$ TERMINATION_SECONDS=$(($TERMINATION_SECONDS+1))
$ patch_string="{\"spec\":{\"template\":{\"spec\":{\"terminationGracePeriodSec
onds\":$TERMINATION_SECONDS}}}}"
$ kubectl patch deploy service-go-v1 -p $patch_string
$ kubectl patch deploy service-go-v2 -p $patch_string
```

在控制平面组件全部升级完成后再进行此实验。

7）观察服务 Envoy 代理升级过程中的服务访问日志：

```
...
Sat Jan 19 04:27:33 UTC 2019 200
```

```
Sat Jan 19 04:27:34 UTC 2019 200
Sat Jan 19 04:27:35 UTC 2019 503
Sat Jan 19 04:27:36 UTC 2019 200
...
Sat Jan 19 04:27:53 UTC 2019 200
Sat Jan 19 04:27:54 UTC 2019 200
Sat Jan 19 04:27:55 UTC 2019 200
...
```

从服务访问日志中可以看出，只有极少数异常服务响应码（503），这表明整个 Envoy 代理升级过程中，网格内的服务几乎不受影响，可以正常提供服务。

8）清理：

```
$ kubectl delete -f istio/upgrade/virtual-service-go-rbac-mtls-retry.yaml
$ kubectl delete ns testing
$ kubectl scale --replicas=1 -n istio-system deployment istio-egressgateway
$ kubectl scale --replicas=1 -n istio-system deployment istio-ingressgateway
$ kubectl scale --replicas=1 -n istio-system deployment istio-pilot
$ kubectl scale --replicas=1 -n istio-system deployment istio-policy
$ kubectl scale --replicas=1 -n istio-system deployment istio-telemetry
```

12.4　使用 Helm 定制部署 Istio

Helm 是 Kubernetes 的包管理工具，我们之前安装 Istio 时使用的 istio-demo.yaml 文件实际上就是根据 Istio 的 Helm 包生成的部署文件。

（1）安装 Helm

下载 Helm 安装包。到 Helm 的发布页面 https://github.com/helm/helm/releases 查找当前可用的版本的包地址，由于这个压缩包在国外服务器上，可能由于网络问题不能顺利下载，可以自己先在网络正常的机器上下载完成后，再上传到实验环境的虚拟机。使用如下命令下载安装包：

```
$ wget https://storage.googleapis.com/kubernetes-helm/helm-v2.12.2-linux-amd64.tar.gz
```

安装 Helm：

```
$ tar xf helm-v2.12.2-linux-amd64.tar.gz
$ sudo mv linux-amd64/ /usr/local/helm-2.12.2
$ sudo ln -sv /usr/local/helm-2.12.2 /usr/local/helm
$ echo 'export PATH=/usr/local/helm:$PATH' | sudo tee /etc/profile.d/helm.sh
$ source /etc/profile.d/helm.sh
$ helm version
Client: &version.Version{SemVer:"v2.12.2", GitCommit:"7d2b0c73d734f6586ed222a567c5d103fed435be", GitTreeState:"clean"}
Error: could not find tiller
```

（2）生成Istio部署文件

根据配置生成Istio的部署文件，使用如下命令生成定制的部署文件[⊖]：

```
$ helm template /usr/local/istio/install/kubernetes/helm/istio  \
--name istio --namespace istio-system \
--set global.hyperkube.hub=registry.cn-shanghai.aliyuncs.com/gcr-k8s \
--set pilot.resources.requests.memory=300Mi  \
--set gateways.istio-ingressgateway.enabled=true \
--set gateways.istio-egressgateway.enabled=true \
--set galley.enabled=true \
--set sidecarInjectorWebhook.enabled=true \
--set global.mtls.enabled=false \
--set prometheus.enabled=true \
--set grafana.enabled=true \
--set tracing.enabled=true \
--set pilot.traceSampling=100 \
--set servicegraph.enabled=true > istio.yaml
```

（3）部署Istio

创建istio-system命名空间：

```
$ kubectl create ns istio-system
```

创建Istio CRD：

```
$ kubectl apply -f /usr/local/istio/install/kubernetes/helm/istio/templates/
crds.yaml
```

查看Istio CRD：

```
$ kubectl get crd
NAME                                         CREATED AT
adapters.config.istio.io                     2018-11-28T06:44:40Z
...
rules.config.istio.io                        2018-11-28T06:44:39Z
servicecontrolreports.config.istio.io        2018-11-28T06:44:40Z
servicecontrols.config.istio.io              2018-11-28T06:44:39Z
serviceentries.networking.istio.io           2018-11-28T06:44:39Z
servicerolebindings.rbac.istio.io            2018-11-28T06:44:40Z
serviceroles.rbac.istio.io                   2018-11-28T06:44:40Z
signalfxs.config.istio.io                    2018-11-28T06:44:39Z
solarwindses.config.istio.io                 2018-11-28T06:44:39Z
stackdrivers.config.istio.io                 2018-11-28T06:44:39Z
statsds.config.istio.io                      2018-11-28T06:44:39Z
stdios.config.istio.io                       2018-11-28T06:44:39Z
templates.config.istio.io                    2018-11-28T06:44:40Z
```

⊖ 可选配置选项参考官方文档：https://istio.io/docs/reference/config/installation-options/。

```
tracespans.config.istio.io                2018-11-28T06:44:40Z
virtualservices.networking.istio.io       2018-11-28T06:44:39Z
```

部署 Istio 相关组件：

```
$ kubectl apply -f istio.yaml
```

查看 Istio 组件状态：

```
$ kubectl get deploy -n istio-system
NAME                     DESIRED   CURRENT   UP-TO-DATE   AVAILABLE   AGE
grafana                  1         1         1            1           61s
istio-citadel            1         1         1            1           60s
istio-egressgateway      1         1         1            1           62s
istio-galley             1         1         1            1           62s
istio-ingressgateway     1         1         1            1           61s
istio-pilot              1         1         1            1           61s
istio-policy             1         1         1            1           61s
istio-sidecar-injector   1         1         1            1           60s
istio-telemetry          1         1         1            1           61s
istio-tracing            1         1         1            1           60s
prometheus               1         1         1            1           60s
servicegraph             1         1         1            1           60s

$ kubectl get job -n istio-system
NAME                          COMPLETIONS   DURATION   AGE
istio-cleanup-secrets         1/1           41s        78s
istio-grafana-post-install    1/1           17s        78s
istio-security-post-install   1/1           16s        78s

$ kubectl get pod -n istio-system
NAME                                     READY   STATUS      RESTARTS   AGE
grafana-546d9997bb-f2scr                 1/1     Running     0          92s
istio-citadel-6955bc9cb7-47cnf           1/1     Running     0          91s
istio-cleanup-secrets-l4shp              0/1     Completed   0          95s
istio-egressgateway-7dc5cbbc56-5gcz4     1/1     Running     0          92s
istio-galley-545b6b8f5b-gbvtx            1/1     Running     0          92s
istio-grafana-post-install-v5j5b         0/1     Completed   0          95s
istio-ingressgateway-7958d776b5-9rgbx    1/1     Running     0          92s
istio-pilot-64958c46fc-9vmlq             2/2     Running     0          92s
istio-policy-5c689f446f-wltfj            2/2     Running     0          92s
istio-security-post-install-xwph2        0/1     Completed   0          95s
istio-sidecar-injector-99b476b7b-lxqcb   1/1     Running     0          91s
istio-telemetry-55d68b5dfb-qjvbm         2/2     Running     0          92s
istio-tracing-6445d6dbbf-7qdqk           1/1     Running     0          91s
prometheus-65d6f6b6c-rsrdr               1/1     Running     0          91s
servicegraph-57c8cbc56f-fqftt            1/1     Running     0          91s

$ kubectl get svc -n istio-system
```

```
NAME                 TYPE              CLUSTER-IP       EXTERNAL-IP    PORT(S)    AGE
...
istio-ingressgateway    LoadBalancer    10.100.103.226    <pending>    80:31380/
TCP,443:31390/TCP,31400:31400/TCP,15011:30675/TCP,8060:30928/TCP,853:31502/
TCP,15030:31492/TCP,15031:30442/TCP   110s
istio-pilot    ClusterIP    10.110.236.54    <none>    15010/TCP,15011/TCP,8080/
TCP,9093/TCP    110s...
zipkin        ClusterIP      10.101.209.252   <none>        9411/TCP         107s
```

当组件全部处于Running或者Completed时再进行实验，由于需要拉取较多镜像，如果网速较慢，可能需要等待很长一段时间。如果没有跟着之前的步骤配置镜像拉取加速，会更加缓慢，强烈建议配置镜像加速。

（4）服务部署测试

部署httpbin服务：

```
$ kubectl apply -f kubernetes/httpbin.yaml

$ kubectl get pod
NAME                            READY        STATUS        RESTARTS        AGE
httpbin-b67975b8f-wjzsg         2/2          Running       0               57s
```

创建gateway暴露服务：

```
$ kubectl apply -f istio/route/gateway-httpbin-http.yaml
```

使用curl访问：

```
$ curl http://11.11.11.111:31380/get
{
    "args": {},
    "headers": {
        "Accept": "*/*",
        "Content-Length": "0",
        "Host": "11.11.11.111:31380",
        "User-Agent": "curl/7.29.0",
        "X-B3-Sampled": "1",
        "X-B3-Spanid": "c21ad57ad952c5ae",
        "X-B3-Traceid": "c21ad57ad952c5ae",
        "X-Envoy-Internal": "true",
        "X-Request-Id": "399cd990-172e-966f-8f1f-ef4a4b150dce"
    },
    "origin": "10.244.0.0",
    "url": "http://11.11.11.111:31380/get"
}
```

浏览器访问。访问地址http://11.11.11.111:31380/get，结果如图12-5所示。

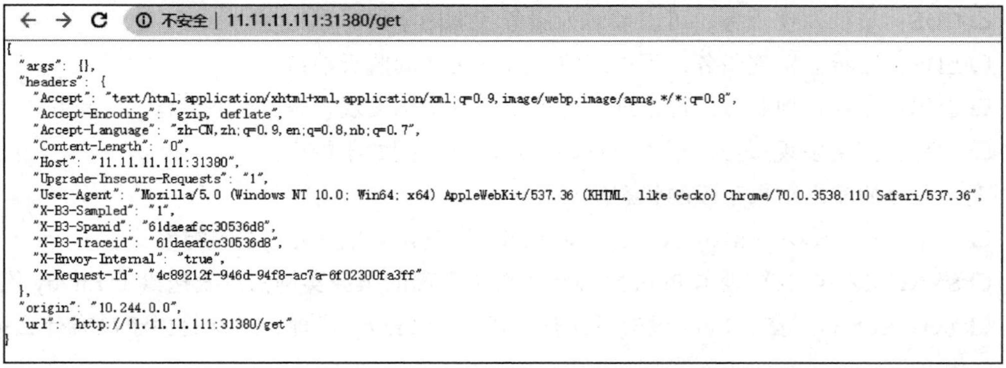

不安全 | 11.11.11.111:31380/get


```
{
  "args": {},
  "headers": {
    "Accept": "text/html,application/xhtml+xml,application/xml;q=0.9,image/webp,image/apng,*/*;q=0.8",
    "Accept-Encoding": "gzip, deflate",
    "Accept-Language": "zh-CN,zh;q=0.9,en;q=0.8,nb;q=0.7",
    "Content-Length": "0",
    "Host": "11.11.11.111:31380",
    "Upgrade-Insecure-Requests": "1",
    "User-Agent": "Mozilla/5.0 (Windows NT 10.0; Win64; x64) AppleWebKit/537.36 (KHTML, like Gecko) Chrome/70.0.3538.110 Safari/537.36",
    "X-B3-Sampled": "1",
    "X-B3-Spanid": "61daeafcc30536d8",
    "X-B3-Traceid": "61daeafcc30536d8",
    "X-Envoy-Internal": "true",
    "X-Request-Id": "4c89212f-946d-94f8-ac7a-6f02300fa3ff"
  },
  "origin": "10.244.0.0",
  "url": "http://11.11.11.111:31380/get"
}
```


图 12-5　浏览器访问

12.5　故障排除

本节主要介绍当网格中出现问题时，应该如何排查问题，以及如何解决这些问题。我们将主要学习故障排除的技巧和方向。

1. 路由不生效

如果创建路由规则后并没有达到预期的效果，可以使用如下的方法来找到路由不生效的问题并解决。

1）查看 Pilot 是否正常：

```
$ kubectl get pod -l app=pilot -n istio-system
NAME                             READY    STATUS     RESTARTS    AGE
istio-pilot-64958c46fc-cdk62     2/2      Running    0           26m
```

2）查看网格整体情况：

```
$ istioctl proxy-status
PROXY              CDS          LDS          EDS          RDS          PILOT           VERSION
istio-egressgateway-7dc5cbbc56-rc5pl.istio-system     SYNCED     SYNCED    ZSYNCED
(100%)     NOT SENT    istio-pilot-64958c46fc-cdk62      1.0.2
istio-ingressgateway-7958d776b5-qfpg7.istio-system     SYNCED     SYNCED     SYNCED
(100%)     NOT SENT     istio-pilot-64958c46fc-cdk62     1.0.2
service-go-v1-7cc5c6f574-sj2mn.default      SYNCED      SYNCED      SYNCED (100%)
SYNCED     istio-pilot-64958c46fc-cdk62     1.0.2
service-go-v2-7656dcc478-f8zwc.default      SYNCED      SYNCED      SYNCED (100%)
SYNCED     istio-pilot-64958c46fc-cdk62     1.0.2
```

如果服务实例没有出现在上述的命令结果里，表明服务实例可能出了故障，没能加入网格内。

上述命令的输出结果字段解释如下：

- ❑ CDS：集群发现服务，可以理解为服务实例的集群。
- ❑ LDS：监听器发现服务，可以理解为服务实例的监听端口。
- ❑ EDS：端点发现服务，可以理解为单个服务实例发现。
- ❑ RDS：路由发现服务，根据条件路由请求到不同的集群中。

上述命令的输出结果字段的状态解释如下：

- ❑ SYNCED：表示Envoy代理已经确认收到了Pilot上一次发送的配置。
- ❑ SYNCED(100%)：表示Pilot已经将集群中全部的服务实例信息发送给了Envoy代理。
- ❑ NOT SENT：表示Pilot没有发送任何配置给Envoy代理，通常是由于没有什么可发送的。
- ❑ STALE：表示Pilot已经发送了更新配置给Envoy代理，但是还没有收到Envoy代理的确认，通常可能是由于Envoy代理和Pilot之前的网络连接出现了问题或者机器性能有问题。

3）检查Envoy代理能否连接到Pilot。

在没有启用mTLS时使用如下命令：

```
$ SERVICE_GO_POD_NAME=$(kubectl get pod -l app=service-go -o jsonpath={.items[0].metadata.name})
$ PILOT_POD_IP=$(kubectl get pod -l app=pilot -n istio-system -o jsonpath={.items[0].status.podIP})
$ kubectl exec $SERVICE_GO_POD_NAME -c istio-proxy -- curl -s http://$PILOT_POD_IP:15003/v1/registration
```

命令的执行结果如下：

```
[
  {
  "service-key": "grafana.istio-system.svc.cluster.local|http",
  "hosts": [
  {
    "ip_address": "10.244.1.7",
    "port": 3000
    }
    ]
  },
...
  {
  "service-key": "service-go.default.svc.cluster.local|http",
  "hosts": [
    {
    "ip_address": "10.244.1.15",
    "port": 80
    },
  {
  "ip_address": "10.244.2.10",
```

```
  "port": 80
    }
    ]
  },
...
  {
  "service-key": "zipkin.istio-system.svc.cluster.local|http",
  "hosts": [
    {
     "ip_address": "10.244.1.10",
     "port": 9411
    }
    ]
  }
]
```

4）查看 Pilot 和 Envoy 的配置文件的差异：

```
$ istioctl proxy-status service-go-v1-7cc5c6f574-sj2mn.default
Clusters Match
Listeners Match
Routes Match
```

正常情况下应该如上述的结果所示，如果出现了一不致，就说明 Pilot 到 Envoy 的配置同步出现了问题。

5）深入 Envoy 配置信息。

查看服务集群配置信息：

```
$ istioctl proxy-config clusters -n default service-go-v1-7cc5c6f574-sj2mn
SERVICE FQDN    PORT      SUBSET     DIRECTION     TYPE
BlackHoleCluster   -   -   -   STATIC
...
jaeger-query.istio-system.svc.cluster.local   16686    -     outbound     EDS
kube-dns.kube-system.svc.cluster.local        53       -     outbound     EDS
kubernetes.default.svc.cluster.local          443      -     outbound     EDS
prometheus.istio-system.svc.cluster.local     9090     -     outbound     EDS
prometheus_stats                              -        -     -            STATIC
service-go.default.svc.cluster.local          80       -     inbound      STATIC
service-go.default.svc.cluster.local          80       -     outbound     EDS
...
```

```
$ istioctl proxy-config clusters service-go-v1-7cc5c6f574-sj2mn --fqdn service-go.default.svc.cluster.local -o json
[
    {
        "name": "inbound|80||service-go.default.svc.cluster.local",
        "connectTimeout": "1.000s",
        "hosts": [
            {
```

```
                "socketAddress": {
                    "address": "127.0.0.1",
                    "portValue": 80
                }
            }
        ],
        "circuitBreakers": {
            "thresholds": [
                {}
            ]
        }
    },
    {
        "name": "outbound|80||service-go.default.svc.cluster.local",
        "type": "EDS",
        "edsClusterConfig": {
            "edsConfig": {
                "ads": {}
            },
            "serviceName": "outbound|80||service-go.default.svc.cluster.local"
        },
        "connectTimeout": "1.000s",
        "circuitBreakers": {
            "thresholds": [
                {}
            ]
        }
    }
]
```

查看服务监听端口配置信息：

```
$ istioctl proxy-config listeners -n default service-go-v1-7cc5c6f574-sj2mn
ADDRESS                 PORT      TYPE
...
10.108.252.151          443       TCP
10.108.252.151          15011     TCP
10.96.0.1               443       TCP
10.96.0.10              53        TCP
0.0.0.0                 9090      HTTP
0.0.0.0                 80        HTTP
...

$ istioctl proxy-config listeners -n default service-go-v1-7cc5c6f574-sj2mn
--port 15001 -o json
[
    {
        "name": "virtual",
        "address": {
            "socketAddress": {
                "address": "0.0.0.0",
                "portValue": 15001
```

```
            }
        },
        "filterChains": [
            {
                "filters": [
                    {
                        "name": "envoy.tcp_proxy",
                        "config": {
                            "cluster": "BlackHoleCluster",
                            "stat_prefix": "BlackHoleCluster"
                        }
                    }
                ]
            }
        ],
        "useOriginalDst": true
    }
]
```

查看路由配置信息：

```
$ istioctl proxy-config routes -n default service-go-v1-7cc5c6f574-sj2mn -o json
...
    "routes": [
        {
            "match": {
                "prefix": "/"
            },
            "route": {
                "cluster":"outbound|80||service-go.default.svc.cluster.local",
                "timeout": "0.000s",
                "maxGrpcTimeout": "0.000s"
            },
...
```

查看启动时的配置信息：

```
$ istioctl proxy-config bootstrap -n default service-go-v1-7cc5c6f574-sj2mn -o
json
{
    "bootstrap": {
        "node": {
            "id": "sidecar~10.244.2.10~service-go-v1-7cc5c6f574-sj2mn.default~default.
                svc.cluster.local",
            "cluster": "service-go",
            "metadata": {
                "INTERCEPTION_MODE": "REDIRECT",
                "ISTIO_PROXY_SHA": "istio-proxy:6953ca783697da07ebe565322d12e9
                                    69280d8b03",
                "ISTIO_PROXY_VERSION": "1.0.2",
                "ISTIO_VERSION": "1.0.3",
```

```
                "POD_NAME": "service-go-v1-7cc5c6f574-sj2mn",
                "app": "service-go",
                "istio": "sidecar",
                "pod-template-hash": "7cc5c6f574",
                "version": "v1"
            },
        "buildVersion": "0/1.8.0-dev//RELEASE"
    },
```

查看服务实例的日志信息：

```
$ kubectl logs -f -n default service-go-v1-7cc5c6f574-sj2mn istio-proxy
...
[2019-01-19 04:48:58.770][11][info][main] external/envoy/source/server/server.
cc:401] all clusters initialized. initializing init manager
[2019-01-19 04:48:58.935][11][info][upstream] external/envoy/source/server/
lds_api.cc:80] lds: add/update listener '10.105.194.10_15011'
[2019-01-19 04:48:58.938][11][info][upstream] external/envoy/source/server/
lds_api.cc:80] lds: add/update listener '10.105.194.10_853'
...
[2019-01-19 04:48:58.990][11][info][upstream] external/envoy/source/server/
lds_api.cc:80] lds: add/update listener '0.0.0.0_15031'
[2019-01-19 04:48:59.003][11][info][upstream] external/envoy/source/server/
lds_api.cc:80] lds: add/update listener '0.0.0.0_8060'
[2019-01-19 04:48:59.010][11][info][upstream] external/envoy/source/server/
lds_api.cc:80] lds: add/update listener '0.0.0.0_8080'
```

2. mTLS 异常

当启用 mTLS 后，服务出现无法正常访问的情况，关闭 mTLS 后服务可以正常访问，可以使用如下的排查步骤来发现问题并解决。

1）查看 Citadel 是否正常：

```
$ kubectl get pod -l istio=citadel -n istio-system
NAME                              READY     STATUS      RESTARTS     AGE
istio-citadel-6955bc9cb7-dsl78    1/1       Running     0            36m
```

2）查看证书和密钥：

```
$ kubectl exec $(kubectl get pod -l app=service-go -o jsonpath={.items[0].
metadata.name}) -c istio-proxy -- ls /etc/certs
cert-chain.pem
key.pem
root-cert.pem

$ kubectl exec $(kubectl get pod -l app=service-go -o jsonpath={.items[0].
metadata.name}) -c istio-proxy -- cat /etc/certs/cert-chain.pem | openssl x509
-text -noout  | grep Validity -A 2
    Validity
```

```
        Not Before: Nov 30 04:53:59 2018 GMT
        Not After : Feb 28 04:53:59 2019 GMT

    $ kubectl exec $(kubectl get pod -l app=service-go -o jsonpath={.items[0].
metadata.name}) -c istio-proxy -- cat /etc/certs/cert-chain.pem | openssl x509
-text -noout  | grep 'Subject Alternative Name' -A 1
            X509v3 Subject Alternative Name:
                URI:spiffe://cluster.local/ns/default/sa/default
```

3）检查 mTLS 配置。

查看默认全局 mTLS 策略：

```
$ kubectl get meshpolicy default -o yaml
apiVersion: authentication.istio.io/v1alpha1
kind: MeshPolicy
metadata:
...
  name: default
...
spec:
  peers:
  - mtls:
      mode: PERMISSIVE
```

由于 Istio 安装时会默认设置全局的 mTLS 策略，所以默认情况下会显示为 CONFLICT 状态，但是由于默认策略模式为 PERMISSIVE，因此服务仍然可以正常访问。

```
$ istioctl authn tls-check service-go.default.svc.cluster.local
HOST:PORT  STATUS   SERVER   CLIENT  AUTHN POLICY  DESTINATION RULE
service-go.default.svc.cluster.local:80  CONFLICT  mTLS  HTTP  default/    -
```

istioctl authn tls-check 命令的输出解释如下：

- HOST:PORT 表示被检查服务的地址和端口。
- STATUS 表示 mTLS 当前的状态，OK 表示正常，CONFLICT 表示 mTLS 的服务端与客户端配置存在冲突，访问异常。
- SERVER 表示服务端使用的协议。
- CLIENT 表示客户端使用的协议。
- AUTHN POLICY 表示使用的 Policy 策略名称，形如“策略名称 / 命名空间名称”，default/default 表示使用的是 default 命名空间的 default 策略。
- DESTINATION RULE 表示使用的 DestinationRule 路由规则名称，形如“路由规则名称 / 命名空间名称”，service-go/default 表示使用的是 default 命名空间的名为 service-go 的 DestinationRule 路由规则。

【实验】

模拟 mTLS 故障，排查问题。

1）创建测试Pod：

```
$ kubectl apply -f kubernetes/dns-test.yaml
```

2）故意设置错误的mTLS策略：

```
$ kubectl apply -f istio/security/mtls-service-go-bad-rule.yaml
```

3）查看认证策略：

```
$ istioctl authn tls-check service-go.default.svc.cluster.local
HOST:PORT   STATUS   SERVER   CLIENT   AUTHN POLICY    DESTINATION RULE
service-go.default.svc.cluster.local:80   CONFLICT   HTTP    mTLS   service-go/
default   service-go/default
```

从上面的输出结果可知，service-go服务mTLS的配置存在冲突，服务端使用了HTTP，而客户端使用了mTLS协议，因此访问会出现异常。

4）访问服务：

```
$ kubectl exec dns-test -c dns-test -- curl -s http://service-go/env
upstream connect error or disconnect/reset before headers
```

5）更新mTLS策略：

```
$ kubectl apply -f istio/security/mtls-service-go-on.yaml
```

6）查看认证状态：

```
$ istioctl authn tls-check service-go.default.svc.cluster.local
HOST:PORT      STATUS     SERVER     CLIENT     AUTHN POLICY     DESTINATION RULE
service-go.default.svc.cluster.local:80     OK     mTLS     mTLS      service-go/
default    service-go/default
```

7）访问服务：

```
$ kubectl exec dns-test -c dns-test -- curl -s http://service-go/env
{"message":"go v2"}
```

8）清理：

```
$ kubectl delete -f istio/security/mtls-service-go-on.yaml
$ kubectl delete -f kubernetes/dns-test.yaml
```

3. RBAC异常

当启用RBAC后，服务出现无法正常访问的情况，关闭RBAC后服务可以正常访问，可以使用如下的排查步骤来发现问题，并解决问题。

1）确保RBAC已经正确开启。

开启RBAC规则：

```
$ kubectl apply -f istio/security/rbac-config-on.yaml
rbacconfig.rbac.istio.io/default created
```

查看 RbacConfig：

```
$ kubectl get rbacconfigs.rbac.istio.io --all-namespaces
NAMESPACE        NAME            AGE
default          default         48s
```

确保只有一个名为 default 的 RbacConfig 配置实例，否则 Istio 会禁用 RBAC 功能，并忽略所有策略。如果有多余的 RbacConfig 配置实例，删除所有多余的 RbacConfig 配置实例。

2）开启 Polit 的 RBAC 调试日志。

另外开一个终端执行如下的命令，该命令不会结束，会持续输出访问日志。设置 RBAC 调试日志完成后，按 Ctrl+C 停止该命令：

```
$ kubectl port-forward $(kubectl -n istio-system get pods -l istio=pilot -o jsonpath='{.items[0].metadata.name}') -n istio-system 19876:9876
```

使用 iptables 开放远程访问：

```
$ sudo iptables -t nat -I PREROUTING -d 11.11.11.111 -p tcp --dport 9876 -j DNAT --to-destination 127.0.0.1:19876
```

通过 ControlZ 控制功能开启 RBAC 调试日志。

访问地址 http://11.11.11.111:9876/scopez/，设置 RBAC 的输出级别为 debug，如图 12-6 所示。

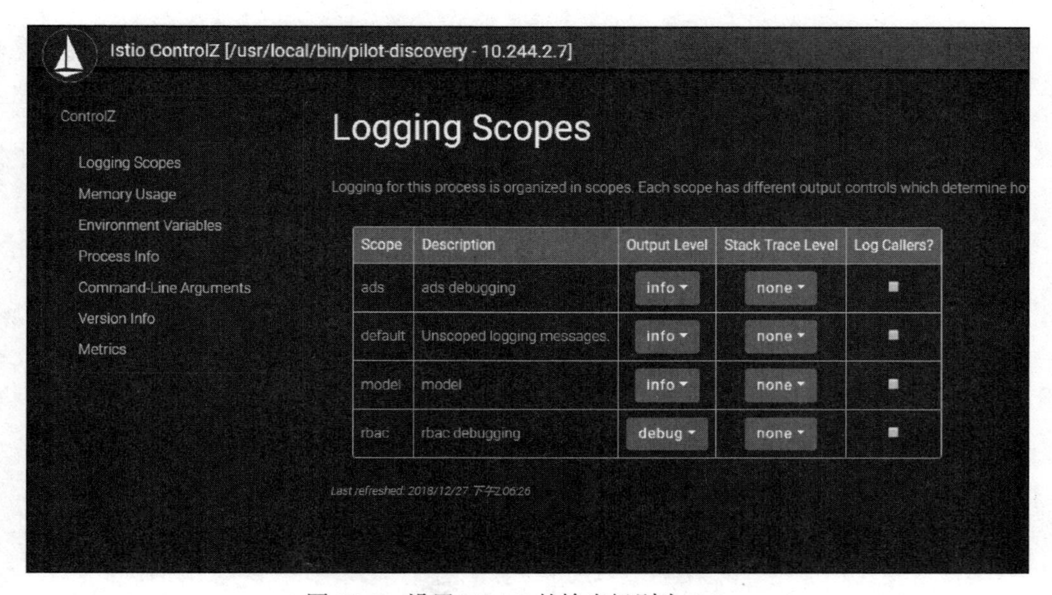

图 12-6　设置 RBAC 的输出级别为 debug

应用服务的 RBAC 策略：

```
$ kubectl apply -f istio/security/rbac-service-go-user-agent-policy.yaml
servicerole.rbac.istio.io/service-viewer created
servicerolebinding.rbac.istio.io/bind-service-viewer created
```

查看 Pilot 的日志输出：

```
$ kubectl logs $(kubectl -n istio-system get pods -l istio=pilot -o jsonpath='{.items[0].metadata.name}') -c discovery -n istio-system | grep rbac
2019-01-19T03:07:07.413353Z  info  registering for apiVersion rbac.istio.io/v1alpha1
2019-01-19T05:00:56.365084Z  info rbac   no service role in namespace default
2019-01-19T05:00:56.365915Z  info   rbac  no service role binding in namespace default
2019-01-19T05:00:56.366458Z  info  rbac  built filter config for service-go.default.svc.cluster.local
2019-01-19T05:00:56.373868Z  info  rbac  no service role in namespace default
2019-01-19T05:00:56.373952Z  info  rbac  no service role binding in namespace default
2019-01-19T05:00:56.374160Z  info  rbac  built filter config for service-go.default.svc.cluster.local
2019-01-19T05:02:43.948543Z  debug  rbac  building filter config for {service-go.default.svc.cluster.local map[pod-template-hash:7656dcc478 version:v2 app:service-go] map[destination.name:service-go destination.namespace:default destination.user:default]}
```

3）确保 Pilot 正确的分发了策略

查看 Envoy 代理的配置：

```
$ kubectl exec  $(kubectl get pods -l app=service-go -o jsonpath='{.items[0].metadata.name}') -c istio-proxy -- curl localhost:15000/config_dump -s
...
{
  "name": "envoy.filters.http.rbac",
  "config": {
    "shadow_rules": {
      "policies": {}
    },
    "rules": {
      "policies": {
        "service-viewer": {
          "permissions": [
            {
              "and_rules": {
                "rules": [
                  {
                    "or_rules": {
                      "rules": [
                        {
```

```
                                "header": {
                                  "exact_match": "GET",
                                  "name": ":method"
                                }
...
          "principals": [
          {
            "and_ids": {
              "ids": [
                {
                  "header": {
                    "name": "User-Agent",
                    "prefix_match": "RBAC-"
                  }
                }
              ]
            }
```

从上述命令结果的 "envoy.filters.http.rbac" 中查看 RBAC 策略 Envoy 代理是否已经收到并应用到配置中。

4）确保 Envoy 代理正确的执行了策略。

设置 Envoy 代理 RBAC 的日志级别为 debug：

```
$ kubectl exec -ti $(kubectl get pod -l app=service-go -o jsonpath={.items[0].
metadata.name}) -c istio-proxy -- curl -X POST 127.0.0.1:15000/logging?rbac=debug
active loggers:
...
    http: info
    http2: info
...
    rbac: debug
    redis: info
```

5）创建测试 Pod：

```
$ kubectl apply -f kubernetes/dns-test.yaml
```

6）访问 service-go 服务：

```
$ kubectl exec dns-test -c dns-test -- curl -s http://service-go/env
RBAC: access denied

$ kubectl exec dns-test -c dns-test -- curl -s -H "User-Agent: RBAC-TEST"
http://service-go/env
{"message":"go v2"}

$ kubectl exec dns-test -c dns-test -- curl -s -H "User-Agent: RBAC-TEST"
http://service-go/env
{"message":"go v1"}
```

7）查看Envoy代理日志：

```
$ kubectl logs $(kubectl get pods -l app=service-go -o jsonpath='{.items[0].metadata.name}') -c istio-proxy
[2019-01-19 04:57:30.512][19][debug][rbac] external/envoy/source/extensions/filters/http/rbac/rbac_filter.cc:65] checking request: remoteAddress: 10.244.1.16:46074, localAddress: 10.244.1.15:80, ssl: none, headers: ':authority', 'service-go'
':path', '/env'
':method', 'GET'
'user-agent', 'curl/7.35.0'
...
}

[2019-01-19 04:58:18.446][19][debug][rbac] external/envoy/source/extensions/filters/http/rbac/rbac_filter.cc:78] shadow denied
[2019-01-19 04:58:18.456][19][debug][rbac] external/envoy/source/extensions/filters/http/rbac/rbac_filter.cc:112] enforced denied

[2019-01-19 05:00:12.409][20][debug][rbac] external/envoy/source/extensions/filters/http/rbac/rbac_filter.cc:65] checking request: remoteAddress: 10.244.1.16:46594, localAddress: 10.244.1.15:80, ssl: none, headers: ':authority', 'service-go'
':path', '/env'
':method', 'GET'
'accept', '*/*'
'user-agent', 'RBAC-TEST'
...
}

[2019-01-19 05:05:25.669][20][debug][rbac] external/envoy/source/extensions/filters/http/rbac/rbac_filter.cc:78] shadow denied
[2019-01-19 05:05:25.669][20][debug][rbac] external/envoy/source/extensions/filters/http/rbac/rbac_filter.cc:108] enforced allowed
[2019-01-19T05:05:25.669Z] "GET /envHTTP/1.1" 200 - 0 19 6 1 "-" "RBAC-TEST" "9f594a94-6be0-924d-83a0-39448f020701" "service-go" "127.0.0.1:80" inbound|80||service-go.default.svc.cluster.local - 10.244.1.15:80 10.244.1.16:46074
```

从日志中可以看出，当'user-agent' = 'RBAC-TEST'请求被允许（enforced allowed），当'user-agent' = 'curl/7.35.0'请求被拒绝（enforced denied）。

8）清理：

```
# 通过ControlZ控制功能恢复RBAC日志级别为info

# 删除iptables规则
$ sudo iptables -t nat -D PREROUTING -d 11.11.11.111 -p tcp --dport 9876 -j DNAT --to-destination 127.0.0.1:19876

# 重置Envoy代理RBAC的日志级别为info
```

```
$ kubectl exec -ti $(kubectl get pod -l app=service-go -o jsonpath={.items[0].
metadata.name}) -c istio-proxy -- curl -X POST 127.0.0.1:15000/logging?rbac=info

$ kubectl delete -f istio/security/rbac-config-on.yaml
$ kubectl delete -f istio/security/rbac-service-go-user-agent-policy.yaml
$ kubectl delete -f kubernetes/dns-test.yaml
```

4. 指标或日志收集异常

这里只演示指标数据收集异常的情况，日志数据收集问题的思路类似，不再演示。

1）基础环境准备。

创建测试 Pod：

```
$ kubectl apply -f kubernetes/fortio.yaml
```

访问服务：

```
$ kubectl exec fortio -c fortio /usr/local/bin/fortio -- load -qps 10 -n 100
-loglevel Error http://service-go/env
05:18:48 I logger.go:97> Log level is now 4 Error (was 2 Info)
Fortio 1.0.1 running at 10 queries per second, 2->2 procs, for 100 calls:
http://service-go/env
Aggregated Function Time : count 100 avg 0.0067368204 +/- 0.003404 min
0.002795267 max 0.025809888 sum 0.673682036
# target 50%      0.00593333
# target 75%      0.00833333
# target 90%      0.0105
# target 99%      0.018
# target 99.9%    0.0257289
Sockets used: 4 (for perfect keepalive, would be 4)
Code 200 : 100 (100.0 %)
All done 100 calls (plus 0 warmup) 6.737 ms avg, 10.0 qps
```

2）查看 Mixer 是否接收到了 Envoy 代理报告请求。

使用如下命令查看 Mixer 接收到 Envoy 代理报告的请求次数：

```
$ TELEMETRY_IP=$(kubectl get svc istio-telemetry -n istio-system -o
jsonpath='{.spec.clusterIP}')
$ curl -s $TELEMETRY_IP:9093/metrics | grep grpc_server_handled_total
# HELP grpc_server_handled_total Total number of RPCs completed on the server,
regardless of success or failure.
# TYPE grpc_server_handled_total counter
grpc_server_handled_total{grpc_code="OK",grpc_method="Report",grpc_
service="istio.mixer.v1.Mixer",grpc_type="unary"} 52
```

3）查看 Mixer 规则是否存在。

使用如下命令查看已经创建的 rule：

```
$ kubectl get rules --all-namespaces
```

```
NAMESPACE                    NAME                          AGE
istio-system                 kubeattrgenrulerule           28d
istio-system                 promhttp                      28d
istio-system                 promtcp                       28d
istio-system                 stdio                         28d
istio-system                 stdiotcp                      28d
istio-system                 tcpkubeattrgenrulerule        28d
```

4）查看 Prometheus 处理程序是否存在。

使用如下命令查看已经创建的 Prometheus 处理程序：

```
$ kubectl get prometheuses.config.istio.io --all-namespaces
NAMESPACE                    NAME                          AGE
istio-system                 handler                       28d
```

5）查看 Mixer 指标收集实例是否存在。

使用如下命令查看已经创建的 metric 实例：

```
$ kubectl get metrics.config.istio.io --all-namespaces
NAMESPACE                    NAME                          AGE
istio-system                 requestcount                  28d
istio-system                 requestduration               28d
istio-system                 requestsize                   28d
istio-system                 responsesize                  28d
istio-system                 tcpbytereceived               28d
istio-system                 tcpbytesent                   28d
```

6）查看是否有配置错误。

使用如下命令查看是存在错误的配置，当存在计数不为 0 的条目时，表示存在配置错误：

```
$ TELEMETRY_IP=$(kubectl get svc istio-telemetry -n istio-system -o jsonpath='{.spec.clusterIP}')
$ curl -s $TELEMETRY_IP:9093/metrics | grep config_error_count
# HELP mixer_config_adapter_info_config_error_count The number of errors encountered during processing of the adapter info configuration.
# TYPE mixer_config_adapter_info_config_error_count counter
...
mixer_config_instance_config_error_count{configID="17"} 1
mixer_config_instance_config_error_count{configID="2"} 0
...
mixer_config_rule_config_error_count{configID="15"} 0
mixer_config_rule_config_error_count{configID="16"} 0
mixer_config_rule_config_error_count{configID="17"} 1
...
```

7）查看 Mixer 日志。

使用如下的命令查看 Mixer 日志有无错误信息。如果需要，也可以像之前开启 Pilot 的 RBAC 的日志输出级别的步骤一样，开启 Mixer 的 debug 级别的输出日志：

```
$ kubectl logs -f -n istio-system $(kubectl get pod -l app=telemetry -n istio-system -o jsonpath='{.items[0].metadata.name}') mixer | egrep 'error|warn'
...
2018-12-28T07:16:30.155044Z     error     failed to evaluate expression for field 'Dimensions[source_service]'; unknown attribute source.service.name
2018-12-28T07:16:30.155882Z     error     Instance not found: instance='mytcpsentbytes.metric'
2018-12-28T07:16:30.156061Z     warn Neither --kubeconfig nor --master was specified.  Using the inClusterConfig.  This might not work.
2018-12-28T07:16:30.265742Z     error     adapters     adapter did not close all the scheduled daemons     {"adapter": "handler.kubernetesenv.istio-system"}
2018-12-28T07:16:37.171114Z     warn input set condition evaluation error: id='9', error='lookup failed: 'destination.service''
2018-12-28T07:16:47.173559Z     warn input set condition evaluation error: id='9', error='lookup failed: 'destination.service''
...
```

8）查看 Mixer 是否发送了指标给 Prometheus 处理程序。

使用如下命令查看 Mixer 发送指标给 Prometheus 处理程序的次数：

```
$ TELEMETRY_IP=$(kubectl get svc istio-telemetry -n istio-system -o jsonpath='{.spec.clusterIP}')
$ curl -s $TELEMETRY_IP:9093/metrics | grep mixer_runtime_dispatch_count
# HELP mixer_runtime_dispatch_count Total number of adapter dispatches handled by Mixer.
# TYPE mixer_runtime_dispatch_count counter
mixer_runtime_dispatch_count{adapter="kubernetesenv",error="false",handler="handler.kubernetesenv.istio-system",meshFunction="kubernetes"} 884
mixer_runtime_dispatch_count{adapter="kubernetesenv",error="true",handler="handler.kubernetesenv.istio-system",meshFunction="kubernetes"} 0
mixer_runtime_dispatch_count{adapter="prometheus",error="false",handler="handler.prometheus.istio-system",meshFunction="metric"} 213
mixer_runtime_dispatch_count{adapter="prometheus",error="false",handler="tcphandler.prometheus.default",meshFunction="metric"} 83
mixer_runtime_dispatch_count{adapter="prometheus",error="true",handler="handler.prometheus.istio-system",meshFunction="metric"} 0
mixer_runtime_dispatch_count{adapter="prometheus",error="true",handler="tcphandler.prometheus.default",meshFunction="metric"} 0
mixer_runtime_dispatch_count{adapter="stdio",error="false",handler="handler.stdio.istio-system",meshFunction="logentry"} 213
mixer_runtime_dispatch_count{adapter="stdio",error="true",handler="handler.stdio.istio-system",meshFunction="logentry"} 0
```

9）查看 Prometheus 配置。

使用如下命令查看 Prometheus 配置文件，确认是否有 Istio 相关的配置：

```
$ PROMETHEUS_IP=$(kubectl get svc prometheus -n istio-system -o jsonpath='{.spec.clusterIP}')
$ curl -s $PROMETHEUS_IP:9090/config | grep -B15 'istio-telemetry;prometheus'
scrape_configs:
```

```
- job_name: istio-mesh
    scrape_interval: 5s
    scrape_timeout: 5s
    metrics_path: /metrics
    scheme: http
    kubernetes_sd_configs:
    - api_server: null
        role: endpoints
        namespaces:
            names:
            - istio-system
    relabel_configs:
      - source_labels: [__meta_kubernetes_service_name, __meta_kubernetes_
endpoint_port_name]
        separator: ;
        regex: istio-telemetry;prometheus
```

12.6 一个请求的完整过程分析

本节展示一个请求从一个服务实例调用另一个服务的过程，分析一下请求是怎么从调用方到服务提供方的。

1. HTTP 协议的请求

以 service-python 服务访问 service-lua 服务为例。

（1）DNS 解析

当 service-python 服务要访问 service-lua 服务时，需要先通过 kube-dns 找到要访问服务的 IP 地址，也就是 service-lua 服务的 Cluster IP，执行如下的命令可以看到 DNS 解析得到的 service-lua 服务的 IP 地址：

```
$ kubectl exec $(kubectl get pod -l app=service-python -o jsonpath='{.
items[0].metadata.name}') -c service-python -- ping -c 1 service-lua
PING service-lua (10.111.228.100): 56 data bytes
64 bytes from 10.111.228.100: seq=0 ttl=64 time=0.052 ms

--- service-lua ping statistics ---
1 packets transmitted, 1 packets received, 0% packet loss
round-trip min/avg/max = 0.052/0.052/0.052 ms
```

（2）调用方发起请求

当 service-python 服务拿到 service-lua 服务的 IP 地址后，发送请求给服务对应的 IP 和端口。由于 Envoy 代理会拦截所有流量，这个请求会最先到达 service-python 服务实例的 Envoy 代理上，查看 Envoy 代理监听了哪些地址和端口：

```
$ istioctl proxy-config listener $(kubectl get pod -l app=service-python -o
```

```
jsonpath='{.items[0].metadata.name}')
    ADDRESS                            PORT                      TYPE
    10.244.1.16                        80                        HTTP
    10.99.91.255                       15011                     TCP
    10.110.93.233                      443                       TCP
    10.99.91.255                       31400                     TCP
    10.105.42.173                      443                       TCP
    10.96.0.1                          443                       TCP
    10.97.150.161                      443                       TCP
    10.99.91.255                       8060                      TCP
    10.101.237.88                      15011                     TCP
    10.96.0.10                         53                        TCP
    10.99.91.255                       853                       TCP
    10.111.27.172                      42422                     TCP
    10.99.91.255                       443                       TCP
    0.0.0.0                            15001                     TCP
    0.0.0.0                            9093                      HTTP
    0.0.0.0                            15010                     HTTP
    0.0.0.0                            15031                     HTTP
    0.0.0.0                            80                        HTTP
    0.0.0.0                            9091                      HTTP
    0.0.0.0                            15030                     HTTP
    0.0.0.0                            15004                     HTTP
    0.0.0.0                            8080                      HTTP
    0.0.0.0                            8060                      HTTP
```

从上面的监听器可以看出，由于 service-python 服务访问 service-lua 服务使用的是 HTTP 协议并且是 80 端口，所以会匹配到 0.0.0.0:80 监听器。

查看具体监听器的信息：

```
    $ istioctl proxy-config listener $(kubectl get pod -l app=service-python -o
jsonpath='{.items[0].metadata.name}') --type HTTP --port 80 --address 0.0.0.0 -o
json
    [
        {
            "name": "0.0.0.0_80",
            "address": {
                "socketAddress": {
                    "address": "0.0.0.0",
                    "portValue": 80
                }
            },
            "filterChains": [
                {
                    "filters": [
                        {
                            "name": "envoy.http_connection_manager",
                            "config": {
                                ...
```

```
                            "rds": {
                                "config_source": {
                                    "ads": {}
                                },
                                "route_config_name": "80"
                            },
                            ...
                        }
                    }
                ]
            }
        ],
    }
]
```

从上面的输出信息可以看出，0.0.0.0:80 的监听器使用了名为 80 的路由规则。

查看路由规则信息如下：

```
$ istioctl proxy-config route $(kubectl get pod -l app=service-python -o
jsonpath='{.items[0].metadata.name}') --name=80 -o json
[
    {
        "name": "80",
        "virtualHosts": [
            ...
            {
                "name": "service-lua.default.svc.cluster.local:80",
                "domains": [
                    "service-lua.default.svc.cluster.local",
                    "service-lua.default.svc.cluster.local:80",
                    "service-lua",
                    "service-lua:80",
                    "service-lua.default.svc.cluster",
                    "service-lua.default.svc.cluster:80",
                    "service-lua.default.svc",
                    "service-lua.default.svc:80",
                    "service-lua.default",
                    "service-lua.default:80",
                    "10.109.146.20",
                    "10.109.146.20:80"
                ],
                "routes": [
                    {
                        "match": {
                            "prefix": "/"
                        },
                        "route": {
                            "cluster": "outbound|80||service-lua.default.svc.
                                        cluster.local",
                            "timeout": "0.000s",
```

```
                "maxGrpcTimeout": "0.000s"
              }
              ...
            }
          ]
        },
        ...
      ],
      "validateClusters": false
    }
]
```

从上面的路由信息可以看出，对于请求 service-lua 服务的流量使用了名为 "outbound|80||service-lua.default.svc.cluster.local" 的服务实例集群。

查看服务实例集群的信息：

```
$ istioctl proxy-config cluster $(kubectl get pod -l app=service-python -o jsonpath='{.items[0].metadata.name}') --fqdn=service-lua.default.svc.cluster.local -o json
[
    {
        "name": "outbound|80||service-lua.default.svc.cluster.local",
        "type": "EDS",
        "edsClusterConfig": {
            "edsConfig": {
                "ads": {}
            },
            "serviceName": "outbound|80||service-lua.default.svc.cluster.local"
        },
        "connectTimeout": "1.000s",
        "circuitBreakers": {
            "thresholds": [
                {}
            ]
        }
    }
]
```

从中可以看出，请求 service-lua 服务的流量最终都转发到了名为 "outbound|80||service-lua.default.svc.cluster.local" 的 endpoint 上。

查看 endpoint 信息：

```
$ istioctl proxy-config endpoint $(kubectl get pod -l app=service-python -o jsonpath='{.items[0].metadata.name}') --cluster "outbound|80||service-lua.default.svc.cluster.local"
ENDPOINT        STATUS      CLUSTER
10.244.1.15:80     HEALTHY     outbound|80||service-lua.default.svc.cluster.local
10.244.2.9:80      HEALTHY     outbound|80||service-lua.default.svc.cluster.local
```

从中可以看出，请求 service-lua 服务的流量最终到达 10.244.1.10:80 和 10.244.2.8:80 地址上，而这两个地址其实就是 Kubernetes 中 service-lua 服务实例的 Pod IP 地址。

查看 service-lua 服务的 Pod IP 地址：

```
$ kubectl get pod -l app=service-lua -o wide
NAME      READY     STATUS     RESTARTS     AGE     IP     NODE     NOMINATED NODE
service-lua-v1-5c9bcb7778-l6qv2     2/2     Running     0     22m     10.244.1.15
lab3    <none>
service-lua-v2-75cb5cdf8-svn5c     2/2     Running     0     22m     10.244.2.9
lab2    <none>
```

调用方的 Envoy 代理在获取到具体的 service-lua 服务实例地址后，对相应的地址发起请求。

（3）服务方接收请求

由于 Envoy 代理会拦截所有流量，这个请求会最先到达 service-lua 服务实例的 Envoy 代理上，查看 Envoy 代理监听了哪些地址和端口：

```
$ istioctl proxy-config listener $(kubectl get pod -l app=service-lua -o
jsonpath='{.items[0].metadata.name}')
ADDRESS                  PORT         TYPE
10.244.1.15              80           HTTP
10.99.91.255             443          TCP
10.99.91.255             8060         TCP
10.111.27.172            42422        TCP
10.96.0.1                443          TCP
10.110.93.233            443          TCP
10.97.150.161            443          TCP
10.99.91.255             31400        TCP
10.101.237.88            15011        TCP
10.96.0.10               53           TCP
10.99.91.255             15011        TCP
10.99.91.255             853          TCP
10.105.42.173            443          TCP
0.0.0.0                  9091         HTTP
0.0.0.0                  8060         HTTP
0.0.0.0                  15010        HTTP
0.0.0.0                  8080         HTTP
0.0.0.0                  80           HTTP
0.0.0.0                  9093         HTTP
0.0.0.0                  15030        HTTP
0.0.0.0                  15031        HTTP
0.0.0.0                  15004        HTTP
0.0.0.0                  15001        TCP
```

从上面的监听器可以看出，请求 service-lua 服务实例的流量会匹配到 10.244.1.15:80 监听器。

查看具体监听器的信息：

```
$ istioctl proxy-config listener $(kubectl get pod -l app=service-lua -o
jsonpath='{.items[0].metadata.name}') --type HTTP --port 80 --address 10.244.1.15
-o json
    ...
    "virtual_hosts": [
        {
            "domains": [
                "*"
            ],
            "name": "inbound|http|80",
            "routes": [
                {
                    ...
                    "route": {
                        "cluster": "inbound|80||service-lua.default.svc.cluster.local",
                        "max_grpc_timeout": "0.000s",
                        "timeout": "0.000s"
                    }
                }
            ]
        }
    ]
    ...
```

从上面的路由规则可以看出，10.244.1.15:80 监听器会使用名为 "inbound|80||service-lua.default.svc.cluster.local" 的服务实例集群。

查看服务实例集群的信息：

```
$ istioctl proxy-config cluster $(kubectl get pod -l app=service-lua -o
jsonpath='{.items[0].metadata.name}') --fqdn=service-lua.default.svc.cluster.local
--direction inbound -o json
    [
        {
            "name": "inbound|80||service-lua.default.svc.cluster.local",
            "connectTimeout": "1.000s",
            "hosts": [
                {
                    "socketAddress": {
                        "address": "127.0.0.1",
                        "portValue": 80
                    }
                }
            ],
            "circuitBreakers": {
                "thresholds": [
                    {}
                ]
```

```
            }
        }
    ]
```

从上面的服务实例集群信息可以看出，请求service-lua服务的流量最终地址为127.0.0.1:80。service-lua服务实例的Envoy代理请求本地址获取响应结果后，返回给服务调用方的Envoy代理，再由调用方的Envoy代理返回数据给服务调用方。

2. TCP协议的请求

以service-go服务访问Redis服务为例。

（1）DNS解析

当service-go服务要访问Redis服务时，需要先通过kube-dns找到要访问服务IP地址，也就是Redis服务的Cluster IP，执行如下的命令可以看到DNS解析得到Redis服务的IP地址：

```
    $ kubectl exec $(kubectl get pod -l app=service-go -o jsonpath='{.items[0].
metadata.name}') -c service-go -- ping -c 1 redis
    PING redis (10.96.84.23): 56 data bytes
    64 bytes from 10.96.84.23: seq=0 ttl=64 time=0.066 ms

    --- redis ping statistics ---
    1 packets transmitted, 1 packets received, 0% packet loss
    round-trip min/avg/max = 0.066/0.066/0.066 ms
```

（2）调用方发起请求

当service-go服务拿到Redis服务的IP地址后，发送请求给服务对应的IP和端口。由于Envoy代理会拦截所有流量，这个请求会最先到达service-go服务实例的Envoy代理上，查看Envoy代理监听了哪些地址和端口：

```
    $ istioctl proxy-config listener $(kubectl get pod -l app=service-go -o
jsonpath='{.items[0].metadata.name}')
    ADDRESS                           PORT                        TYPE
    10.244.2.8                        80                          HTTP
    10.99.91.255                      443                         TCP
    10.96.0.1                         443                         TCP
    10.97.150.161                     443                         TCP
    10.105.42.173                     443                         TCP
    10.99.91.255                      31400                       TCP
    10.99.91.255                      8060                        TCP
    10.96.0.10                        53                          TCP
    10.110.93.233                     443                         TCP
    10.111.27.172                     42422                       TCP
    10.101.237.88                     15011                       TCP
    10.99.91.255                      15011                       TCP
    10.99.91.255                      853                         TCP
```

```
0.0.0.0                          15030                    HTTP
0.0.0.0                          15010                    HTTP
0.0.0.0                          8080                     HTTP
0.0.0.0                          8060                     HTTP
0.0.0.0                          15031                    HTTP
0.0.0.0                          9091                     HTTP
0.0.0.0                          15004                    HTTP
0.0.0.0                          9093                     HTTP
0.0.0.0                          80                       HTTP
0.0.0.0                          15001                    TCP
10.96.84.23                      6379                     TCP
```

从上面的监听器可以看出，由于 service-go 服务访问 Redis 服务使用的是 TCP 协议并且是 6379 端口，所以会匹配到 10.96.84.23:6379 监听器。

查看具体监听器的信息：

```
$ istioctl proxy-config listener $(kubectl get pod -l app=service-go -o jsonpath='{.items[0].metadata.name}') --type TCP --port 6379 --address 10.96.84.23 -o json
...
"filters": [
    {
        "name": "envoy.tcp_proxy",
        "config": {
            ...
            "cluster": "outbound|6379||redis.default.svc.cluster.local",
            "stat_prefix": "outbound|6379||redis.default.svc.cluster.local"
        }
    }
]
```

从上面的监听器信息可以看出，对于请求 Redis 服务的流量使用了名为 "outbound|6379||redis.default.svc.cluster.local" 的服务实例集群。

查看服务实例集群的信息：

```
$ istioctl proxy-config cluster $(kubectl get pod -l app=service-go -o jsonpath='{.items[0].metadata.name}') --fqdn=redis.default.svc.cluster.local -o json
[
    {
        "name": "outbound|6379||redis.default.svc.cluster.local",
        "type": "EDS",
        "edsClusterConfig": {
            "edsConfig": {
                "ads": {}
            },
            "serviceName": "outbound|6379||redis.default.svc.cluster.local"
        },
```

```
        "connectTimeout": "1.000s",
        "circuitBreakers": {
            "thresholds": [
                {}
            ]
        }
    }
]
```

从上面的服务实例集群信息可以看出，请求 Redis 服务的流量最终都转发到了名为 "outbound|6379||redis.default.svc.cluster.local" 的 endpoint 上。

查看 endpoint 信息：

```
$ istioctl proxy-config endpoint $(kubectl get pod -l app=service-go -o jsonpath='{.items[0].metadata.name}') --cluster "outbound|6379||redis.default.svc.cluster.local"
ENDPOINT            STATUS      CLUSTER
10.244.2.12:6379    HEALTHY     outbound|6379||redis.default.svc.cluster.local
```

可以看出，请求 Redis 服务的流量最终到达 10.244.2.12:6379 地址上，而这个地址其实就是 Kubernetes 中 Redis 服务实例的 Pod IP 地址。

查看 Redis 服务实例的 Pod IP 地址：

```
$ kubectl get pod -l app=redis -o wide
NAME                READY   STATUS    RESTARTS   AGE   IP           NODE   NOMINATED NODE
redis-v1-7d56d758f5-5bx5w  2/2  Running  0  22m   10.244.2.12   lab2   <none>
```

调用方的 Envoy 代理在获取到具体的服务实例地址后，对相应的地址发起请求。

（3）服务方接收请求

由于 Envoy 代理会拦截所有流量，这个请求会最先到达 Redis 服务实例的 Envoy 代理上，查看 Envoy 代理监听了哪些地址和端口：

```
$ istioctl proxy-config listener $(kubectl get pod -l app=redis -o jsonpath='{.items[0].metadata.name}')
ADDRESS                   PORT          TYPE
10.244.2.12               6379          TCP
10.111.27.172             42422         TCP
10.96.0.10                53            TCP
10.110.93.233             443           TCP
10.99.91.255              443           TCP
10.99.91.255              8060          TCP
10.97.150.161             443           TCP
10.99.91.255              15011         TCP
10.96.84.23               6379          TCP
10.99.91.255              31400         TCP
10.99.91.255              853           TCP
10.101.237.88             15011         TCP
```

```
10.96.0.1                         443                      TCP
10.105.42.173                     443                      TCP
0.0.0.0                           9093                     HTTP
0.0.0.0                           15031                    HTTP
0.0.0.0                           15010                    HTTP
0.0.0.0                           80                       HTTP
0.0.0.0                           15004                    HTTP
0.0.0.0                           8080                     HTTP
0.0.0.0                           8060                     HTTP
0.0.0.0                           15030                    HTTP
0.0.0.0                           9091                     HTTP
0.0.0.0                           15001                    TCP
```

从上面的监听器可以看出，请求 Redis 服务实例的流量会匹配到 10.244.2.12:6379 监听器。

查看具体监听器的信息：

```
$ istioctl proxy-config listener $(kubectl get pod -l app=redis -o jsonpath='{.items[0].metadata.name}') --type TCP --port 6379 --address 10.244.2.12 -o json
...
{
    "name": "envoy.tcp_proxy",
    "config": {
        ...
        "cluster": "inbound|6379||redis.default.svc.cluster.local",
        "stat_prefix": "inbound|6379||redis.default.svc.cluster.local"
    }
}
...
```

从中可以看出，10.244.2.12:6379 监听器会使用名为 "inbound|6379||redis.default.svc.cluster.local" 的服务实例集群。

查看服务实例集群的信息：

```
$ istioctl proxy-config cluster $(kubectl get pod -l app=redis -o jsonpath='{.items[0].metadata.name}') --fqdn=redis.default.svc.cluster.local --direction inbound -o json
[
    {
        "name": "inbound|6379||redis.default.svc.cluster.local",
        "connectTimeout": "1.000s",
        "hosts": [
            {
                "socketAddress": {
                    "address": "127.0.0.1",
                    "portValue": 6379
                }
```

```
                }
            ],
            "circuitBreakers": {
                "thresholds": [
                    {}
                ]
            }
        }
    ]
```

从中可以看出，请求Redis服务的流量最终地址为127.0.0.1:6379。Redis服务实例的Envoy代理请求本地地址获取响应结果后，返回给服务调用方的Envoy代理，再由Envoy代理返回数据给服务调用方。

12.7 本章小结

部署是使用Istio时最简单的环节，在部署上线以后，维护Istio变成了最重要的一项任务，如何定制安装，如何在不中断业务的情况下升级Istio，出现故障时如何排除，这些都是我们要重点关注的问题。通过了解一个请求的完整过程，我们可以更加深入地理解Istio，为以后的问题排查和深入学习打下基础。

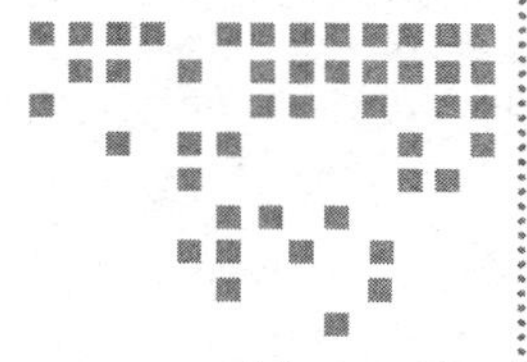

第13章 Chapter 13

杂　项

本章介绍一些 Istio 提供的其他功能，主要包括：使用 CORS 来解决跨域问题，如何进行 URL 重定向和重写，TCP 路由和 TLS 路由的配置。还介绍了在 Istio 集群中的服务健康检查问题，以及生产环境中如何暴露 ingressgateway。最后介绍了 Mixer 与 Adapter 模型，这对理解前面章节实验中使用到的 yaml 配置文件会有很大的帮助。

【实验前的准备】

进行本章实验前，需要先执行如下的前置步骤。

1）下载实验时用到的源码仓库：

```
$ sudo yum install -y git

$ git clone https://github.com/mgxian/istio-lab
Cloning into 'istio-lab'...
remote: Enumerating objects: 252, done.
remote: Counting objects: 100% (252/252), done.
remote: Compressing objects: 100% (177/177), done.
remote: Total 779 (delta 157), reused 166 (delta 74), pack-reused 527
Receiving objects: 100% (779/779), 283.37 KiB | 243.00 KiB/s, done.
Resolving deltas: 100% (451/451), done.

$ cd istio-lab
```

2）开启 default 命名空间的自动注入功能：

```
$ kubectl label namespace default istio-injection=enabled
namespace/default labeled
```

3）部署用于测试的服务：

```
$ kubectl apply -f service/go/service-go.yaml
$ kubectl get pod
NAME                              READY     STATUS     RESTARTS    AGE
service-go-v1-7cc5c6f574-xdgxz    2/2       Running    0           40s
service-go-v2-7656dcc478-cf2mf    2/2       Running    0           40s
```

13.1 CORS

跨域资源共享（CORS）是一种机制，它使用额外的 HTTP 头来告诉浏览器让运行在一个域名上的 Web 应用可以访问来自不同源服务器上的指定资源。当一个资源从与该资源本身所在的服务器不同的域、协议或端口请求另外一个资源时，资源会发起一个跨域 HTTP 请求。

比如，站点 http://domain-a.com 的某 HTML 页面通过 <img> 的 src 请求 http://domain-b.com/image.jpg。出于安全原因，浏览器限制跨域 HTTP 请求。可以通过配置 CORS 头来解除这种限制。更深入的内容可以参考如下两篇文章：

- http://www.ruanyifeng.com/blog/2016/04/cors.html
- https://developer.mozilla.org/zh-CN/docs/Web/HTTP/AccesscontrolCORS

Istio 提供了 CORS 的配置，我们可以在 VirtualService 中配置 CORS。

服务 service-go 配置 CORS 示例如下：

```
 1 apiVersion: networking.istio.io/v1alpha3
 2 kind: VirtualService
 3 metadata:
 4   name: service-go-cors
 5 spec:
 6   hosts:
 7   - service-go
 8   http:
 9   - route:
10     - destination:
11         host: service-go
12     corsPolicy:
13       allowOrigin:
14       - http://www.will.com
15       allowMethods:
16       - POST
17       - GET
18       allowHeaders:
19       - X-Custom-Header
20       exposeHeaders:
21       - X-Expose-Header
22       maxAge: 24h
23       allowCredentials: false
```

第 12 ～ 23 行定义了服务 service-go 的 CORS 设置，allowOrigin 表示允许跨域的域名，allowMethods 表示允许跨域的 HTTP 方法，allowHeaders 表示在请求时允许携带的请求头，exposeHeaders 表示允许暴露的请求响应头，maxAge 表示跨域预检查的缓存时间，allowCredentials 表示是否允许发送 Cookie。

【实验】

1）创建测试 Pod：

```
$ kubectl apply -f kubernetes/dns-test.yaml
```

2）创建 service-go 服务的 CORS 规则：

```
$ kubectl apply -f istio/miscellaneous/virtual-service-go-cors.yaml
```

3）简单请求访问测试：

```
$ kubectl exec dns-test -c dns-test -- curl -sv -H "Origin: http://www.will.
com" http://service-go/env
> GET /env HTTP/1.1
> User-Agent: curl/7.35.0
> Host: service-go
> Accept: */*
> Origin: http://www.will.com
>
{"message":"go v2"}< HTTP/1.1 200 OK
< content-type: application/json; charset=utf-8
< date: Sat, 19 Jan 2019 05:54:22 GMT
< content-length: 19
< x-envoy-upstream-service-time: 18
< server: envoy
< access-control-allow-origin: http://www.will.com
< access-control-expose-headers: X-Expose-Header
<
{ [data not shown]

$ kubectl exec dns-test -c dns-test -- curl -sv -H "Origin: http://not-
allowed-origin.will.com" http://service-go/env
> GET /env HTTP/1.1
> User-Agent: curl/7.35.0
> Host: service-go
> Accept: */*
> Origin: http://not-allowed-origin.will.com
>
{"message":"go v2"}< HTTP/1.1 200 OK
< content-type: application/json; charset=utf-8
< date: Sat, 19 Jan 2019 05:55:22 GMT
< content-length: 19
< x-envoy-upstream-service-time: 1
< server: envoy
```

```
<
{ [data not shown]
```

4）非简单请求访问测试。

预检请求：

```
$ kubectl exec dns-test -c dns-test -- curl -sv -X OPTIONS -H "Origin: http://www.will.com" -H "Access-Control-Request-Method: GET" -H "Access-Control-Request-Headers: X-Custom-Header" http://service-go/env
> OPTIONS /env HTTP/1.1
> User-Agent: curl/7.35.0
> Host: service-go
> Accept: */*
> Origin: http://www.will.com
> Access-Control-Request-Method: GET
> Access-Control-Request-Headers: X-Custom-Header
>
< HTTP/1.1 200 OK
< access-control-allow-origin: http://www.will.com
< access-control-allow-methods: POST,GET
< access-control-allow-headers: X-Custom-Header
< access-control-max-age: 24h0m0s
< access-control-expose-headers: X-Expose-Header
< date: Sat, 19 Jan 2019 05:56:18 GMT
< server: envoy
< content-length: 0
<

$ kubectl exec dns-test -c dns-test -- curl -sv -X OPTIONS -H "Origin: http://not-allowed-origin.will.com" -H "Access-Control-Request-Method: GET" -H "Access-Control-Request-Headers: X-Custom-Header" http://service-go/env
> OPTIONS /env HTTP/1.1
> User-Agent: curl/7.35.0
> Host: service-go
> Accept: */*
> Origin: http://not-allowed-origin.will.com
> Access-Control-Request-Method: GET
> Access-Control-Request-Headers: X-Custom-Header
>
< HTTP/1.1 404 Not Found
< date: Sat, 19 Jan 2019 05:57:32 GMT
< server: envoy
< content-length: 0
< x-envoy-upstream-service-time: 1
<
```

正常请求：

```
$ kubectl exec dns-test -c dns-test -- curl -sv -H "Origin: http://www.will.com" -H "X-Custom-Header: value" http://service-go/env
> GET /env HTTP/1.1
```

```
> User-Agent: curl/7.35.0
> Host: service-go
> Accept: */*
> Origin: http://www.will.com
> X-Custom-Header: value
>
{"message":"go v2"}< HTTP/1.1 200 OK
< content-type: application/json; charset=utf-8
< date: Sat, 19 Jan 2019 05:57:58 GMT
< content-length: 19
< x-envoy-upstream-service-time: 1
< server: envoy
< access-control-allow-origin: http://www.will.com
< access-control-expose-headers: X-Expose-Header
<
{ [data not shown]

$ kubectl exec dns-test -c dns-test -- curl -sv -H "Origin: http://not-allowed-origin.will.com" -H "X-Custom-Header: value" http://service-go/env
> GET /env HTTP/1.1
> User-Agent: curl/7.35.0
> Host: service-go
> Accept: */*
> Origin: http://not-allowed-origin.will.com
> X-Custom-Header: value
>
{"message":"go v2"}< HTTP/1.1 200 OK
< content-type: application/json; charset=utf-8
< date: Sat, 19 Jan 2019 05:58:40 GMT
< content-length: 19
< x-envoy-upstream-service-time: 17
< server: envoy
<
{ [data not shown]
```

从上面的请求测试可以得出如下的结论：当使用允许的域名进行跨域请求时，返回的响应头中都会有浏览器跨域请求需要的 Access-Control-Allow-Origin 等字段。当使用了其他不被允许的域名进行跨域资源请求时，请求的响应中并不包含浏览器跨域请求资源时需要的响应头。

5）清理：

```
$ kubectl delete -f kubernetes/dns-test.yaml
$ kubectl delete -f istio/miscellaneous/virtual-service-go-cors.yaml
```

13.2 URL 重定向

URL 重定向（Redirect）也称为 URL 转发，是一种当实际资源，如单个页面、表单或者整个 Web 应用，被迁移到新的 URL 时，保持（原有）链接可用的技术。一般情况下，服务会返回一个重定向响应，响应码一般为 301 或者 302，重定向的 URL 放在响应头的 Location 字段中。

Istio 提供的 URL 重定向功能，会返回一个响应码为 301 的重定向响应。

服务 service-go 配置 URL 重定向示例如下：

```
1 apiVersion: networking.istio.io/v1alpha3
2 kind: VirtualService
3 metadata:
4   name: service-go-redirect
5 spec:
6   hosts:
7   - service-go
8   http:
9   - match:
10     - uri:
11         exact: /v1/env
12     redirect:
13       uri: /env
14       authority: service-go
15   - route:
16     - destination:
17         host: service-go
18         subset: v1
19 ---
20 apiVersion: networking.istio.io/v1alpha3
21 kind: DestinationRule
22 metadata:
23   name: service-go
24 spec:
25   host: service-go
26   subsets:
27   - name: v1
28     labels:
29       version: v1
30   - name: v2
31     labels:
32       version: v2
```

第 1 ～ 18 行 VirtualService 的定义表明，当访问服务 service-go 的 /v1/env 链接时，会重定向到服务 service-go 的 /env 链接。

【实验】

1）创建测试 Pod：

```
$ kubectl apply -f kubernetes/dns-test.yaml
```

2）创建 service-go 服务的 redirect 规则：

```
$ kubectl apply -f istio/miscellaneous/virtual-service-go-redirect.yaml
```

3）服务访问测试：

```
$ kubectl exec dns-test -c dns-test -- curl -sI http://service-go/v1/env
HTTP/1.1 301 Moved Permanently
location: http://service-go/env
date: Sat, 19 Jan 2019 06:01:33 GMT
server: envoy
transfer-encoding: chunked
```

从上面的请求测试可以看出，当访问 service-go 服务的 /v1/env 链接时，链接被重定向为 /env，重定向后的 /env 请求被转发到了 service-go 服务的 v1 版本。

4）清理：

```
$ kubectl delete -f kubernetes/dns-test.yaml
$ kubectl delete -f istio/miscellaneous/virtual-service-go-redirect.yaml
```

13.3 URL 重写

URL 重写（Rewrite）是一种在请求真正发送到后端服务之前改写请求 URI 的技术。

服务 service-go 配置 URL 重写示例如下：

```
 1 apiVersion: networking.istio.io/v1alpha3
 2 kind: DestinationRule
 3 metadata:
 4   name: service-go
 5 spec:
 6   host: service-go
 7   subsets:
 8   - name: v1
 9     labels:
10       version: v1
11   - name: v2
12     labels:
13       version: v2
14 ---
15 apiVersion: networking.istio.io/v1alpha3
16 kind: VirtualService
17 metadata:
18   name: service-go-rewrite
19 spec:
20   hosts:
21   - service-go
22   http:
23   - match:
24     - uri:
25         exact: /v1/env
```

```
26      rewrite:
27        uri: /env
28      route:
29      - destination:
30          host: service-go
31          subset: v1
32    - route:
33      - destination:
34          host: service-go
35          subset: v2
```

第 15 ～ 35 行定义的 VirtualService 表示，当请求服务 service-go 的 /v1/env 链接时，在转发到后端服务之前重写为访问后端服务的 /env 链接。

【实验】

1）创建测试 Pod：

```
$ kubectl apply -f kubernetes/dns-test.yaml
```

2）创建 service-go 服务的 rewrite 规则：

```
$ kubectl apply -f istio/miscellaneous/virtual-service-go-rewrite.yaml
```

3）服务访问测试：

```
$ kubectl exec dns-test -c dns-test -- curl -s http://service-go/v1/env
{"message":"go v1"}

$ kubectl exec dns-test -c dns-test -- curl -s http://service-go/env
{"message":"go v2"}
```

从上面的请求测试可以看出，当访问 service-go 服务的 /v1/env 链接时，链接被重写为 /env，并把请求转发到了 service-go 服务的 v1 版本，当访问 service-go 服务的其他链接时，把请求转发到了 service-go 服务的 v2 版本。

4）清理：

```
$ kubectl delete -f kubernetes/dns-test.yaml
$ kubectl delete -f istio/miscellaneous/virtual-service-go-rewrite.yaml
```

13.4 TCP 路由

TCP 路由（TCPRoute）与之前 HTTPRoute 类似，根据 TCP 请求的信息，转发请求到后端不同服务上。

目前版本不支持使用权重来进行多路由转发，而 1.1 版本中已经支持。

Redis 服务器的 TCP 路由示例如下：

```
apiVersion: networking.istio.io/v1alpha3
kind: VirtualService
metadata:
  name: redis-proxy
spec:
  hosts:
  - redis-proxy
  tcp:
  - match:
    - port: 16379
    route:
    - destination:
      host: redis
      port:
      number: 6379
```

上面代码定义的 VirtualService 表明，当以 TCP 协议请求 redis-proxy 的 16379 时，转发请求到 Redis 服务的 6379 端口。

【实验】

1）部署 Redis client：

```
$ kubectl apply -f kubernetes/redis-cli.yaml
```

2）部署 Redis server：

```
$ kubectl apply -f kubernetes/redis-server.yaml
$ kubectl get pod
NAME                            READY      STATUS      RESTARTS      AGE
redis-v1-7d56d758f5-tkp6n       1/1        Running     0             11s
```

3）创建 Redis server 的 TCP 路由规则：

```
$ kubectl apply -f istio/miscellaneous/virtual-service-tcp-route.yaml
```

4）创建不包含后端 Pod 的 redis-proxy Service，用于 DNS 域名解析：

```
$ kubectl apply -f kubernetes/redis-proxy-service.yaml
```

5）服务访问测试：

```
$ REDIS_CLI_POD=$(kubectl get pod -l app=redis-cli -o jsonpath={.items..
metadata.name})
$ kubectl exec $REDIS_CLI_POD -c redis-cli -- redis-cli -h redis-proxy -p
16379 info cpu
# CPU
used_cpu_sys:0.423661
used_cpu_user:0.490072
used_cpu_sys_children:0.000783
used_cpu_user_children:0.001674
```

从上面的请求测试可以看出，当访问 redis-proxy 服务的 16379 端口时，请求转发到了 Redis 服务的 6379 端口。

6）清理：

```
$ kubectl delete -f kubernetes/redis-cli.yaml
$ kubectl delete -f kubernetes/redis-server.yaml
$ kubectl delete -f kubernetes/redis-proxy-service.yaml
$ kubectl delete -f istio/miscellaneous/virtual-service-tcp-route.yaml
```

13.5 TLS 路由

TLS 路由（TLSRoute）用于配置未终止的 TLS/HTTPS 流量。一般情况下配合 Gateway 或者 ServiceEntry 来使用。

Nginx 服务器的 TLS 路由示例如下：

```
 1 apiVersion: networking.istio.io/v1alpha3
 2 kind: Gateway
 3 metadata:
 4   name: https-gateway
 5 spec:
 6   selector:
 7     istio: ingressgateway # use Istio default gateway implementation
 8   servers:
 9   - port:
10       number: 443
11       name: https
12       protocol: HTTPS
13     tls:
14       mode: PASSTHROUGH
15     hosts:
16     - "*"
17 ---
18 apiVersion: networking.istio.io/v1alpha3
19 kind: VirtualService
20 metadata:
21   name: my-nginx
22 spec:
23   hosts:
24   - "*"
25   gateways:
26   - https-gateway
27   tls:
28   - match:
29     - port: 443
30       sniHosts:
31       - "foo.my-nginx.com"
32     route:
```

```
33      - destination:
34          host: my-nginx.foo.svc.cluster.local
35          port:
36            number: 443
37    - match:
38      - port: 443
39        sniHosts:
40        - "bar.my-nginx.com"
41      route:
42      - destination:
43          host: my-nginx.bar.svc.cluster.local
44          port:
45            number: 443
```

第1～16行定义了名为https-gateway的网关，监听443端口，使用HTTPS协议，TLS模式选择PASSTHROUGH，表示不终止TLS。

第18～45行定义了名为my-nginx的VirtualService，当sni host为foo.my-nginx.com时，路由请求到my-nginx.foo.svc.cluster.local服务的443端口，当sni host为bar.my-nginx.com时，路由请求到my-nginx.bar.svc.cluster.local服务的443端口。

【实验】

1）创建HTTPS证书：

```
$ openssl req -x509 -nodes -days 365 -newkey rsa:2048 -keyout /tmp/nginx.key
-out /tmp/nginx.crt -subj '/CN=*.my-nginx.com/O=my-nginx'

$ kubectl create ns foo
$ kubectl create ns bar

$ kubectl create secret tls nginxsecret -n foo --key /tmp/nginx.key --cert /
tmp/nginx.crt
$ kubectl create secret tls nginxsecret -n bar --key /tmp/nginx.key --cert /
tmp/nginx.crt

$ kubectl get secret nginxsecret -n foo
NAME            TYPE                  DATA      AGE
nginxsecret     kubernetes.io/tls     2         20s

$ kubectl get secret nginxsecret -n bar
NAME            TYPE                  DATA      AGE
nginxsecret     kubernetes.io/tls     2         10s
```

2）创建Nginx的HTTPS配置：

```
$ kubectl create configmap nginxconfigmap -n foo --from-file=istio/
miscellaneous/https-default.conf

$ kubectl create configmap nginxconfigmap -n bar --from-file=istio/
miscellaneous/https-default.conf
```

```
$ kubectl get configmap nginxconfigmap -n foo
NAME                   DATA      AGE
nginxconfigmap         1         20s

$ kubectl get configmap nginxconfigmap -n bar
NAME                   DATA      AGE
nginxconfigmap         1         10s
```

3）创建测试Pod：

```
$ kubectl apply -f kubernetes/dns-test.yaml
```

4）部署nginx服务：

```
$ kubectl apply -f <(istioctl kube-inject -f kubernetes/my-nginx.yaml) -n foo

$ kubectl apply -f <(istioctl kube-inject -f kubernetes/my-nginx.yaml) -n bar

$ kubectl get pod -n foo
NAME                         READY     STATUS     RESTARTS    AGE
my-nginx-97744d9bd-2z8v5     2/2       Running    0           48s

$ kubectl get pod -n bar
NAME                         READY     STATUS     RESTARTS    AGE
my-nginx-97744d9bd-t4mdd     2/2       Running    0           54s
```

5）关闭nginx服务的mTLS：

```
$ kubectl apply -f istio/security/mtls-my-nginx-disable.yaml -n foo
$ kubectl apply -f istio/security/mtls-my-nginx-disable.yaml -n bar
```

6）服务访问测试：

```
$ kubectl exec dns-test -c dns-test -- curl -sk https://my-nginx.foo/
Hello from nginx in foo

$ kubectl exec dns-test -c dns-test -- curl -sk https://my-nginx.bar/
Hello from nginx in bar
```

7）创建TLS路由规则：

```
$ kubectl apply -f istio/miscellaneous/virtual-service-tls-route.yaml
```

8）服务访问测试：

```
$ curl -sk --resolve foo.my-nginx.com:31390:11.11.11.112 https://foo.my-nginx.
com:31390/
Hello from nginx in foo

$ curl -sk --resolve bar.my-nginx.com:31390:11.11.11.112 https://bar.my-nginx.
```

```
com:31390/
    Hello from nginx in bar
```

从上面的请求测试可以看出，当请求的 sni host 为 foo.my-nginx.com，请求转发到了 foo 命名空间的 my-nginx 服务；当请求的 sni host 为 bar.my-nginx.com，请求转发到了 bar 命名空间的 my-nginx 服务。

9）清理：

```
$ kubectl delete -f kubernetes/dns-test.yaml
$ kubectl delete ns foo
$ kubectl delete ns bar
$ kubectl delete -f istio/miscellaneous/virtual-service-tls-route.yaml
```

13.6 mTLS 迁移

在服务迁移到服务网格中时，可能需要逐渐地迁移，这时候可能会由于网格内的服务启用了 mTLS，导致网格外的服务调用网格中的服务失败。为了解决这种问题，可以把网格中服务的 mTLS 模式设置为 PERMISSIVE，这样网格中的服务既可以接收 mTLS 加密流量，又可以接收没有经过 mTLS 加密的普通流量。等到所有服务都迁移到网格中，再把网格中服务的 mTLS 模式设置为 STRICT，使网格内的服务只接受 mTLS 加密的流量。这样可以保证服务之间的通信安全。

配置服务 httpbin 接收两种类型的流量示例：

```
 1 apiVersion: authentication.istio.io/v1alpha1
 2 kind: Policy
 3 metadata:
 4   name: httpbin
 5 spec:
 6   targets:
 7   - name: httpbin
 8   peers:
 9   - mtls:
10       mode: PERMISSIVE
11 ---
12 apiVersion: networking.istio.io/v1alpha3
13 kind: DestinationRule
14 metadata:
15   name: httpbin
16 spec:
17   host: "httpbin.default.svc.cluster.local"
18   trafficPolicy:
19     tls:
20       mode: ISTIO_MUTUAL
```

第 10 行设置 httpbin 服务的服务端 mTLS 模式为 PERMISSIVE，表示既可以接收

mTLS 加密流量，又可以接收没有经过 mTLS 加密的普通流量。

【实验】

1）创建测试 Pod：

```
$ kubectl apply -f kubernetes/dns-test.yaml
$ kubectl create ns legacy
$ kubectl apply -f kubernetes/dns-test.yaml -n legacy
$ kubectl get pod
NAME                          READY     STATUS     RESTARTS     AGE
dns-test                      2/2       Running    0            25s

$ kubectl get pod -n legacy
NAME                          READY     STATUS     RESTARTS     AGE
dns-test                      1/1       Running    0            24s
```

2）部署 httpbin 服务：

```
$ kubectl apply -f kubernetes/httpbin.yaml
$ kubectl get pod -l app=httpbin
NAME                          READY     STATUS     RESTARTS     AGE
httpbin-b67975b8f-vx4gb       2/2       Running    0            81s
```

3）创建 httpbin 服务的 mTLS 规则为 PERMISSIVE 模式的路由：

```
$ kubectl apply -f istio/security/mtls-httpbin-both.yaml
```

4）服务访问测试：

```
$ kubectl exec dns-test -c dns-test -- curl -s http://httpbin.default:8000/
headers
{
  "headers": {
    "Accept": "*/*",
    "Content-Length": "0",
    "Host": "httpbin.default:8000",
    "User-Agent": "curl/7.35.0",
    "X-B3-Sampled": "1",
    "X-B3-Spanid": "98d45df0bd84370d",
    "X-B3-Traceid": "98d45df0bd84370d",
    "X-Request-Id": "794f2d8a-38e1-9037-835c-ed29e06e62f0"
  }
}

$ kubectl exec dns-test -c dns-test -n legacy -- curl -s http://httpbin.
default:8000/headers
{
  "headers": {
    "Accept": "*/*",
    "Content-Length": "0",
    "Host": "httpbin.default:8000",
```

```
        "User-Agent": "curl/7.35.0",
        "X-B3-Sampled": "1",
        "X-B3-Spanid": "a720bcba6c04c95c",
        "X-B3-Traceid": "a720bcba6c04c95c",
        "X-Request-Id": "96431a94-84f5-9ef5-9192-7d2323e5686e"
    }
}
```

5）创建 httpbin 服务的 mTLS 规则为 STRICT 模式：

```
$ kubectl apply -f istio/security/mtls-httpbin-strict.yaml
```

6）服务访问测试：

```
$ kubectl exec dns-test -c dns-test -- curl -s http://httpbin.default:8000/
headers
{
    "headers": {
        "Accept": "*/*",
        "Content-Length": "0",
        "Host": "httpbin.default:8000",
        "User-Agent": "curl/7.35.0",
        "X-B3-Sampled": "1",
        "X-B3-Spanid": "eea1d828dbd22bd9",
        "X-B3-Traceid": "eea1d828dbd22bd9",
        "X-Request-Id": "feea08c3-c97f-9200-862b-df8297f83a7b"
    }
}

$ kubectl exec dns-test -c dns-test -n legacy -- curl -s http://httpbin.
default:8000/headers
command terminated with exit code 56
```

从上面的服务访问测试可以得出如下的结论：当 httpbin 服务使用 PERMISSIVE 模式的 mTLS 时，普通请求和经过 mTLS 加密的请求都会成功。当 httpbin 服务使用 STRICT 模式的 mTLS 时，只有使用 mTLS 加密的请求会成功，而普通请求则会失败。

7）清理：

```
$ kubectl delete ns legacy
$ kubectl delete -f kubernetes/dns-test.yaml
$ kubectl delete -f kubernetes/httpbin.yaml
$ kubectl delete -f istio/security/mtls-httpbin-strict.yaml
```

13.7　EnvoyFilter

EnvoyFilter 用于自定义 Envoy 代理的配置，使用时必须小心，如果出现错误，可能导致整个服务网格出现问题。

修改访问 httpbin 服务的请求信息示例如下：

```
 1 apiVersion: networking.istio.io/v1alpha3
 2 kind: EnvoyFilter
 3 metadata:
 4   name: httpbin
 5 spec:
 6   workloadLabels:
 7     app: httpbin
 8   filters:
 9   - listenerMatch:
10       portNumber: 8000
11       listenerType: SIDECAR_INBOUND
12     filterName: envoy.lua
13     filterType: HTTP
14     filterConfig:
15       inlineCode: |
16         function envoy_on_request(request_handle)
17           request_handle:headers():add("X-Foo", "bar")
18         end
19         function envoy_on_response(response_handle)
20           body_size = response_handle:body():length()
21           response_handle:headers():add("X-Response-Body-Size", tostring
                                              (body_size))
22         end
```

第 1 ～ 22 行定义了名为 httpbin 的 EnvoyFilter。

第 6 ～ 7 行的定义表明此规则只对包含 app 标签值为 httpbin 的 pod 生效。

第 9 ～ 11 行的定义表明只处理端口地址为 8000 的进入 pod 的请求。

第 12 ～ 13 行定义所使用过滤器的类型和名称。

第 14 ～ 22 行表明，对于符合上述条件的请求，在请求后端服务时添加字段名为 X-Foo、值为 bar 的请求头，在得到后端服务实例的响应后，添加 X-Response-Body-Size 响应头，并且值为服务响应内容的字节大小。

【实验】

1）创建测试 Pod：

```
$ kubectl apply -f kubernetes/dns-test.yaml
```

2）部署 httpbin 服务：

```
$ kubectl apply -f kubernetes/httpbin.yaml

$ kubectl get pod -l app=httpbin
NAME                           READY     STATUS      RESTARTS     AGE
httpbin-b67975b8f-c7jp7        2/2       Running     0            61s
```

3）创建 httpbin 服务的 EnvoyFilter：

```
$ kubectl apply -f istio/miscellaneous/envoy-filter.yaml
```

4）服务访问测试：

```
$ kubectl exec dns-test -c dns-test -- curl -sv http://httpbin:8000/headers
> GET /headers HTTP/1.1
> User-Agent: curl/7.35.0
> Host: httpbin:8000
> Accept: */*
>
{
    "headers": {
        "Accept": "*/*",
        "Content-Length": "0",
        "Host": "httpbin:8000",
        "User-Agent": "curl/7.35.0",
        "X-B3-Sampled": "0",
        "X-B3-Spanid": "bbaccdc8c132ea85",
        "X-B3-Traceid": "bbaccdc8c132ea85",
        "X-Foo": "bar",
        "X-Request-Id": "337f2180-6ec4-4ff9-a0e3-0cf11c3da91c"
    }
}
< HTTP/1.1 200 OK
< server: envoy
< date: Sat, 19 Jan 2019 06:37:19 GMT
< content-type: application/json
< access-control-allow-origin: *
< access-control-allow-credentials: true
< content-length: 323
< x-envoy-upstream-service-time: 19
< x-response-body-size: 323
<
{ [data not shown]
```

从上面的服务访问测试可以看出，所有进入 httpbin 服务的请求都添加了值为 bar 的 X-Foo 请求头字段，请求的响应中包含了字段名为 x-response-body-size、值为响应内容字节数的响应头。

5）清理：

```
$ kubectl delete -f kubernetes/dns-test.yaml
$ kubectl delete -f kubernetes/httpbin.yaml
$ kubectl delete -f istio/miscellaneous/envoy-filter.yaml
```

13.8　添加请求头

在请求发送给后端服务实例时，给请求添加请求头。

【实验】

1）创建测试 Pod：

```
$ kubectl apply -f kubernetes/dns-test.yaml
```

2）部署 httpbin 服务：

```
$ kubectl apply -f kubernetes/httpbin.yaml
$ kubectl get pod -l app=httpbin
NAME                          READY     STATUS      RESTARTS    AGE
httpbin-b67975b8f-vdstk       2/2       Running     0           7s
```

3）创建 httpbin 服务的添加请求头路由规则：

```
$ kubectl apply -f istio/miscellaneous/virtual-service-httpbin-append-headers.yaml
```

4）服务访问测试：

```
$ kubectl exec dns-test -c dns-test -- curl -s http://httpbin:8000/headers
{
  "headers": {
    "Accept": "*/*",
    "Content-Length": "0",
    "Host": "httpbin:8000",
    "User-Agent": "curl/7.35.0",
    "X-B3-Sampled": "1",
    "X-B3-Spanid": "5961ccd986a78b3a",
    "X-B3-Traceid": "5961ccd986a78b3a",
    "X-Custom-1": "1",
    "X-Custom-2": "2",
    "X-Request-Id": "4748e668-b49c-989f-9047-803a6654ec5d"
  }
}
```

从上面的服务访问测试可以看出，当访问 httpbin 服务时，Envoy 代理自动给请求添加了 X-Custom-1 和 X-Custom-2 请求头。

5）清理：

```
$ kubectl delete -f kubernetes/dns-test.yaml
$ kubectl delete -f kubernetes/httpbin.yaml
$ kubectl delete -f istio/miscellaneous/virtual-service-httpbin-append-headers.yaml
```

13.9 在 Gateway 上使用 HTTPS

在 Istio Ingressgateway 上配置 TLS，实现以 HTTPS 的方式接收外部请求。

【实验一】 在 Gateway 上使用单个证书。

1）创建 HTTPS 证书：

```
$ openssl req -x509 -nodes -days 365 -newkey rsa:2048 -keyout /tmp/httpbin.key
-out /tmp/httpbin.crt -subj '/CN=*.httpbin.will/O=httpbin'

$ kubectl create -n istio-system secret tls istio-ingressgateway-certs --key /
tmp/httpbin.key --cert /tmp/httpbin.crt

$ kubectl get secret istio-ingressgateway-certs -n istio-system
NAME                           TYPE                DATA   AGE
istio-ingressgateway-certs     kubernetes.io/tls   2      39s
```

2）部署 httpbin 服务：

```
$ kubectl apply -f kubernetes/httpbin.yaml

$ kubectl get pod -l app=httpbin
NAME                        READY   STATUS    RESTARTS   AGE
httpbin-b67975b8f-mfxsk     2/2     Running   0          9s
```

3）创建 Gateway 暴露服务：

```
$ kubectl apply -f istio/route/gateway-httpbin-https.yaml
```

4）使用 curl 访问：

```
$ curl -k https://11.11.11.111:31390/get
{
  "args": {},
  "headers": {
    "Accept": "*/*",
    "Content-Length": "0",
    "Host": "11.11.11.111:31390",
    "User-Agent": "curl/7.29.0",
    "X-B3-Sampled": "1",
    "X-B3-Spanid": "238abb319357989f",
    "X-B3-Traceid": "238abb319357989f",
    "X-Envoy-Internal": "true",
    "X-Request-Id": "a006cacd-5adb-90d6-9884-f43377562d10"
  },
  "origin": "10.244.0.0",
  "url": "https://11.11.11.111:31390/get"
}
```

5）使用浏览器访问。

访问地址 https://11.11.11.111:31390/get，得到如图 13-1 所示的结果。

第一次访问时可能会遇到如图 13-2 所示的安全问题，点击“高级”，然后选择继续前往即可。

← → C ▲ 不安全 | https://11.11.11.111:31390/get


```
{
  "args": {},
  "headers": {
    "Accept": "text/html,application/xhtml+xml,application/xml;q=0.9,image/webp,image/apng,*/*;q=0.8",
    "Accept-Encoding": "gzip, deflate, br",
    "Accept-Language": "zh-CN,zh;q=0.9,en;q=0.8,nb;q=0.7",
    "Cache-Control": "max-age=0",
    "Content-Length": "0",
    "Host": "11.11.11.111:31390",
    "Upgrade-Insecure-Requests": "1",
    "User-Agent": "Mozilla/5.0 (Windows NT 10.0; Win64; x64) AppleWebKit/537.36 (KHTML, like Gecko) Chrome/70.0.3538.110 Safari/537.36",
    "X-B3-Sampled": "1",
    "X-B3-Spanid": "7a0590ffa4a4c39c",
    "X-B3-Traceid": "7a0590ffa4a4c39c",
    "X-Envoy-Internal": "true",
    "X-Request-Id": "4bf9773d-0b1c-9a03-bf25-7c6092187093"
  },
  "origin": "10.244.0.0",
  "url": "https://11.11.11.111:31390/get"
}
```


图 13-1 使用浏览器访问

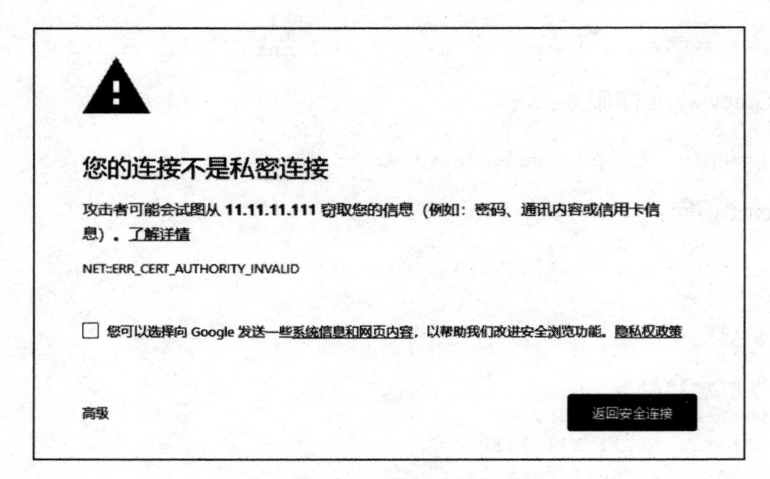

图 13-2 安全问题

【实验二】 在 Gateway 上使用多个证书。在上面的实验基础上进行。

1）删除之前实验创建的 Gateway：

```
$ kubectl delete -f istio/route/gateway-httpbin-https.yaml
```

2）创建 service-go 服务的 HTTPS 证书：

```
$ openssl req -x509 -nodes -days 365 -newkey rsa:2048 -keyout /tmp/service-go.key -out /tmp/service-go.crt -subj '/CN=*.service-go.will/O=service-go'

$ kubectl create -n istio-system secret tls istio-ingressgateway-service-go-certs --key /tmp/service-go.key --cert /tmp/service-go.crt

$ kubectl get secret istio-ingressgateway-service-go-certs -n istio-system
```

```
NAME                                    TYPE                DATA    AGE
istio-ingressgateway-service-go-certs   kubernetes.io/tls   2       9s
```

3）部署 service-go 服务：

```
$ kubectl apply -f service/go/service-go.yaml

$ kubectl get pod -l app=service-go
NAME                            READY   STATUS    RESTARTS   AGE
service-go-v1-7cc5c6f574-hl4vn  2/2     Running   0          32s
service-go-v2-7656dcc478-nzwmb  2/2     Running   0          32s
```

4）使用 Helm 生成支持多证书的 Ingressgateway 部署文件：

```
helm template /usr/local/istio/install/kubernetes/helm/istio \
-x /usr/local/istio/install/kubernetes/helm/istio/charts/gateways/templates/
deployment.yaml \
--name istio-ingressgateway --namespace istio-system \
--set gateways.istio-egressgateway.enabled=false \
--set gateways.istio-ingressgateway.secretVolumes[0].name=ingressgateway-certs \
--set gateways.istio-ingressgateway.secretVolumes[0].secretName=istio-
ingressgateway-certs \
--set gateways.istio-ingressgateway.secretVolumes[0].mountPath=/etc/istio/
ingressgateway-certs \
--set gateways.istio-ingressgateway.secretVolumes[1].name=ingressgateway-ca-
certs \
--set gateways.istio-ingressgateway.secretVolumes[1].secretName=istio-
ingressgateway-ca-certs \
--set gateways.istio-ingressgateway.secretVolumes[1].mountPath=/etc/istio/
ingressgateway-ca-certs \
--set gateways.istio-ingressgateway.secretVolumes[2].name=ingressgateway-
service-go-certs \
--set gateways.istio-ingressgateway.secretVolumes[2].secretName=istio-
ingressgateway-service-go-certs \
--set gateways.istio-ingressgateway.secretVolumes[2].mountPath=/etc/istio/
ingressgateway-service-go-certs \
> istio-ingressgateway.yaml
```

5）部署新的 Ingressgateway：

```
$ kubectl apply -f istio-ingressgateway.yaml
$ kubectl get pod -n istio-system -l app=istio-ingressgateway
NAME                                   READY   STATUS    RESTARTS   AGE
istio-ingressgateway-85dbc5947-bfq5x   1/1     Running   0          49s
```

6）创建 Gateway 暴露服务：

```
$ kubectl apply -f istio/route/gateway-httpbin-service-go-https.yaml
```

7）服务访问测试：

```
$ curl -sk --resolve test.httpbin.will:31390:11.11.11.111 -H "Host: test.
```

```
httpbin.will" https://test.httpbin.will:31390/get
    {
      "args": {},
      "headers": {
        "Accept": "*/*",
        "Content-Length": "0",
        "Host": "test.httpbin.will",
        "User-Agent": "curl/7.29.0",
        "X-B3-Sampled": "1",
        "X-B3-Spanid": "7f4a96d1d3218cd3",
        "X-B3-Traceid": "7f4a96d1d3218cd3",
        "X-Envoy-Internal": "true",
        "X-Request-Id": "d51dca60-5bb6-9e07-b69b-84e42297dfb5"
      },
      "origin": "10.244.0.0",
      "url": "https://test.httpbin.will/get"
    }

    $ curl -sk --resolve test.service-go.will:31390:11.11.11.111 -H "Host: test.
service-go.will" https://test.service-go.will:31390/env
    {"message":"go v2"}
```

8）清理：

```
$ kubectl delete -f istio/route/gateway-httpbin-service-go-https.yaml
$ kubectl delete -f kubernetes/httpbin.yaml
$ kubectl delete secret istio-ingressgateway-certs -n istio-system
$ kubectl delete secret istio-ingressgateway-service-go-certs -n istio-system
$ kubectl delete -f service/go/service-go.yaml
```

13.10 在HTTPS服务上开启mTLS

当Envoy代理跟随使用HTTPS协议的服务部署时，不管服务是否启用mTLS，代理会自动降级为4层协议处理，这就意味着Envoy不会终止原始HTTPS流量。这也是Istio可以处理使用HTTPS协议服务的原因。

【实验】

1）创建my-nginx服务的HTTPS证书：

```
    $ openssl req -x509 -nodes -days 365 -newkey rsa:2048 -keyout /tmp/nginx.key
-out /tmp/nginx.crt -subj '/CN=*.my-nginx.com/O=my-nginx'

    $ kubectl create ns foo
    $ kubectl create secret tls nginxsecret -n foo --key /tmp/nginx.key --cert /
tmp/nginx.crt

    $ kubectl get secret nginxsecret -n foo
```

```
NAME           TYPE                DATA   AGE
nginxsecret    kubernetes.io/tls   2      10s
```

2）创建 Nginx 的 HTTPS 配置：

```
$ kubectl create configmap nginxconfigmap -n foo --from-file=istio/
miscellaneous/https-default.conf

$ kubectl get configmap nginxconfigmap -n foo
NAME             DATA   AGE
nginxconfigmap   1      10s
```

3）创建测试 Pod：

```
$ kubectl apply -f kubernetes/dns-test.yaml
$ kubectl get pod
NAME                          READY        STATUS     RESTARTS    AGE
dns-test                      2/2          Running    0           7s
```

4）部署 nginx 服务，不使用 Envoy 代理：

```
$ kubectl apply -f kubernetes/my-nginx.yaml -n foo
$ kubectl get pod -n foo
NAME                          READY        STATUS     RESTARTS    AGE
my-nginx-97744d9bd-sj2mn      1/1          Running    0           34s
```

5）服务访问测试：

```
$ kubectl exec dns-test -c dns-test -- curl -sk https://my-nginx.foo/
Hello from nginx in foo
```

6）部署 Nginx 服务，使用 Envoy 代理：

```
$ kubectl apply -f <(istioctl kube-inject -f kubernetes/my-nginx.yaml) -n foo
$ kubectl get pod -n foo
NAME                           READY       STATUS     RESTARTS    AGE
my-nginx-5c7d4cf55b-tkp6n      2/2         Running    0           46s
```

7）关闭 Nginx 服务的 mTLS：

```
$ kubectl apply -f istio/security/mtls-my-nginx-disable.yaml -n foo
```

8）服务访问测试：

```
$ kubectl exec dns-test -c dns-test -- curl -sk https://my-nginx.foo/
Hello from nginx in foo
```

9）开启 Nginx 服务的 mTLS：

```
$ kubectl apply -f istio/security/mtls-my-nginx-enable.yaml -n foo
```

10）服务访问测试：

```
$ kubectl exec dns-test -c dns-test -- curl -sk https://my-nginx.foo/
Hello from nginx in foo
```

从上面的测试可以看出，当 my-nginx 服务不在网格内时，可以正常访问。当 my-nginx 服务在网格内时，关闭和启用 mTLS 的情况下都能正常访问 my-nginx 服务。

11）清理：

```
$ kubectl delete ns foo
$ kubectl delete -f kubernetes/dns-test.yaml
```

13.11 网格中的服务健康检查

部署在 Kubernetes 集群中的服务，一般情况下会使用 liveness 和 readiness 进行服务的健康检查。当服务部署在 Istio 网格中时，服务的健康检查要做适当的改变，主要分为如下两种情况：

- ❑ 服务关闭 mTLS：当服务没有启用 mTLS 时，Command 和 HTTP 请求以及 TCP 端口检查类型的服务健康检查都可以正常使用。
- ❑ 服务开启 mTLS：当服务启用 mTLS 时，Command 类型的服务健康检查和 TCP 端口检查类型都能正常使用，HTTP 请求不能正常使用。

TCP 端口检查时，只会检查端口是否开启，由于 Envoy 代理会根据服务的配置监听相应的端口，这会导致不管 Pod 有没有出现问题，TCP 端口检查一定会成功，所以 TCP 端口检查类型的健康检查机制并不能真正检测出服务实例的真实健康情况，不推荐使用此种类型的健康检查。

【实验一】 服务关闭 mTLS 时的健康检查。

1）创建测试 Pod：

```
$ kubectl apply -f kubernetes/dns-test.yaml
```

2）关闭 service-go 服务的 mTLS：

```
$ kubectl apply -f istio/security/mtls-service-go-off.yaml
```

3）部署使用 Command 类型健康检查的 service-go 服务：

```
$ kubectl apply -f kubernetes/service-go-liveness-command.yaml
$ kubectl get pod -l app=service-go
NAME                              READY     STATUS    RESTARTS   AGE
service-go-v1-5d98689766-6rcv2    2/2       Running   0          64s
```

4）部署使用 HTTP 请求类型健康检查的 service-go 服务：

```
$ kubectl apply -f kubernetes/service-go-liveness-http.yaml
```

```
$ kubectl get pod -l app=service-go
NAME                                  READY     STATUS     RESTARTS     AGE
service-go-v1-67dffc6768-tbftf        2/2       Running    0            9s
```

5）部署使用TCP端口类型健康检查的service-go服务：

```
$ kubectl apply -f kubernetes/service-go-liveness-tcp.yaml
$ kubectl get pod -l app=service-go
NAME                                  READY     STATUS     RESTARTS     AGE
service-go-v1-6ff45f7cbc-mf7jf        2/2       Running    0            54s
```

6）清理：

```
$ kubectl delete -f istio/security/mtls-service-go-off.yaml
$ kubectl delete -f kubernetes/service-go-liveness-tcp.yaml
```

【实验二】 服务开启mTLS时的健康检查。

1）开启service-go服务的mTLS：

```
$ kubectl apply -f istio/security/mtls-service-go-on.yaml
```

2）部署使用Command类型健康检查的service-go服务：

```
$ kubectl apply -f kubernetes/service-go-liveness-command.yaml
$ kubectl get pod -l app=service-go
NAME                                  READY     STATUS     RESTARTS     AGE
service-go-v1-5d98689766-t4mdd        2/2       Running    0            7s
```

3）部署使用HTTP请求类型健康检查的service-go服务：

```
$ kubectl apply -f kubernetes/service-go-liveness-http.yaml
$ kubectl get pod -l app=service-go
NAME                                READY   STATUS              RESTARTS    AGE
service-go-v1-67dffc6768-sjqmk      1/2     CrashLoopBackOff    6           4m19s
```

4）查看service-go服务Pod的events事件：

```
    SERVICE_GO_POD=$(kubectl get pod -l app=service-go -o jsonpath={.items..
metadata.name})
    $ kubectl describe pod $SERVICE_GO_POD | grep -A50 Events
    Events:
        Type       Reason       Age        From              Message
        ----       ------       ----       ----              -------
    ...
        Normal     Started      14m        kubelet, lab3     Started container
        Normal     Created      14m        kubelet, lab3     Created container
        Warning    Unhealthy    14m        kubelet, lab3     Liveness probe failed:
Get http://10.244.2.13:80/status: read tcp 10.244.2.1:51260->10.244.2.13:80: read:
connection reset by peer
         Warning    Unhealthy    14m    kubelet, lab3    Liveness probe failed:Get
http://10.244.2.13:80/status: read tcp 10.244.2.1:51268->10.244.2.13:80: read:
```

```
connection reset by peer
    Warning   Unhealthy    14m      kubelet, lab3      Liveness probe
  ...
```

5）部署使用 TCP 端口类型健康检查的 service-go 服务：

```
$ kubectl apply -f kubernetes/service-go-liveness-tcp.yaml
$ kubectl get pod -l app=service-go
NAME                              READY     STATUS      RESTARTS     AGE
service-go-v1-6ff45f7cbc-4cbbj    2/2       Running     0            2m23s
```

6）清理：

```
$ kubectl delete -f istio/security/mtls-service-go-on.yaml
$ kubectl delete -f kubernetes/service-go-liveness-tcp.yaml
```

13.12 Envoy 代理 Ingressgateway

由于 Ingressgateway 是使用 NodePort 的方式暴露在 31380 和 31390 端口，当外部访问集群内暴露的服务时，需要在 URL 中添加端口才能正常访问。由于 Gateway 并不允许在 hosts 中指定带端口的域名，导致我们在创建 Gateway 时只能指定 * 泛域名，这就会造成我们不能使用 host 来区分服务的功能。为了解决这个问题，通常可以创建一个代理服务来转发流量到 Ingressgateway 中，一般情况下，这个代理服务会监听 Web 常用的 80 和 443 端口。在使用代理服务的情况下，后端的 Ingressgateway 还可以创建多个副本来做高可用。

由于 Istio 使用 Envoy 作为服务网格的代理，因此我们也可以直接使用 Envoy 作为我们的代理服务器，代理后端的 Ingressgateway。通过给 Kubernetes 集群中的节点设置特定的标签，如 edgenode=true，来标记这些节点作为集群中的边界节点，在这些节点上部署 Envoy 代理，接收外部流量。

虽然我们使用了 Envoy 作为前端代理 Ingressgateway，接收外部流量，但是我们仍然需要对外部调用方提供唯一的 IP 地址。我们需要在 Envoy 前面再加一层代理，比如，可以使用 Keepalived 与 LVS 配合，保证当集群中的节点出现问题，导致 Ingressgateway 部分实例出现故障或者 Envoy 代理部分实例出现故障时，不会改变集群外的流量入口 IP 地址，进而减少对服务调用方的影响。由于篇幅问题，本书不进行此部分的实验。

【实验】

1）创建 Ingressgateway 的 Envoy 代理服务器配置文件：

```
$ kubectl create configmap front-envoy -n istio-system --from-file=front-envoy.
yaml=istio/miscellaneous/istio-ingressgateway-proxy.yaml

$ kubectl get configmap front-envoy -n istio-system
```

```
NAME                  DATA           AGE
front-envoy           1              10s
```

2）给需要部署 Envoy 代理服务器的边界节点打标签：

```
$ kubectl get no --show-labels
NAME   STATUS   ROLES    AGE   VERSION   LABELS
lab1   Ready    master   12d   v1.12.2   beta.kubernetes.io/arch=amd64,beta.
kubernetes.io/os=linux,kubernetes.io/hostname=lab1,node-role.kubernetes.io/master=
lab2   Ready    <none>   12d   v1.12.2   beta.kubernetes.io/arch=amd64,beta.
kubernetes.io/os=linux,kubernetes.io/hostname=lab2
lab3   Ready    <none>   12d   v1.12.2   beta.kubernetes.io/arch=amd64,beta.
kubernetes.io/os=linux,kubernetes.io/hostname=lab3

$ kubectl label nodes lab1 edgenode=true

$ kubectl get no --show-labels
NAME   STATUS   ROLES    AGE   VERSION   LABELS
lab1   Ready    master   12d   v1.12.2   beta.kubernetes.io/arch=amd64,beta.
kubernetes.io/os=linux,edgenode=true,kubernetes.io/hostname=lab1,node-role.
kubernetes.io/master=
lab2   Ready    <none>   12d   v1.12.2   beta.kubernetes.io/arch=amd64,beta.
kubernetes.io/os=linux,kubernetes.io/hostname=lab2
lab3   Ready    <none>   12d   v1.12.2   beta.kubernetes.io/arch=amd64,beta.
kubernetes.io/os=linux,kubernetes.io/hostname=lab3
```

3）部署 Envoy 代理服务器：

```
$ kubectl apply -f service/envoy/envoy.yaml -n istio-system
$ kubectl get pod -n istio-system -l app=front-envoy
NAME                           READY     STATUS    RESTARTS   AGE
front-envoy-5df48f5dc7-tkp6n   1/1       Running   0          102s
```

4）部署 httpbin 服务：

```
$ kubectl apply -f kubernetes/httpbin.yaml
$ kubectl get pod -l app=httpbin
NAME                           READY     STATUS    RESTARTS   AGE
httpbin-b67975b8f-6rcv2        2/2       Running   0          92s
```

5）创建 Gateway 暴露服务：

```
$ kubectl apply -f istio/miscellaneous/gateway-httpbin-use-host.yaml
```

6）服务访问测试：

```
$ curl -I http://11.11.11.111/
HTTP/1.1 404 Not Found
date: Wed, 09 Jan 2019 03:16:40 GMT
server: envoy
transfer-encoding: chunked
```

7）配置 hosts 解析：

```
11.11.11.111 httpbin.will
```

8）使用 curl 访问：

```
$ curl http://httpbin.will/get
{
  "args": {},
  "headers": {
    "Accept": "*/*",
    "Content-Length": "0",
    "Host": "httpbin.will",
    "User-Agent": "curl/7.29.0",
    "X-B3-Sampled": "1",
    "X-B3-Spanid": "3be937ed7ed61086",
    "X-B3-Traceid": "3be937ed7ed61086",
    "X-Envoy-Internal": "true",
    "X-Request-Id": "51c91552-2e83-98bb-853f-f31ff7923c42"
  },
  "origin": "10.244.0.0",
  "url": "http://httpbin.will/get"
}
```

9）使用浏览器访问。

访问地址 http://httpbin.will/get，得到的结果如图 13-3 所示。

```
← → C  ⓘ 不安全 | httpbin.will/get
{
  "args": {},
  "headers": {
    "Accept": "text/html,application/xhtml+xml,application/xml;q=0.9,image/webp,image/apng,*/*;q=0.8",
    "Accept-Encoding": "gzip, deflate",
    "Accept-Language": "zh-CN,zh;q=0.9,en;q=0.8,nb;q=0.7",
    "Content-Length": "0",
    "Host": "httpbin.will",
    "Upgrade-Insecure-Requests": "1",
    "User-Agent": "Mozilla/5.0 (Windows NT 10.0; Win64; x64) AppleWebKit/537.36 (KHTML, like Gecko) Chrome/70.0.3538.110 Safari/537.36",
    "X-B3-Sampled": "1",
    "X-B3-Spanid": "92747cf1637e8894",
    "X-B3-Traceid": "92747cf1637e8894",
    "X-Envoy-Internal": "true",
    "X-Request-Id": "c0d9b556-b4a8-9ffd-b2c6-86480d37b88b"
  },
  "origin": "10.244.0.0",
  "url": "http://httpbin.will/get"
}
```

图 13-3　实验结果

从上面的服务访问测试可以看出，当创建了使用指定域名的 Gateway 时，访问时必须使用指定的域名，当使用其他域名访问时，会直接返回 404 响应码。

10）清理：

```
$ kubectl delete -f istio/miscellaneous/gateway-httpbin-use-host.yaml
$ kubectl delete -f kubernetes/httpbin.yaml
$ kubectl delete -f service/envoy/envoy.yaml -n istio-system
$ kubectl delete configmap front-envoy -n istio-system
$ kubectl label nodes lab1 edgenode-
```

13.13 Mixer 与 Adapter 模型

Mixer 组件用于遥测数据收集和服务的访问权限控制。Mixer 组件工作机制如图 13-4 所示。

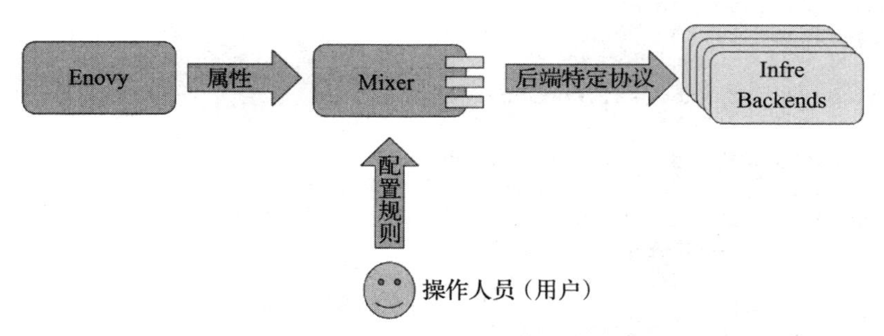

图 13-4 Mixer 组件工作机制图（图片来源：Istio 官方网站）

Mixer 配置基于 adapter 和 template：

- adapter：封装对后端组件的操作。内置的 adapter 及配置可以参考如下官方文档：https://istio.io/docs/reference/config/policy-and-telemetry/adapters/。
- template：定义传递给后端 adapter 组件的数据格式。内置的 template 及配置可以参考如下官方文档：https://istio.io/docs/reference/config/policy-and-telemetry/templates/。

Mixer 配置模型包括如下参数：

- rule：用来绑定 handler 和 instance，把 instance 定义的数据发送给指定的 handler。
- handler：指定后端 adapter。
- instances：指定传递给后端的 template 数据格式，可以指定多个。

配置示例如下：

```
1 apiVersion: config.istio.io/v1alpha2
2 kind: logentry
3 metadata:
4   name: newlog
5   namespace: istio-system
6 spec:
7   severity: '"info"'
8   timestamp: request.time
9   variables:
10     source: source.labels["app"] | source.workload.name | "unknown"
11     user: source.principal | "unknown"
12     destination: destination.labels["app"] | destination.service.name |
               "unknown"
13     response_code: response.code | 0
14     response_size: response.size | 0
15     latency: response.duration | "0ms"
16   monitored_resource_type: '"UNSPECIFIED"'
```

```
17 ---
18 apiVersion: config.istio.io/v1alpha2
19 kind: fluentd
20 metadata:
21   name: fluentdhandler
22   namespace: istio-system
23 spec:
24   address: "fluentd-es.logging:24224"
25 ---
26 apiVersion: config.istio.io/v1alpha2
27 kind: rule
28 metadata:
29   name: newlogtofluentd
30   namespace: istio-system
31 spec:
32   match: "true"
33   actions:
34    - handler: fluentdhandler.fluentd
35      instances:
36      - newlog.logentry
```

第1～16行定义的logentry是一个template，用于定义日志收集时的日志格式。

第18～24行定义的fluentd是一个用于存储日志到fluentd的adapter。

第26～36行定义的rule表示把logentry收集到的日志数据发送给fluentd来处理。

13.14 本章小结

本章介绍了前面没有提到但是可能会非常有用的功能。比如使用Envoy代理Ingressgateway，可能对于在非云平台部署Istio的用户来说非常重要。理解了Mixer与Adapter模型之后，对前面很多功能的配置方式应该会有更深入的理解，在使用Istio时也会更加得心应手。